THE ANATOMY OF A SCIENTIFIC INSTITUTION:

THE PARIS ACADEMY OF SCIENCES, 1666–1803

The Anatomy of a Scientific Institution

The Paris Academy of Sciences, 1666-1803

by Roger Hahn

UNIVERSITY OF CALIFORNIA PRESS

BERKELEY, LOS ANGELES, LONDON 1971

UNIVERSITY OF CALIFORNIA PRESS
BERKELEY AND LOS ANGELES, CALIFORNIA
UNIVERSITY OF CALIFORNIA PRESS, LTD.
LONDON, ENGLAND
COPYRIGHT © 1971, BY
THE REGENTS OF THE UNIVERSITY OF CALIFORNIA

TRANSLATION RIGHTS RESERVED FOR ROGER HAHN
ISBN: 0–520–01818–4
LIBRARY OF CONGRESS CATALOG CARD NUMBER: 70–130795

PRINTED IN THE UNITED STATES OF AMERICA
DESIGNED BY DAVE COMSTOCK

To Henry Guerlac
Scholar, teacher, and friend

Contents

Preface

EVERY historian of science, whatever his special interest, has acknowledged the importance of the Parisian Académie Royale des Sciences as the central theater in which science's intricate plot was unraveled during the Enlightenment. Curiously enough, however, no serious study of that institution has been written since the 1880's, despite the universal recognition of its place in the annals of science.

This study is meant first as a response to this gross lacuna. It is therefore based upon extensive archival research and drawn together from the mass of scholarly literature on French science written in the intervening period. These sources are discussed in Appendix I ("A Bibliographical Note"), to which the specialist is referred for details. Beyond that, this study is an essay into the history of a scientific institution, conceived as something more than the sum of the accomplishments of its illustrious members, however famous they were as individual creators. Any institution that has a continuous history extending into three turbulent centuries of French history develops a special flavor, a character of its own which transcends the acts recorded in its proceedings. It is this elusive spirit that is sought here.

The Académie Royale des Sciences, building upon its early successes, will be seen to have established a set of traditions and to have carved out important functions in the world of science and the state. A delineation of these traditions will be the central issue of the first three chapters. Chapters 4 and 5 will explore the pre-Revolutionary forces that challenged the patterns established by the Academy in the early years of its existence. The next three chapters are devoted to the overwhelming effects of the French Revolution, which successfully called into question all of the values for which the institution had been established, and which eventually led to its demise in 1793. The rest of the book is concerned with the re-emergence of the Academy as part of the Institut National, this time in a new political and scientific setting and with a significantly modified role, which was not substantially altered during the rest of the nineteenth century.

Taken as a whole, this study illustrates the extent to which scientific institutions are shaped by the requirements of both science and society. In writing about scientific institutions as something more than a chronicle of discoveries or the social history of its members, there is a need to recognize that they are the main instrument for the formulation and the transmittal of scientific norms, and that these norms must harmonize with the dominant nonscientific modes of the time. The juxtaposition of the Academy with three different societal settings —the Old Regime, Revolutionary France, and the France of the Napoleonic era—will illustrate this concretely far better than a theoretical discussion could accomplish. More than any other entity, the scientific institution is the anvil on which the often conflicting values of science and society are shaped into a viable form. The institution's success is in large part determined by its ability to make the peculiar enterprise of science palatable to society while at the same time maintaining the advancement of the discipline itself.

This book is also intended to fill a gap in the monographic study of French science, flanked in historical time by Harcourt Brown's *Scientific Organizations in Seventeenth Century France* and Maurice Crosland's *The Society of Arcueil*. What is needed most now to fill out the sequence are similar studies on France's important scientific institutions of the nineteenth century, the Ecole Polytechnique, the Muséum d'Histoire Naturelle, the Faculté des Sciences, and the Collège de France. Only then, perhaps, will it be possible to fathom that special genius and the peculiar limitations that characterize the French scientific mind.

ROGER HAHN

Aptos, California
July 1969

Acknowledgments

I WILL try in the footnotes and in the Bibliographical Note (Appendix I) to make specific acknowledgment to authors whose writings have been indispensable for this book. Institutions and individuals who extended help in other but less visible ways deserve special recognition in this note.

The National Science Foundation permitted me to work in European archives by awarding two fellowships, in 1959–1960 and 1964–1965. Though countless research centers, libraries, archives, and their curators, opened their doors generously, and are thus really entitled to individual thanks, limitations of space do not allow this luxury. A special note of gratitude must nonetheless be registered for Monsieur and Madame Pierre Gauja and their staff at the Académie's archives. Of critical importance also were the Centre Universitaire International and the Centre Alexandre Koyré, which offered a welcomed hospitality while I worked in Paris. Suzanne Delorme, Maurice Daumas, and René Taton were always ready to donate their time, knowledge, and encouragement for the project. A separate mention of Arthur Birembaut's assistance in archival research is also in order. For all this, I am eternally grateful.

In this country, my teachers, friends, and students have given equal support, even when they were critical of the enterprise. I am grateful to Susan Calkin, Gerald Cavanaugh, Seymour Chapin, Maurice Crosland, Vernard Foley, Thomas Kuhn, Larry Loewinger, Charles Paul, and Rhoda Rappaport, each of whom read part of the typescript and offered valuable suggestions. Barbara Boonstoppel prepared the index. Earlier in my research, Denis I. Duveen generously opened his rich private collection for my use. Later, at Berkeley, my colleagues helped by providing a stimulating and demanding intellectual atmosphere. I have especially profited from conversations with Irwin Scheiner and Wolfgang Sauer, who made important suggestions at critical points in my writing. Thanks also must be offered to the Institute of International Studies and the Humanities Institute of the University of California, Berkeley, which provided material assistance for the completion of this book. The typing chores were carried out with effectiveness and good humor by Bojana Ristich. At the University of California Press, Grant Barnes, Barbara Zimmerman, and Jerome Fried were the thoughtful and efficient editors who saw the book through production.

A special paragraph must be reserved for my family and for my mentor at Cornell University. My parents, wife, and children have been patient and sympathetic beyond expectation; their optimism has supplied vital moral support. Henry Guerlac, who initially persuaded me to work on a part of this topic for my doctoral dissertation, has been a constant source of help and inspiration. Without them, I would not have had the fortitude to complete this study.

R. H.

Abbreviations

AdS	Paris, Académie des Sciences, Archives
AdS, Lav	Paris, Académie des Sciences, Archives, Lavoisier papers
AdS, Reg	Paris, Académie des Sciences, Archives, Registre des procès-verbaux des séances
AIHS	*Archives Internationales d'Histoire des Sciences*
AN	Paris, Archives Nationales
AP	*Archives Parlementaires*, lère série (1787–1799)
APS	Philadelphia, American Philosophical Society
AS	*Annals of Science*
BHVP	Paris, Bibliothèque Historique de la Ville de Paris
BI	Paris, Bibliothèque de l'Institut
BM	London, British Museum
BMHN	Paris, Bibliothèque Centrale du Muséum d'Histoire Naturelle
BN	Paris, Bibliothèque Nationale
BS	Paris, Bibliothèque de la Sorbonne
Cassini, "Mes Annales"	Manuscript by Jean-Dominique Cassini in Clermont (Oise), Bibliothèque Municipale, MS 40
(x) Congrès International	*Actes du (x)ème Congrès International d'Histoire des Sciences*
CIPAL	*Procès-verbaux du Comité d'Instruction Publique de l'Assemblée Législative*, ed. J. Guillaume.

CIPCN

Procès-verbaux du Comité d'Instruction Publique de la Convention Nationale, ed. J. Guillaume.

DBF

Dictionnaire de Biographie Française

Enseignement et Diffusion

Enseignement et Diffusion des Sciences en France au XVIIIè Siècle, ed. R. Taton.

HARS

Histoire de l'Académie Royale des Sciences

Institut, Reg

Procès-verbaux des Séances de l'Académie Tenues depuis la Fondation de l'Institut jusqu'au Mois d'Août 1835

JHI

Journal of the History of Ideas

JT

Manuscript "Journal du Trésor de l'Académie, 1792–1793," Ithaca, N.Y., Cornell University Library, MS 114

Lalande, "Collection"

Manuscript of Académie Royale des Sciences, "Collection de ses Règlemens et Déliberations par Ordre de Matière," annotated by J. J. Lalande, Florence, Biblioteca Medicea-Laurenziana, Ashburnham-Libri No. 1700

RHS

Revue d'Histoire des Sciences et de Leurs Applications

Chapter 1

Initiating a Tradition

IT HAS taken decades of scholarly effort to establish what one of the earliest historians of the Royal Academy of Sciences already sensed in the first years of the eighteenth century. Bernard de Fontenelle once remarked that the "scientific renaissance of true philosophy" had necessarily promoted a new Age of Academies all over Europe.[1] The accuracy of his observation has made it a standard theme, reiterated ever since by commentators on the Scientific Revolution. All agree that wherever the spirit of scientific inquiry penetrated, new patterns of social organization sprang up, and that by the middle of the eighteenth century the Academy had become the major symbol for the advancement of learning of the enlightened world.[2]

For a long time this symbol was considered merely an epiphenomenal consequence of the intellectual upheavals that culminated in the establishment of Newtonian science. Deeper study has revealed, however, that the relationship of thought to social organization was in fact considerably more complex. The

1. *HARS* (1666–1699), I, 5. For a key to abbreviations, see pp. xiii–xiv. For full bibliographical citations, consult pp. 374–418.
2. Daumas, "Esquisse d'une Histoire de la Vie Scientifique," pp. 77–99; Ben-David, *Minerva*, IV, 49–50.

transmutation of the techniques, concepts, and methodologies of science was both dependent upon the emergence of new patterns of association among men of science and a stimulus to their further elaboration. Because of this fundamental inter-dependence, the organizational revolution must now be con-sidered an essential component of the Scientific Revolution rather than its by-product. Their relationship was concurrent and symbiotic rather than sequential.

Like the accompanying intellectual transformation, the or-ganizational revolution of the seventeenth century has pro-voked the general historian's curiosity because of its apparent irrelevance to the deep religious, dynastic, and national con-flicts that preoccupied most Western Europeans at that time.[3] As a consequence, the Scientific Revolution is generally en-visioned as a single historical movement of such magnitude and pervasiveness that it transcended the internal dissensions plaguing Europe and flourished everywhere regardless of local issues. In this view, a Kepler, a Galileo, a Descartes, a Huygens, or a Newton is represented in terms of the universal import of his accomplishments rather than for the particular manner in which he was tied to the local culture from which he may have derived considerable sustenance. Similarly, historians have pre-ferred to focus on the shared beliefs that tied together learned circles in Italy, England, Germany, or France instead of the linguistic, religious, or cultural barriers that might have im-peded their intercommunication.

Insofar as the emergence of academies reflected the needs of this general scientific movement, the emphasis upon the similarities among newly constituted scientific organizations is meaningful and significant. By this emphasis, the international character of the movement is underscored and the rather ab-stract nature of scientific thought is endowed with concrete meaning by highlighting generalized social practices. As ama-teur circles developed into well-established learned societies, they delimited a scientific community and gradually defined professional norms. It was through the practices and activities

3. I am deliberately discounting the vast but generally unenlightening litera-ture on Protestantism and the rise of modern science, which has not been successfully integrated into general treatments of the Scientific Revolution.

of these academies that the ideals of the new science most vivid-
ly found public expression.

The precise manner in which professional scientific habits
were gradually established remains to be studied in detail. We
do know that by the time permanent scientific societies were
founded in the 1660's, at least three characteristics directly
related to the demands of the international movement were
already visible. These became the trade-mark of all groups
seriously concerned with the promotion of science. Each organization, by its very nature, had a communal instinct first to
share information within its membership, and secondly to make
its collective findings generally available to other interested
circles; hence the existence of recording and corresponding
secretaries, and the tendency to compose proceedings. Each
learned society had its organizer, its "intelligencer," and ultimately its published *Saggi*, *Mémoires*, or *Transactions*. Yet if in
our eyes communality and publicity seem crucial, the members
themselves placed an even higher value upon a third activity:
experimentation. The mottoes of these seventeenth-century
groups glorified the value of visual demonstration for confirm-
ing scientific truths. Thus the typical academy was apt to
pride itself as much on its laboratory equipment as upon its
secretary or journal. Such distinctive and universal features underscore the common organizational response to a set of profound needs.

Fontenelle was therefore well-advised in stating that the
age of "true philosophy" had called the new academies into
existence in all parts of Western Europe. Yet he was careful to
note that that his Royal Academy of Sciences, in contradistinc-
tion to the London Royal Society, was established to satisfy an
entirely different need as well. As its title indicates, it was more
than just another scientific academy; it was also a royal institution. Fontenelle knew all too well that his institution was
fashioned to satisfy another more parochial force as well, the
French State. He was sufficiently realistic to proclaim that the
Academy had two acknowledged masters, science and the
crown.

To penetrate behind the Academy's development, one
must come to grips with the significance of this dual allegiance.

It was at once the molding force behind the learned society's internal practices, the source of its strength and prominence in the Old Regime, and ultimately the cause of its profound transformation during the Revolutionary period. Our discussion therefore turns first to a detailed examination of the nature of this double loyalty to science and the state and, in a more encompassing sense, to an international community of learning and to French culture.

Although the Academy of Sciences held its first official meeting on 22 December 1666, there is a complex prehistory of scientific organization to be found in the story of the numerous private gatherings held in and around Paris from the early decades of the century.[4] Whatever their idiosyncrasies, these voluntary associations were initiated to satisfy the increasing curiosity about nature's secrets evident in many urban centers of Western Europe in the seventeenth century. As private organizations, they had no obligations to the Crown. Their existence depended principally upon the needs of the general scientific movement. For our purposes, the history of the Royal Academy must begin with the transformation of these amateur circles into a legally recognized learned society attached to the state. As will become evident, that transformation involved more than the creation of a new organ of government. It was no less than the means by which the French virtuosi were metamorphosed into professional scientists.

Three events, spanning almost half a century, punctuate this profound change. The first was the aforementioned inaugural meeting, held in the private library of Louis XIV on rue Vivienne.[5] It was celebrated not only by the opening of a register of minutes of the meeting, but also by a medal struck in honor of the occasion. This medal, symbolically bearing on one side the handsome effigy of the King, and on the other an enthroned Minerva surrounded by symbols of astronomy, anat-

4. H. Brown, *Scientific Organizations*, ch. I–VI, VIII, XI; Taton, *Origines*, pp. 12–29; Gauja, pp. 1–46.
5. *HARS* (1666–1699), I, 14–16; Maindron, *Académie*, pp. 1–7; Schiller, *RHS*, XVII, 100–104; Gauja, pp. 46–51. The illustration on the title page and figure 1 are contemporary drawings of the King's Library.

omy, and chemistry, was the first public documentation of the founding of the Academy. The second episode came in 1699 when the Academy was "renovated"—to follow Fontenelle's terminology. In that year, the Academy was presented with the first set of regulations spelling out its duties, its relations to the Crown, the composition of its membership, and the rules for the conduct of its business. A second medal was struck a few months later when the Academy moved into more spacious quarters in the king's "apartments" at the Louvre. It was there, with considerable pomp duly reported in the press, that the assembly's first public meeting was held on 29 April 1699.[6] Finally, in 1713 the Crown issued letters-patent, properly registered by the Paris Parlement. This ordinance "confirmed the establishment" of the Academy, provided it with "a stable and solid form," and insured that its existence was once and for all recorded "firmly and stably."[7] At this point, the extended process of the creation of the Academy as a royal institution came to an end.

By scrutinizing the circumstances surrounding these events and analyzing official documents produced on these occasions, one can obtain a clear notion of the Academy's legal functions. Yet such an analysis, as commonly given by historians, provides a misleading and incomplete picture of the development of the institution's actual operations, for these public pronouncements were merely the expression of academic practices already in use. They had been elaborated by the dynamic interplay of factors that originally gave birth to the Academy. Therefore, in order to shed light upon the structure and functions of the institution, these events must be taken principally as mere indicators of the more fundamental forces that shaped the Academy.

Prior to the establishment of the Royal Academy, the amateurs of the New Philosophy did not form a well-defined community. Like their literary brethren early in the century, they had no fixed organization, but moved rather freely from one self-constituted group to another. Each private circle had its own character and reputation. Some, like Le Pailleur's hu-

6. *HARS* (1699), pp. 1–16.
7. Aucoc, *Lois*, pp. lx–lxi.

manistic Academy, which encompassed poetry, music, and mathematics, were informal and relaxed in atmosphere.[8] Another, Bourdelot's Academy, was built around weekly "conversations," very much on the model of Renaudot's defunct Bureau d'Adresse. Rohault's Cartesian Wednesdays, attended by large groups, including would-be *femmes savantes*, were filled with demonstrations of a propagandistic nature.[9] Thévenot's Academy, on the other hand, was more intimate and devoted its efforts to experiments rather than rhetoric. Meetings at Montmor's Academy, taken as occasions to set competing philosophies in direct opposition, often degenerated into sophistic wrangling.[10] Whatever its peculiar flavor, each circle was eager to play host to famous foreigners like Steno, Huygens, or Wren, and to welcome news of sister circles abroad. They not only followed the exploits of the Accademia del Cimento in Florence and the Royal Society in London, but were deeply impressed by what could be accomplished through royal support. There can be no doubt that the Tuscan and English examples played a role in bringing about the foundation of the Académie Royale des Sciences.

The foreign examples were important in supplying models often imitated by the Parisians, or simply in stimulating royal action by playing on dynastic jealousies.[11] But the force impelling them to seek new institutional forms came from problems indigenous to the French scene. Within each group, a persisting set of weaknesses forced consideration of alternate forms of association. Most annoying, although probably not most serious, was the unpredictable and often ephemeral nature of the voluntary associations. Le Pailleur's Academy expired in 1654 with the patron's unexpected death. The existence of Bourdelot's Academy was interrupted by its founder's hasty departure to Queen Christina's court in 1652. Thévenot was forced to close his doors for lack of funds for experimentation. Being attached to mortal persons rather than the eternal Crown, the private societies suffered along with their patrons' fortunes. It was the price that had to be paid for the spontaneous and unsystematic

8. Mesnard, *RHS*, XVI, 6–10.
9. Mouy, pp. 108–113.
10. Bigourdan, *Premières Sociétés*, pp. 14–19.
11. George, *AS*, III, 390; Auzout as cited in Wolf, pp. 3–4.

character of these seventeenth-century circles. Indeed, the lack of rigid structure was something of an advantage for groups displaying such a variety of interests and functions.

In the last analysis, however, it was less the daily vicissitudes of having a patron than inadequate finances that impeded scientific development in the seventeenth century. The financial problem became the central issue in the soliciting of governmental assistance by the more dedicated members of the learned circles. On one level, their needs stemmed from the cost of constructing improved instruments (especially the expensive ones for astronomical observations) and of purchasing raw materials to carry on chemical and biological experiments. Without substantial sums, it was also impossible to initiate large-scale enterprises such as scientific expeditions. But behind these specific monetary demands lay a conscious strategy for the advancement of science, a policy which was the work of no more than a handful of professionally minded scientists who deliberately set out to transform the relatively unproductive nature of amateur scientific circles.[12]

Their program was twofold. It called for the deliberate abandonment of verbal dispute in favor of visual demonstration, and for the creation of salaried scientific professionals who would devote their full time to the enterprise.[13] Such demands had already manifested themselves in Montmor's Academy in 1663, when the secretary, Samuel Sorbière, gave a devastatingly candid criticism of his circle's practices. He spoke sarcastically of those who attended the meetings "to kill time and acquire a reputation" rather than to contribute to the advancement of knowledge. He referred bitterly to disputes that were even worse than the wrangling of scholastics. Experimentation was infinitely more bearable and desirable, yet Sorbière realistically pointed out that

12. These scientists, referred to by Taton as "Montmoriens dissidents," include Adrien Auzout, Pierre Petit, Christiaan Huygens, and Melchisédech Thévenot, most of whom became academicians. Despite his delayed entrance into the Academy in 1685, Thévenot is known to have been invited to join the group in 1666. See McKeon, *RHS*, XVIII, 2.

13. In 1733, Fontenelle had said, "Le règne des mots et des termes est passé; on veut des choses": *HARS* (1666–1699), I, 2. Ricken, pp. 96–117, discusses the tradition behind this and similar phrases in France, which suggests an indigenous rather than a Baconian origin.

to build an arsenal of machines to carry out all sorts of experiments is impossible. . . . Think of the space needed for observation of the stars, and of the size of the apparatus necessary for a forty-foot telescope. . . . Was not Tycho Brahe forced to build his Uraniborg, a castle not so much for lodgings as for the making of celestial observations?

Truly, gentlemen, only kings and wealthy sovereigns or a few wise and rich republics can undertake to erect a physical academy where there would be constant experimentation. A special structure must be built to order; a number of artisans must be hired; and considerable funds are necessary for other expenses.[14]

Another patron of science, Thévenot, tried to create a more professional atmosphere free from verbal bickering by providing his group with ample means for experimentation. To that end, he even hired a chemical demonstrator and limited attendance to serious scientists. But, within a few years, when the expenses proved too high, he was forced to give up his courageous plan.

The emerging spirit of this new professional scientific community is best embodied in the plans for a Compagnie des Sciences et des Arts, probably elaborated in 1664 or early in 1665 by Thévenot and the astronomers Auzout and Petit.[15] Here, spelled out in twenty-six concise paragraphs, are the demands of the growing scientific community: laboratories, observatories, systematic data-gathering expeditions, permanent translators, foreign correspondents, etc. Here too is one of the earliest attempts to prescribe and limit the function of academic officers and to establish a system of subcommittees of experts. Pervading all these proposals was an implicit definition of professional activities and standards which would of necessity exclude the dilettante amateur.

It was not surprising that the scientists turned to the Crown and its chief minister, Colbert, for guidance in this matter. Throughout his life, Colbert had displayed a keen interest in scientific research, even when its dividends could not be directly measured in utilitarian terms. His admiration for scientific enterprise sprang from a profound faith in the rational and precise habits of mind that were epitomized by and learned

14. Bigourdan, *Premières Sociétés*, p. 18.
15. Huygens, IV, 325–329, XXII, 620 n. 74.

from the study of science itself. In his work as an administrator, he displayed a love for detail, accuracy, and method characteristic of both the new bureaucratic mentality and the pursuit of science.[16] The promotion of science was therefore a natural complement to his own personal predilections.

The minister welcomed the individual advances made by Sorbière, Auzout, and the poet and scientific amateur Chapelain on behalf of the new science. No one, least of all Colbert, failed to note how beneficial this enterprise might prove for the state and for the glory of its ambitious sovereign. Neither the practical benefits of an advisory council of scientific experts as outlined in the projected Compagnie des Sciences et des Arts, nor the utilitarian promises their research held out for navigation, warfare, and architecture escaped Colbert's calculations. Even more crucial for him and for Louis XIV was the opportunity to fulfill the dream of centralizing all the cultural activities of the realm around the monarch.[17] Science, as much as literature and the arts, was meant to bring brilliance to the Crown as well as to bask in the dazzling glory of the Roi Soleil.

Moreover, the creation of an Academy of Sciences for the glory and the utility of the state fitted in well with many other of the Crown's aspirations. Mercantilist policies would find support in careful scientific inquiries made in the technological sphere. Technical education could be encouraged to promote trade and productivity. In addition, the existence of a new Royal Academy could continue the process of cultural regimentation and maximize the loyalty of its most creative subjects by making their livelihood dependent upon the State.[18] Colbert was all the more anxious to show his ability in this domain in order to surpass the achievements of his recently deposed ministerial rival, Foucquet. Colbert wanted to be remembered as the *Mecenas des gens de lettres*, and to that end he had already initiated a talent search for men of letters, artists, and scientists who would welcome government subsidy.[19]

16. King, pp. 100–115.
17. Clément, V, xxx–c; George, *AS*, III, 372, 387, 398.
18. Barrière, *Vie Intellectuelle*, p. 127, and Condorcet's unpublished memoir, "Sur les Académies," in APS, MS. 506 C 75, p. lv.
19. Charles Perrault used this phrase in his poem "Responce a un Poeme de Monsieur Quinault, où Apollon se plaint que le Mecenas des gens de Lettres refuse d'estre loüé," in *Recueil*, pp. 311, 316.

The emerging pattern of governmental paternalism, itself a curious mixture of self-interest and self-glorification so characteristic of Louis XIV's absolutist government, made the marriage between the aspiring professional scientists and the state all the more propitious. Both parties were anxious to enter into a partnership that would satisfy their particular needs. Scientists, by accepting government pensions, tacitly indicated their willingness to devote their efforts to the "national good" and to forgo the freedom and pleasures of dilettantism. In return, the state was willing to commit itself to a considerable outlay of funds to staff the Academy and provide it with necessary equipment, if it could reap the benefits of scientific inquiry.

This promising partnership based on mutual self-interest was not realized without difficulty. Negotiations took almost three years, from Sorbière's original demands in 1663 until the Academy's register of minutes was opened. In the interim, a sense of insecurity to which the scientists were unaccustomed prevailed. Contradictory rumors about the king's plans circulated among contemporaries who, as mere citizens, were not privy to the monarch's decisions. Every sign was grasped as an indication of his intentions. Thus Huygens' appointment by the Crown was generally regarded as the most favorable portent for the success of the Academy's creation. But even Huygens was kept in the dark about the plans for the new group he expected to head. When he was awarded a yearly pension of 6,000 livres and brought to Paris in April 1666, his mandate to form an Academy was by no means clear.[20] Now that royal munificence was involved, the knot of scientific activists began to realize how much its future depended upon the private decisions of the king and his councilors. Failing royal approval, there would be no alternative but to revert to dated practices, a clear admission of failure. Thus, to obtain his desired ends, the new scientist knew he would be forced to abdicate considerable freedom of action by casting his lot with the king.

The three years' delay was not in fact a product of the ruler's desire to make his power more manifest by shrouding

---

20. Huygens, VI, 23. There were earlier intimations made to Huygens by Carcavi regarding "quelques autres plus grands desseins." Huygens, VI, 16, and V, 375, 472, 475.

his intentions. It was a consequence of divergent opinions over the most sensible manner in which to constitute this new royal institution. The principle of a royal academy was agreed upon very soon after Sorbière's demands. Only the interrelated questions of its functional role and personnel remained, but these were fraught with difficulties. In a certain sense, they were not formally resolved until the renovation of 1699, and then as a result of the practice of academic life rather than on the basis of royal policy.

The prospective nature of the learned society's organization was caught between two divergent ideals, both of which held some attraction for the Academy's champions. On the one hand, there was the proposal for a Compagnie des Sciences et des Arts, calculated to appeal to the utilitarian concerns of the monarchy. On the other, there was Charles Perrault's schema for a General Academy, which played upon the ruler's paternalistic dream of controlling French culture. Both plans explicitly included the pursuit of science as a major activity, but toward different ends. For this reason, each implied a different criterion for selection to membership. The choice of either plan would result in only the partial satisfaction of the goals of both the professional scientist and the government in their new partnership.

Behind the plan for the Company was a Baconian desire to study nature and its products for the sake of aiding man to control his environment. This utilitarian bent called for a definition of membership which corresponded better to the idea of a scientist as a specialist than as a natural philosopher:

The Company will be composed of the most learned persons available in all the true sciences, such as geometry, mechanics, optics, astronomy, geography, etc., in physics, medicine, chemistry, anatomy, etc., or in the practice of the arts such as architecture, fortification, sculpture, painting, drawing, the channelling and raising of water, metallurgy, agriculture, navigation, etc. . . .[21]

Professional scientists concerned with the practical arts were to form the core of the new institution. In addition, inventors, travelers, and linguists would also be admitted into the Com-

21. This and the next quotation are from Huygens, IV, 328.

pany, provided they were especially useful to its particular
ends. The group was in one sense very broad, allowing for
members of different occupations and social classes. In another
sense, it was rigidly narrow, excluding more traditional humanistic pursuits and public affairs:

In the meetings, there will never be a discussion of the mysteries
of religion or the affairs of state; and if there is at times talk of
metaphysics, morals, history, or grammar, it will only be in passing
and in relation to physics or to exchanges among men.

Such explicit exclusion of subject matter was attractive to the
government. It gave promise of providing the state with highly
competent technical advisors who would not meddle in political
affairs. For the professional scientist, the plan also met the needs
for a narrower and more specialized institution to deal with
the advancement of science than was possible in the amateur
groups. Yet this solution was narrowly technocratic in a way
that did not square with the grandiose plans of the Roi Soleil.

Charles Perrault, later famous as the author of fairy tales,
put forward a scheme of much loftier vision.[22] In 1666, he proposed a General Academy which, while encompassing all subjects of erudition, excluded the less noble practical arts that
were central to the alternate plans for a Company. In Perrault's
Academy, there were to be sections on belles-lettres, history,
philosophy, and mathematics; and each member was expected
to be versed in a wide variety of subjects even though he was
assigned to one discipline. The plans for a General Academy
called for liberally educated members of society who possessed
a breadth of appreciation rather than narrow technical competence. How seriously this project was considered is evidenced
by the fact that the plans were detailed enough to specify where
and when the sections of this General Academy would meet.[23]
It is most likely that it had the support of Carcavi, one of Colbert's most trusted advisors on scientific matters, who was
named head of the King's Library, where this General Academy was to have assembled.

22. Clément, V, 512–513.
23. *HARS* (1666–1699), I, 6.

For Colbert, this conception was both attractive and time-ly. It gave bureaucratic expression to the humanistic dream of the Crown as patron and regulator of the "Empire of Letters." Indeed, Colbert had already begun to map out an ambitious program to enlist the most talented people of France into the service of their monarch. One by one, all the cultural assets of the nation were to be bought out with favors from the state, and thus securely brought under its protective wing. For some years already, individual grants had been doled out to selected men of letters, and even to foreigners who might thus be at-tracted to France.[24] By virtue of his appointment as Comp-troller of Royal Buildings in 1662, Colbert had the opportunity to become a "Minister of Culture," a role he quietly assumed on every occasion. After Chancellor Séguier's death in 1672, Col-bert had arranged to have the Académie Française ask for royal "protection," and he granted it the right to meet in the King's Library. At another opportune moment, he rescued the Acad-emy of Painting and Sculpture from certain death, and its membership was thereby deeply indebted to him. Moreover, to advise him on general matters of erudition, he had created an inner council of learned men in 1662, including the very instiga-tor of the idea of a General Academy, Charles Perrault.[25] The possibility of founding a major cultural center was therefore both real and tantalizing.

Both these structural conceptions—one primarily utilitarian and the other more broadly cultural—had a strong appeal for Colbert. The glorification of the monarch was as important for him as the execution of his mercantilist policies, and he would undoubtedly have preferred a solution that satisfied both ends. Circumstances in part solved the dilemma for him, for Perrault's grandiose scheme proved unworkable. To organize the entire structure of an Empire of Letters on the rational basis suggested by a General Academy presupposed that Colbert had an entire-ly free hand in the cultural life of France. In practice, he was limited, not so much by the king's personal whims as by already established cultural groups which developed an independent

24. Collas, pp. 389–483.
25. Charles Perrault, *Mémoires*, pp. 34–42.

esprit de corps and resisted all attempts to have their prerogatives diminished.[26] The plan for a General Academy, reminiscent of the Museum of Alexandria, was abandoned, and not
until the holocaust of the French Revolution would it be
feasible once more to attempt a complete restructuring of the
intellectual life of France.

Colbert was thus forced to relinquish any dream he might
have cherished of becoming Minister of Culture through the
creation of a General Academy. But he did not completely
adopt the alternative utilitarian plan that remained. The Academy of Sciences, when it was eventually set up, was clearly
more than a consultative assembly designed to answer the
Crown's queries on technological problems. It was also conceived as an engine for the glorification of the Roi Soleil
through the advancement of science, hopefully outshining
Leopold's Accademia del Cimento and Charles's Royal Society.
In the noble design of the Paris Observatory, in the ample financial provisions made for experiments, and above all in the very
choice of membership, Colbert showed that he had not totally
shelved Perrault's cultural conception of the Academy.

All fifteen original members of the Academy of Sciences
could easily have fitted in Perrault's General Academy, had it
been founded. Among them were men of considerable cultural
attainments and erudition, such as Pierre Carcavi and Christiaan
Huygens, the polymath Marin Cureau de La Chambre, the
Oratorian theologian Jean-Baptiste Du Hamel, the magistrate
Bernard Frénicle de Bessy, and Charles Perrault's equally famous brother, the architect and physician Claude Perrault.
Though many were intimately concerned with the practical
applications of scientific studies, none was selected for his
technical prowess alone. Indeed, one of the academicians, the
mathematician Gilles Personne de Roberval, was well known
for his incompetence in mechanical matters! A tacit understanding almost seems to have existed that a sound liberal education rather than apprenticeship in a trade was the proper
qualification for admission. Hence, from the outset, there was
a certain social and intellectual bond in the Royal Academy

26. *HARS* (1666–1699), I, 6–7; Hirschfield, pp. 18–20.

which ran counter to the hopes of the framers of the Company. Artisans were clearly excluded.[27]

Such exclusiveness was not based upon a sense of snobbery. For Colbert and his advisors, the determining factor for eligibility was competence in scientific matters. They took to heart the recommendations of Chapelain to appoint "those who are proficient in experimentation, or those who can extract all useful matters from it, who have the clarity of mind to gather it, and finally those who possess the different talents which could make the Royal Academy as sound as it is useful."[28] But since such talents generally presupposed a good education, men of lower social rank for whom the possibilities of formal schooling were almost nonexistent were ruled out. If the Academy seemed aristocratic, that resulted not so much from class consciousness as from the fact that avenues of learning were open principally to the wealthy and the titled.

Certainly Colbert did not hesitate to exclude men of high social standing whose interest in science was superficial or tied to ends other than the advancement of knowledge. Chapelain warned him of men with fine social reputations, more concerned with cabals than with making new discoveries. Much to their chagrin, many of the amateurs who had adorned the private circles of Montmor and Bourdelot were consequently left out of the new professional Royal Academy.

Excluded also were Cartesians and Jesuits, not because of any doctrinal differences as one writer has suggested, but because they were considered partisans of rigid philosophic and religious creeds, teachers of the faith, or public demonstrators, rather than open-minded seekers after truth.[29] It was their alleged inability to shed their prejudices that kept them out. Chapelain, in a Baconian frame of mind, once more put his finger on the matter when he wrote that "the principal object [of the Academy] . . . is to banish all prejudices from science, basing everything on experiments, to find in them something certain, to dismiss all chimeras and to open an easy path to truth

27. Bertrand, pp. 74–75; Chapin, *French Historical Studies*, V, 389.
28. Clément, V, 513–514. Collas, pp. 383–388, provides plausible evidence that this unidentified text was in fact authored by Chapelain.
29. Hirschfield, pp. 34–37.

for those who will continue this practice."[30] Considering the difficulties inherent in the creation of such an organization, the pressures to which Colbert must have been subjected from various interest groups and individuals, and the somewhat divergent aims Colbert accepted in fashioning the Academy, it is remarkable that the process of creation took less than four years. In that short period, in addition to performing other and more demanding tasks concerning the rest of the kingdom, Colbert managed to forge a new scientific organization built on the convergence of interests of the new science and the absolutist state.

By the end of 1666, some of the most essential elements of the new Royal Academy had emerged. It was to be composed of two sections: a mathematical one including all the "exact sciences"; and a physical one concerned with the more "experimental sciences," such as physics, chemistry, anatomy, and botany. Though Wednesdays were reserved for the discussion of mathematical questions and Saturdays for physical problems, it was decided that all members would attend and participate in each session. The meetings were to be held in the King's Library, at least until proper laboratory facilities were completed. Sessions were closed to the public; although a register of minutes was kept, it remained the private property of the Academy until extracts were printed. A few aspiring "students," each attached to a pensioned academician, were also selected to attend meetings, but they were present only to listen, watch, and learn, rather than to participate. Presumably they were admonished to keep silent about the proceedings and promised elevation to the rank of regular membership with a stipend when they "matured."

Thus, in the Academy's first years, an air of secrecy prevailed. Outsiders curious about its doings—especially Henry Oldenburg, the secretary of the rival Royal Society of London —were anxious for detailed news of this promising enterprise. Henri Justel, the French Protestant virtuoso who was on the fringe of the scientific community, could write two years after the Academy's foundation: "Our Society is still meeting, but it has produced nothing as yet, which makes people talk who

30. Clément, V, 514.

imagine that great discoveries are made while sleeping and without thinking."[31]

While the scientific world waited, the Academy was making significant progress. Its minutes, which have yet to be examined systematically by historians of science, indicate constant activity and a conscious effort to attack the study of nature methodically. At the outset, each academician was asked every year to draw up lists of problems which were worthy of the society's time and which could also be treated experimentally and in common. These lists were to serve as long-range guides to the group's activities, and for a few years they were followed, with only relatively insignificant interruptions. Occasionally the Academy was consulted on some new substance, project, or invention, or the making of a new map of the kingdom; or it was asked to experiment on the strength of gunpowder or the range of artillery pieces. The minister, while profiting from the expert knowledge of his salaried employees, always kept their higher purpose in mind, so that, while directing the Academy toward problems of potential military or industrial benefit to the nation, he also permitted the academicians to examine the same problems from a generalized viewpoint. He judiciously provided them with a balanced fare of both "pure" and "applied" science.

While Colbert lived, the Academy flourished. Its membership was increased by a new, talented group of professional scientists. Joining Huygens from abroad were the astronomers Giovanni Domenico Cassini and Ole Rømer, enticed to France by offers of sizable pensions. From within the nation, Edme Mariotte, Nicolas-François Blondel, and Philippe de La Hire swelled the ranks of mathematicians, while Denis Dodart, Joseph-Guichard Du Verney, and Jean Marchant joined the society as naturalists. The choice of this additional personnel confirmed the policies adopted by Colbert when he made his original appointments to the Academy.

During his administration, there were also significant developments in publishing practices. Partial announcements of the Academy's deliberations were given in the newly founded *Journal des Savants*, and starting with a text by Huygens pub-

31. Oldenburg, V, 207.

lished in 1667 the Academy began to share its findings with the public. By 1676 there were enough separately printed reports to collect them in a *Recueil de Plusieurs Traitez de Mathématique de l'Académie.*[32] Moreover, under the society's sponsorship, scientific expeditions were outfitted and sent out to Cayenne, to the island of Hveen, and to the Canary Islands.[33] An agreement with Jesuit Fathers traveling to the Far East was also made to provide the learned society with more data of potential scientific interest.[34] Colbert also encouraged the Academy by subsidizing the use of the Jardin du Roi as a natural-science laboratory for its members.[35]

But the most impressive and tangible evidence of the Academy's importance was the building of an observatory. This had been one of the specific requests made by Auzout in his plea for the creation of the Academy. The structure was not conceived simply as the center of the Academy's astronomical pursuits, but as the building that would house all its activities. In the original plan, the observatory building was to be the site of meetings; it was also to have chemical laboratories, a room for displaying models of machines, and space to contain all the specimens of natural history collected by or for the society. In short, what later became solely the Observatory had been destined to be the very site of the Academy.[36]

The plans for its construction matched the Crown's grandiose expectations for its function. Selected for the architectural design was the academician Claude Perrault, who had already conceived plans for the famous colonnade of the Louvre. A parcel of land south of Paris and dominating the countryside was purchased in 1667 and on it was built the imposing structure which still stands with only minor changes. The building was an impressive measure of the importance Colbert attached

32. Plantefol, pp. 64–70.
33. McKeon, *RHS*, XVIII, 5 n. 23; Olmsted, *Isis*, and *Proceedings APS*; Chapin, *French Historical Studies*, V, 377–378.
34. *HARS* (1666–1699), I, 415; Tachard; Fournier, pp. 41–42. In 1692, the Academy published the *Observations Physiques et Mathématiques . . . Envoyées des Indes et de la Chine . . . par les Pères Jésuites.*
35. Schiller, *RHS*, XVII, 111–112.
36. Charles Perrault, *Mémoires*, pp. 219–221; Wolf, pp. 1–112. For various reasons the building of the Observatory never fulfilled this dream. See Cassini, p. 184, and Chapin, "Astronomy," pp. 15–16.

Figure 1—Meeting of the Academy in the Bibliothèque du Roi. Engraving published in 1686, taken from Tachard, p. 1, courtesy of the University of California Los Angeles Library.

R. Sewtjus.

Figure 2–Louis XIV and Colbert on an imaginary visit to the Academy. Engraving by Sébastien Leclerc published in 1671 as the frontispiece to Claude Perrault, *Mémoires*, courtesy of the University of California Berkeley Library.

to his venture into science. It was also the visible evidence of a major cultural achievement in the Age of Louis XIV. Accordingly, another of the famous medals which record the important events of the reign of the Roi Soleil was struck, and a number of royal visits were arranged. In 1677, the Dauphin made an official tour of the Observatory, and in 1682 the King himself made his visit.[37] The building became the effective symbol of the Crown's persistent efforts on behalf of science. Hence its completion in 1672 was taken by many as evidence that the Academy was firmly established. It was indeed a major commitment, and gave academicians a much needed sense of confidence. But its completion was only one step in the long process of fashioning the habits of scientific behavior that came to be known as the academic tradition.

It is customary to speak of a gradual decline in the Academy's activities after 1683, and then to extol the renovation of 1699 as a turning point in the society's history.[38] According to this view, Louvois's succeeding Colbert as the Academy's effective director and his subsequent demands that it pay greater attention to questions of practical importance signaled a reversal for the flourishing institution. Continuing this argument, only when Louvois was in turn replaced by Pontchartrain in 1691, and when the effective leadership of the Academy passed into the hands of his nephew, the Abbé Bignon, was the Academy supposed to have recovered its former stature and, with a new set of regulations, to have become the guiding spirit behind the science of the Enlightenment. Colbert and Bignon are usually represented as farsighted statesmen, while Louvois by contrast is portrayed as a myopic administrator bent upon seeing immediate by-products of the royal institution's efforts. The 1699 regulations are also seen as marking the abandonment of the Baconian ideal of collective activity and anonymous publication in favor of recognizing of individual contributions, thereafter

37. *HARS* (1666-1699), I, 241, 348; Wolf, pp. 116-119. This visit by Louis XIV followed shortly after one he had made in December 1681 to his Library, where the Academy held its meetings. The King's visit of the Academy pictured in figure 2 is a fictitious one set in the Jardin du Roi with the partially completed Observatory in the background.
38. Maury, *Académie des Sciences*, pp. 37-41; Bertrand, pp. 39-41.

published by the Academy under the scientist's name. This reversal of academic policy is usually offered as another significant factor in the learned society's revival during the early years of the eighteenth century.

This standard version, which owes much to Bignon's friend Fontenelle, is more a caricature than an actual description of the Academy's evolution during the last three decades of the seventeenth century. Some fragmentary evidence supports the notion of a decline in activity in the late 1690's, although in fact no acceptable criteria for measuring scientific activity have ever been devised.[39] But on the surface of things, it is quite evident that the parade of ministers and the legal alteration of academic practices in 1699 were not fundamentally responsible for the changing fortunes of the institution. When the definitive history of this period is written, several other and more fundamental factors will have to be taken into account, not the least of which was the changing tone of French political and cultural life, which expressed itself in the revocation of the Edict of Nantes and renewed military adventures. During the 1680's and 1690's, the activities of the royal society were significantly affected by the departure from Paris of Protestant foreigners such as Huygens, Rømer, Leibniz, and Hartsoeker, all of whom had participated in the life of the Academy. Another factor in the apparent decline in scientific activity was the wars, which attenuated communication with foreign centers and temporarily left Paris in a state of sterile isolation.

In the long run, however, these vicissitudes were not as significant as the changing patterns of academic practices within the Academy itself. During the decades preceding the 1699 renovation, the Academy was gradually fashioning a productive working relationship with its parents, the Crown and science. As the standard version suggests, it is true that the state made increasing demands upon the Academy and that collective publications were discontinued. Both of these changes, which were legitimized (rather than initiated) by articles in the

39. Bertrand, pp. 43–44; H. Brown, *Scientific Organizations*, p. 260; Hirschfield, p. 64 n. 12; and a letter of Etienne-François Geoffroy to Hans Sloane, dated 18 April 1709, cited in Jacquot, *Le Naturaliste Sloane*, p. 17.

1699 regulations, are best envisaged as the response of a still malleable institution to the double set of demands placed upon it by the scientific community and the nation. It was just that response that gradually transformed Colbert's semiprivate band of scientific professionals into France's acknowledged arbiter of scientific and technological activity.

From the outset, Colbert and the academicians were perfectly aware of the role the society was to assume as consultant to the Crown. At first, the Academy seems to have been directed toward long-range projects, such as helping to solve the problem of determining longitude at sea, the mapping of France's territory, the establishment of hydraulic theory useful for fountain construction, and the composition of treatises of mechanics relevant to military uses.[40] Gradually, the Academy was also called upon to provide answers of a more particular nature, and to make pronouncements upon the merits of specific technical proposals offered to Louis XIV. The Academy's printed *Mémoires*, for example, refer to the examination of several projects to make saltwater potable, of a metal mirror to focus light rays for the production of high temperatures, and of a series of machines invented to perform human tasks mechanically.[41]

In the early years of the Academy's existence, whenever the matter seemed sufficiently important, intricate arrangements were made to assess individual projects. In 1668 there was the case of André Reusner, who claimed to have solved the longitude problem, and for whom a special committee of the Academy was set up, meeting with Colbert in his personal library. Elaborate procedures were followed, including the presence of an admiral of the French Navy at the hearing and the keeping of careful records of the interrogation. The academic members of the commission raised a variety of objections to Reusner's scheme, and he was eventually discredited.[42] But as the scientists became more accustomed to their role as critics and judges, they streamlined procedures so that, by the early

---

40. The habit, referred to earlier, of annually setting down the list of scientific projects to which they would devote themselves was still practiced as late as February 1699. See AdS, Reg 28 Feb. 1699, fol. 132v–147r.
41. *HARS* (1666–1699), I, 50, 144, 200, 320, 428; Plantefol, pp. 71–74.
42. Huygens, XXII, 218–226.

eighteenth century, it became customary to deal with each case by constituting an ad hoc committee. Instead of keeping elaborate minutes on which to base decisions, the committee would prepare in writing a carefully considered discussion, followed by a verdict suggested for ratification by the entire Academy. Under normal circumstances, only this committee report was meticulously composed and preserved. The committee system was functioning as early as 1678 and quickly became the standard operating practice for carrying on much of the society's business.[43]

The largest number of specific projects examined by the Academy involved inventions and machines. Colbert and Louvois both were beset by petitioners demanding rewards for supposedly new and useful devices, ranging from pumps to machines for cleaning harbors and for milling grain.[44] An accurate count of the number of these projects or actual machines —or even of the requests for recompense—is impossible to obtain from extant ministerial records, since often many would be sent directly to the Academy without first passing through government channels. As knowledge of the Academy's consultative function was spread among inventors, the Crown was frequently by-passed until after the Academy had given a favorable report, when its conclusions could be used to secure a reward from the government. There were, of course, a large number of projects that never got past the Academy.

As the Academy became accustomed to its official function, its committees evolved criteria of judgment, which generally centered on an assessment of the project's novelty and its potential utility. At first the language used to state the Academy's conclusions varied, but from at least 1685 on it included the word "approuvé," as if the learned society possessed some kind of veto power.[45] Even before a legal meaning was given to the phrase "approuvé par l'Académie," it became a standard feature in all favorable reports. This gradual evolution of consultative practices culminated in Article 31 of the 1699 regulations,

---

43. Huygens, XXII, 256.
44. *HARS* (1666–1699), I, 144, 200; *Machines*, I, 5–148; McCloy.
45. *HARS* (1666–1699), I, 448. There, the exact wording is "La Compagnie les a fort approuvées."

which fixed the Academy's advisory function and its legal prerogatives:

The Academy will examine, if the King so rules, all machines for which a privilege has been requested from His Majesty. It will certify whether or not they are new and useful, and the inventor whose work has been approved will be held responsible for leaving the Academy a model [of his invention].[46]

It was in this concise statute that the position of the institution as the officially sanctioned arbiter in technological matters was established. Though formulated to apply only to machines referred to the Academy by the Crown, it came to be applied to all technological projects brought directly to the learned society's attention, and later even to purely scientific matters. Gradually, the institution was being recognized as the arbiter for the scientific community as well as for the government.

The passage from judgment in technology to judgment in science is at first difficult to understand. Not only was the substantive relationship between science and technology quite tenuous in the seventeenth century, but the needs of the technician and natural philosopher appeared to be somewhat different. Whereas the former required the Academy's approval to secure financial gain, the latter could have little expectation of making capital out of his study of nature. Yet the role of the Academy vis-à-vis both groups was increasingly confused, both by academicians and outsiders.

For one thing, the technician found it was possible to reap the benefit of the learned society's judgment even without obtaining a privilege or stipend from the king. He could put the Academy's name to good use by advertising his product or invention as one guaranteed by a group of impartial government experts. Even his personal prestige was raised by being coupled with such a renowned and official group. The clever inventor hardly ever missed the opportunity to display the words *approuvé par l'Académie* to advertise himself and his product. The scientist's reaction was in many ways similar. He too cherished a favorable review of the discoveries he submitted to the Academy, especially when it led to their publication. The stamp

46. Aucoc, *Lois*, p. lxxxix.

of approval from such an authoritative group of professionals was taken as a significant mark of his standing in the scientific community as well as a guarantee of the value of his work. Functionally, the artisan and the savant came to expect similar benefits from submitting their accomplishments to the learned body.

Within the Academy itself, there was also a tendency to consider judgments on technological and scientific matters as a single academic activity. Operationally, the process was always the same. Mechanical projects and scientific papers alike were referred to the same kind of ad hoc committees for critical examination before the matter was sent to a regular meeting for final action. The committee-referral system and the usage of *approuvé* became so much a part of the society's normal practices that academicians made little distinction between the realms of science and technology. It was as if the content of the subject under consideration had been made irrelevant by the bureaucratic procedure and all that counted was the rendering of a considered collective judgment.

It is within the context of the emergent image of the Academy as the arbiter of science and technology that the gradual abandonment of anonymous publication by the Academy must be examined. Curious as it may at first seem, the decision to recognize the individuality of discoveries reinforced the tendency toward considering the Academy's principal collective function as that of supreme judge. It may even have been a contributing factor in this evolution. Practice indicated that communal efforts were more properly applicable to the evaluation of new ideas than to creation. In terms of the role of the learned society, this discovery was a revelation of fundamental significance. To see its implications, we must turn back for a moment to the ideals espoused by the early promoters of scientific societies, from which the ideology of communality was originally taken.

Following Bacon, these men advocated collective activity in science, not only for its cumulative effect upon the advancement of learning, but also because it gave promise of discovering truth with greater certainty. In their minds individual investigators were prone to misjudgment due to personal prejudice, whereas a cooperative enterprise gave greater guarantee of philosophical unbias.[47] The clash of opinions and common scrutiny provided additional assurance of the objectivity of the discovery. The Accademia del Cimento, for example, published its collective findings without citing individual contributors, thereby placing emphasis on the discoveries rather than the discoverers themselves. In reporting the eternal "truths of nature," would it not be foolish and vain to record the names of mere mortals who happened to have stumbled across a right answer?

Views of this sort abounded in the seventeenth century, not only in Bacon's widely read discussions in the *Advancement of Learning*, in the defense of the Royal Society's existence by Glanvill and Sprat, but in France as well.[48] In 1666, Huygens, for example, specifically referring to the program of the new Parisian Academy, urged that "the principal occupation of this Assembly and the most useful must be, in my opinion, to work in natural history somewhat in the manner suggested by Verulam [Bacon] . . ."[49] Both Du Hamel and Gallois spoke highly of the Lord Chancellor, whose experimental method they praised. Even the confirmed Cartesian, Claude Gadroys, who had little regard for the empiricism of the Baconian approach, expected that the collective work of the Academy would bring greater yields than the fruits of any single individual, no matter how gifted.[50]

Perhaps the most eloquent exposition of the virtue of academic cooperation was given by Claude Perrault in the introductory letter to a work dedicated to the Academy:

It is the advantage stemming from being in a Society, and especially a Society like yours, where all things are examined and discussed with such care, that everything emerging from it bears an irreproachable character. I do not know whether or not I am wrong, but I have always looked upon the works which are fashioned in your Company and which have passed before your eyes to be of a special kind. The books that we have were almost all written by

47. This was considered by Bacon as a part of the "idols of the cave." See R. F. Jones, p. 48, and Rossi, pp. 165–166.
48. Penrose, pp. 94–103; Luxembourg, pp. 28–31; Van Leeuwen, ch. I.
49. Huygens, VI, 95–96.
50. Preface to Gadroys.

individuals, and human frailty does not allow a single man to be free from error. Such is not the case for works of your Company, where all the facts brought forth are verified by a great many eyes, all clairvoyant, where arguments are discussed by a great many minds, all enlightened . . . It is on such works that one can, as on solid foundations, build the edifice of science without fear. Everywhere else there is peril.[51]

In its first decades of existence, the Academy had acted on the ideal of collective activity. Most of the publications of the society from 1667 to 1692 excluded or underplayed the individual authors of observations and discoveries for the benefit of the company as a whole. The most impressive of these were the large folio volumes of the *Mémoires pour Servir à l'Histoire Naturelle des Animaux* (1671), which, in its illustrations, portrayed the Academy working as a group.[52] Other volumes, like the *Recueil de Plusieurs Traitez de Mathématique de l'Académie* (1676) or the *Mémoires pour Servir à l'Histoire des Plantes* (1676), named the academicians principally responsible for their composition, but claimed that ownership of the ideas belonged to the Academy as a whole.[53]

Although in theory the Academy opted in favor of anonymous publication, the practice posed many difficulties. Some of the work undertaken for the learned society was not carried out on its premises at the King's Library, usually for lack of proper laboratory facilities. Since that work was not usually witnessed by the entire company, its accuracy could not easily be controlled by the group. The individual academician's skill as an experimenter and his trustworthiness had to be accepted in lieu of communal verification. In such instances, it was suitable for the work to appear under the individual's name. A case in point is the analysis of mineral waters by Samuel Cottereau Du Clos. The Academy agreed to print his results in a miscellany of other observations, in the fourth volume of a *Recueil de l'Académie*. Later this work was published as a separate vol-

51. Claude Perrault, *Recueil*. The same sentiments, though expressed with less eloquence, had been stated earlier by Perrault in the 1671 preface of his *Mémoires pour Servir à l'Histoire Naturelle des Animaux*.
52. Watson.
53. Bertrand, p. 45, asserts that individual names were, by common agreement, left out of the prefaces. In the *Description de Quelques Plantes Nouvelles*, however, they are specifically mentioned.

ume, bearing Du Clos's title as an academician, thus implicitly giving him the approval of the company.[54]

But what if an academician published a work of his own without first presenting it to the entire body for examination? Would not his membership in the Academy easily be construed as a sanction of its contents by the group as a whole? Herein lay a potential danger for the organization's reputation. Indications are that academic etiquette demanded the title "academician" be left out of any book written by a member of the society if the work had not been cleared by the entire company beforehand. Many of Du Hamel's works published after 1672 fall in this category. It is as if the appellation "academician" belonged to the corporate group rather than to the individual.

To preserve the purity of the company's reputation was not the only consideration, however. Seventeenth-century scientists were no less concerned with the establishment of priority than their modern counterparts, for their personal reputations were also at stake. Thus, after announcing a new discovery or idea in the Academy's closed meeting, an individual often sent his findings to the one periodical that could guarantee quick publication, the *Journal des Savants*. Under the rubric of "extracts from the register of meetings of the gentlemen in the assembly at the King's Library" the findings of individual academicians were presented to the world at large.[55] But in the face of such individual actions, how could communal integrity be preserved?

It was to resolve this difficult problem that on 18 August 1688, the Academy decided that any member wishing to publish matters discussed in the society's meetings under his own name would have to permit his manuscript to be scrutinized by the Academy as a body.[56] The purpose was not so much to censor the writings as to separate contributions made by the author from collaborative ones. In this way, everyone's interests

54. In other cases when there was a disagreement between experiments carried out privately and those done in the Academy's presence, the official publications remained silent. See Schiller, *Clio Medica*, I, 34.
55. Sergescu, *AIHS*, I, 63; Plantefol, pp. 69–70. Mlle Suzanne Delorme, of the Centre de Synthèse in Paris, showed me the results of her unpublished study of these extracts in 1965.
56. Bertrand, pp. 45–46.

were safeguarded. The individual scientist was able to obtain his due, and the Academy protected itself from plagiarism by one of its members. The new technique devised to implement this decision was to demand that once the manuscript had been examined by an ad hoc committee and purged of any communal property, the company would grant its stamp of approval. After 1688, books written by academicians generally carried a statement above the censor's approbation similar to the following:

Extracts of the registers of the Royal Academy of Sciences. "The Company, on the report of M. l'Abbé Gallois, who had been asked to examine a book by M. Rolle, of this Company, and bearing the title of *Traité d'Algèbre* . . . , has judged the work worthy of being printed."[57]

By following this procedure, the Academy was at once protecting itself and giving the public a certificate of authenticity.

What is significant in this intricate evolution of practices hammered out to safeguard everyone's interest is the way in which the society unwittingly redefined its function in the scientific community. In effect, it abandoned the group's original commitment to communal investigation in favor of a role as adjudicator of individually authored products. Thus the Academy found itself increasingly redirecting its communal efforts to deciding whether or not the scientific work was "worthy of publication," once more edging toward the position of supreme judge.

We see now that this change in academic practices antedates the 1699 regulation by at least a decade, and that it came about as the result of concrete problems rather than as an abrupt transformation of ideology. In this instance, just as in the case of the gradual evolution of the Academy's precise function as government consultant, the new habits adopted by the company were embodied in the 1699 regulations. Article 30 states:

The Academy will examine all works that academicians propose to have published; it will give its approval only after a complete read-

57. In the copy of Rolle that I examined, the censor's approval was on the last page, dated 16 July 1689. The Academy had approved the book three days earlier. Similar procedures were obviously followed for Varignon and Tournefort.

ing in the meetings, or at least only after an examination is made by those the Company has designated to prepare a report; and no academician shall use the title academician in his writings unless that work has been approved by the Academy.[58]

By the terms of this article, then, the current practice was sanctioned and legalized. Once more, the 1699 regulations marked the end of a trial period that resulted in codification of a set of academic traditions rather than establishment of new practices.

The direction of the new academic practices was clearly to reinforce the society's adjudicatory position. In still another way, the resolutions taken by the body in 1688 forced the Academy into the role of arbiter. The proponents of learned societies who had at first lauded communal activities made little distinction between discovery and verification. The experience of several decades of academic life revealed that, for most kinds of scientific research, the creative act of discovery was the product of a single mind, even when the individual was led to a problem by the prodding of his colleagues. Even observations were recognized as contributions of individuals, regardless of the fact that the instrument or expedition that made them possible was often the property of the Academy or the consequence of a group decision. The notion of communal creation had to be set aside when it was realized that the most important ingredient of discovery is personal.

All that remained for the group acting as a group was to request, collect, sift, and publish information. This was what the Academy had implicitly agreed to do in 1688 when it abandoned the practice of communal publication. It gave proper credit to the individual academician while retaining the right of review. As a body, the Academy might approve or disapprove; it could verify or falsify claims; it could repeat experiments, compare different observations, and even solicit new information—but it found it impossible to carry out collectively the essential steps of creation.[59] This realization, like so many others, was codified by Article 20 of the 1699 regulations, which reads:

58. Aucoc, *Lois,* p. lxxxix.
59. Brunet, *AIHS,* I, 36–39. Mariotte refers to the necessity of scientific collaboration for this very reason.

Experience having shown too many difficulties in works pursued in common by the Academy, each academician will choose a particular object of study, and by his accounts to the assembly, he will try to enrich all those composing the Academy through his wisdom [*lumières*], and to profit from their remarks.[60]

A great many other less important matters relating to the society's internal procedures were also incorporated into the Academy's 1699 regulations. Most of them sanctioned patterns which were already in operation and which buttressed the Academy's position as a tribunal of science and technology. The ease with which this stance was accepted is all the more understandable when set in the context of Old Regime society, the subject to be explored in the next two chapters. But before proceeding to examine this important historical dimension, it is first essential to indicate how the academicians as jurors learned to deal with one of the most characteristic features in the internal life of science, conflict.

For the original promoters of academies, the very form of the new institutions was sufficient to dissolve differences. These utopian spokesmen imagined professional scientific societies to be dedicated to an unostentatious pursuit of truth rather than the glory of public debates. It was the general expectation that the shedding of words for deeds, and the abandonment of authorities in favor of direct experimentation, would yield immediate and tangible results. Such progress, it was thought, was insured by collective action which would neutralize individual prejudice and bring about a consensus based upon unambiguous evidence from nature. Fontenelle, with his knack for turning a phrase, insisted that "physics holds the secret of shortening countless arguments that rhetoric makes infinite."[61]

In practice, the appeal to nature and to collective wisdom did not automatically bring about the expected agreement. The nature of both mankind and science conspired to frustrate these dreams. The annals of science of the late seventeenth century are just as full of controversies as they are today, even among professionals working with allegedly neutral observations and

60. Aucoc, *Lois*, p. lxxxvii.
61. Sprat, pp. 83–86, 104–115; Rossi, pp. 23, 166–177. The quotation appears in *Digression sur les Anciens et Modernes*, p. 164 of the Shackleton edition.

without personal prejudices. Although the early scientist was not able to explain satisfactorily why perception is necessarily subjective, he quickly became aware that agreement was not guaranteed by the academic structure. Faced with this realization, he might well have concluded that techniques for the enforcement of a consensus had to be developed, eventually resulting in the adoption of an academic dogma about the truth of a particular proposition. This might have pushed the Academy into becoming the official carrier of some scientific ideology, such as the popular Cartesian philosophy.

An official ideology did emerge. But it was not in the form of a "party line," explicitly following an Aristotelean, Cartesian, or Leibnizian world view.[62] Nor did the Academy find it necessary to be committed to any specific scientific theory, simply because it had once been found reasonable and then "approved."[63] Indeed, the founders of the Academy had explicitly rejected doctrinal dogmatism as contrary to the progress of science. Fontenelle ably summed up their sentiments, stating that "above all, in the Academy no system should dominate to the exclusion of others."[64] He added that the real spirit of the discovery of truth was dynamic in character, more akin to a process of unfolding in time than a static, immediate, and total discovery at any one time. The Academy reserved the right to change its mind in the face of new evidence.

Yet it is insufficient to characterize its position simply as openmindedness and flexibility. What might be termed academic or official science was expressed in social and philosophical terms rather than by decisions on substantive issues. There was no longer room for the dilettante, the virtuoso, the eccentric, the undisciplined mind. Ostentation, vanity, and pomposity were to be replaced by earnest unaffectedness.[65] Simplicity, hard work, and unadorned logic were to characterize the new academician. Charlatans were consequently to be dismissed

62. Papillon, pp. 689–720.
63. A good example is the Academy's flexible attitude toward preformation, discussed by Roger, *Sciences de la Vie*, pp. 163–453, in which the nature of debates between Perrault, Tauvry, Hartsoeker, and Mery is set out.
64. *HARS* (1666–1699), I, 16.
65. Roger, *Sciences de la Vie*, pp. 173–174, refers to the prohibition of the "débraillé intellectuel."

with merciless dispatch. Such was the policy adopted by the Academy in practice, concurring completely with the initial decision to exclude from its ranks all but those willing to become scientific professionals. In this, the Academy was establishing behavioral norms that actually defined the scientific community.

In philosophical terms, the Academy gradually moved toward a position of phenomenological positivism. This was almost inevitable, given the substantial number of controversies that had already ensued among academicians of good faith. The examples are legion. Mariotte, Frénicle, Buot, Claude Perrault, and Huygens could not agree upon the nature of gravitation. Du Verney and Perrault failed to interpret similar anatomical evidence on the function of the ear in the same fashion. Du Clos's explanation of the increase of weight in calcination and his treatment of the process of coagulation did not bring about any consensus either.[66] To decide by a majority vote which position was correct was absurd, for all the scientists firmly believed that truth was determined by nature and not by the will of men. In practice, the most that the company could assert as being true was that portion of evidence about which there was no disagreement. Generally it was the undigested observation of some phenomenon that was thereby considered as true "positively." For the rest, the greatest service the Academy could perform was to publish the opinion of each of its members, no matter how many contradictions might appear.

The better part of wisdom, if the Academy was to have a long and useful existence, was to resist the temptation to judge where judgment was premature. Fontenelle, who grasped the realities of the life of the Academy better than any of his contemporaries, became an eloquent spokesman for this philosophical position. While discussing academic disagreements over the circulation of sap in plants, he remarked that

The Academy was the natural judge between these two views. But since a large part of wisdom consists in making no judgment, it declared that the subject was not yet sufficiently clear. It is necessary to wait until there is a fairly large number of experiments and facts before extracting a generalization. There is a common

66. *HARS* (1666–1699), I, 22, 395; Huygens, XIX, 628–645.

tendency to rush to formulate general principles, and the mind runs toward systems. But we must not always trust in the [merit of] this ardor.[67]

Fontenelle's position has generally been pictured as a sort of latter-day Pyrrhonism, modeled somewhat after Montaigne's skepticism.[68] Behind it, however, lies the more fundamental belief that the edifice of science would not be built in a day. Progress depended upon the slow accumulation of unassailable facts. In this task, skepticism was not an end in itself, but only a tool necessary to create an absolutely secure foundation. In this sense, Fontenelle's views were more akin to those of Mersenne, Gassendi, and Descartes than to Montaigne's.[69] But unlike the Cartesians, Fontenelle coupled phenomenological positivism with the dream of a new and secure knowledge that would emerge only as time permitted it to unfold.

Fontenelle's ardor and commitment to scientific progress, so characteristic of the new breed of academicians, found its clearest expression in his discussion of the "renovation" of the Academy. There, he said,

We are forced to look upon present-day science, at least physics, as if it were in its cradle. Hence the Academy is only at the stage of gathering an ample store of well-founded observations and facts, which will one day become the basis for a System. For Systematic Physics must refrain from building its edifice until Experimental Physics is able to furnish it with the necessary materials.

Societies—and only societies protected by princes—can succeed in amassing these materials. Neither the wisdom [*lumières*], nor the care, nor the life, nor the faculties of any individual can suffice. There are too many experiments, too many different types of experiments...

Until now the Academy of Sciences has attacked nature only piecemeal. It has adopted no general system, for fear of falling into one of those hastily fashioned systems that the human mind is all too eager to adopt. . . . Today we ascertain one fact, tomorrow another, to which it bears no relation. We do not refrain from hazarding conjectures on causes—but they are only conjectures.[70]

67. *HARS* (1666–1699), I, 62.
68. Marsak, *Fontenelle*, p. 32.
69. Historically, Fontenelle's views are a development of the philosophy of "constructive skepticism" discussed by Popkin, ch. VII, and by Van Leeuwen, ch. I. On this score it is misleading to refer to its origins either in Montaigne or in Bacon.
70. *HARS* (1699), pp. xviii–xix.

Facts could be ascertained, approved, recorded, and accumulated. Logic and procedure might even be scrutinized, especially in the mathematical sciences. But no final judgment could yet be rendered on speculation or on interpretation of the evidence.

It is a remarkable sign of the Academy's sound judgment that it was able to recognize and distinguish fact from conjecture, to accept the former and sanction it, and to suspend corporate judgment on the latter. It is a sign of its wisdom and vitality that it felt sufficiently confident in the future of science and its own role in the progressive enterprise to vent members' personal disagreements about nature in public. The controversy over the interpretation of the blind spot, involving Mariotte, Pecquet, and Perrault, was prominently placed in the Academy's *Recueil* of 1676. Disagreements between Tauvry and Mery on circulation of the blood in the foetus and other minor controversies were also displayed publicly.[71] The Academy's deft handling of this matter, striking a compromise between dogmatism and dilettantism, clearly enhanced its stature in the scientific community. In forging a set of principles for action, the academicians had begun to establish a viable tradition for their new institution.

71. *HARS* (1666–1699), I, 102–103; *HARS* (1699), pp. 25–30; Durry, I, part I, pp. 217–218, for the blind-spot controversy. Chapter VIII of Hirschfield's dissertation is also devoted to this subject. For the Tauvry-Mery controversy, see *HARS* (1699), pp. 25–38, and Roger, *Sciences de la Vie*, pp. 171–173.

Chapter 2

A Republic of Science

THE ACADEMIC practices developed in the late seventeenth century and their codification in 1699 constitute only a small part of the reasons behind the Academy's resounding success in the eighteenth century. As is well known and has repeatedly been documented, the meetings of the Academy were the scene for the announcement or discussion of almost every important discovery of that century, and the society's membership list reads like a roster of all the significant French scientists of the era.[1] Election to the Academy quickly became the most coveted prize for the ambitious savant who, if he acquired the title *membre de l'Académie*, cherished it as proudly as if it were a title of nobility. To be an academician, or to have one's views discussed favorably by the Academy, was in fact to receive the supreme accolade of that loosely knit community of scientists known as the "Republic of Science."[2]

In order to appreciate more fully the reasons for the learned society's meteoric rise to fame, it is necessary to understand how this Republic came to assume such a prodigious importance in the Enlightenment, and why academies could provide the Republic with an ideal institutional framework. This will lead us to consider both the broader historical context in which

1. A list of members is given in Appendix II, along with the important biographical literature existing for each scientist.
2. The term is borrowed and adapted from Jouvenel. His article is based on Polanyi's views of the structure and nature of the community of scientists, given in *Science, Faith and Society*, pp. 42–62. Another valuable analysis is given by the sociologist Hagstrom.

academies of all sorts proliferated, and to examine those aspects of French culture that promoted the high standing quickly attained by this particular academy of sciences. Such a widening of our vision is all the more necessary in that it will suggest reasons for the Academy's eventual decline toward the turn of the eighteenth century.

Among the broader forces that must be considered in such an analysis, the most significant is that "cultural revolution" of Western civilization culminating in the Enlightenment. Writers like Cassirer, Hazard, Smith, Znaniecki, Krauss, and Gay, who have described the movement from different perspectives, have nonetheless uncovered a number of similar themes. For our purposes, the most important by-product of this revolution was the new standing given to knowledge in our society. The Enlightenment enthroned a new image of *scientia* as the universal solvent of ignorance and prejudice. In the hyperbolic language of the period, one could speak of the blinding light of reason banishing the barbarous dark ages forever, or of the power gained from the new knowledge breaking the shackles that had enchained secular society. Whatever the metaphor, it was clear that knowledge obtained through a deft combination of skepticism and common-sense rationalism, stripped of the millstones of tradition, was the key to the new era. Once the errors of the past had been systematically eradicated, mankind could begin to build anew, basing modern civilization upon solid and unalterable foundations.

In such an enterprise, the role of the man of learning was completely transformed. As the carrier of a new ideology, the intellectual was being asked to assume an almost superhuman task. Unconsciously, he appropriated a set of ideals that had earlier served the clergy. Striving for the discovery of supreme and eternal truths, he expected himself to display the same reverent allegiance at the altar of knowledge that his religious counterpart gave to his God. He assumed all the qualities of moderation and serenity that were associated with his dedication to the pursuit of truth. The intellectual, or *philosophe*, was to be a man "immune from the popular preoccupation and errors, stripped of the vanities of the world; . . . who prefers a private life to the fracas of the world."[3] Such a conception made him stand out from among his fellow men and elicited the definition of *philosophe* as "a man who ... places himself above the duties and ordinary obligations of civil and religious life. ... who denies himself nothing, allows himself to be constrained by nothing, who leads the life Philosophy."[4]

To some, this characterization of the intellectual might once have suggested a hermit or cenobite, withdrawn into his "ivory tower" for the sake of higher purposes. Whatever appeal this image of the monk or even of the Renaissance humanist may have had before the seventeenth century, it had largely disappeared by the end of the cultural revolution. The intellectual now lived in his *saeculum*, among men, and his efforts were community-directed.

There were actually two societies to which he felt an obligation. One was composed of his fellow intellectuals, the other of society at large. The intellectual was profoundly convinced that for knowledge to be meaningful, it must be diffused throughout mankind.[5] In his eyes, modernity itself was predicated upon the transformation of habits of action and thought, to be achieved by the enlightenment of large groups of society. Even among cultural relativists like Montesquieu and Voltaire, the distinguishing feature of Europe was its addiction to the pursuit of knowledge and the advancement of science, arts, and letters. It was *scientia* that provided Western man with his polish, his urbanity, his "civilité," or, to adopt eighteenth-century terminology, his civilized character.[6] This self-conscious realization offered the intellectual an image of himself as a part of society in which he fulfilled a crucial utilitarian function. In this context, utilitarianism was not conceived merely with reference to the physical needs of mankind. It had a global connotation which, beyond its promise for political, social, and economic reform, encompassed an even more funda-

3. From the 1732 edition of the *Dictionnaire de Trévoux*, as quoted in Dieckmann, *Philosophe*, p. 73.
4. From the 1694 edition of the *Dictionnaire de l'Académie Française*, as quoted in Dieckmann, *Philosophe*, p. 72.
5. Dieckmann, "The Concept of Knowledge," p. 80.
6. Febvre, pp. 9–16; Benveniste, pp. 47–54; Pflaum, pp. 5–12.

mental moral regeneration of Western society. In 1681, Leibniz could conceive of learning as the panacea for all of mankind's ills.[7]

A new role for the intellectual was thus evolving, one in which he found himself no longer the educator of a single class but the instructor of mankind. Education, based upon the fruits of the new learning, was now conceived as the instrument for the total spiritual reformation of all of society. In 1703, the Italian essayist Muratori was able to envisage the manner in which the study of knowledge itself would establish and propagate a universal sense of good taste, which in turn could enable all mankind to discern eternal truth and to follow the path of moral rectitude.[8] In short, the man of learning, through his intellectual pursuits, was destined to become the most useful of all the citizens in his society.

Not satisfied with the task of being missionary to all of humanity, the intellectual longed even more for the psychic rewards which only a community of peers could bestow. Consequently he developed a sense of allegiance to another society, more restricted in number and different in character: the "Republic of Letters." The term itself, or close variants, has a long history running back to the ancients. In the seventeenth century, it was sufficiently well known to figure in the title of books and journals by Niceron, Bayle, Le Clerc, and others. By the middle of the eighteenth century, the concept of a group of intellectuals—a *société des gens de lettres*—was already the object of essays and critical books.[9] In the last half of the century, efforts to write the history of civilization in terms of chronological annals of this Republic of Letters were commonplace.[10] No doubt Condorcet's *Esquisse d'un Tableau Historique des Progrès de l'Esprit Humain* was a late product of this popular conception which had swept Western Europe.

For our purposes, several features of this intellectual community are noteworthy. As has been suggested, its civilizing

7. Couturat, pp. 135–137.
8. Muratori, pp. 256–261.
9. Duclos, *Considérations*, ch. X; D'Alembert, "Réflexions" and "Essai sur la Société." See also the works cited by Barnes and Kirschstein.
10. Gusdorf, *Sciences Humaines*, I, 59–86.

role was directed toward society at large. Nevertheless, the members of the Republic of Letters possessed special qualities that set them apart. Individually, they were often recognizable by their indifference or even their critical attitude toward social conventions and traditional values. As a group, they constituted a caste with quite peculiar characteristics. Charles Duclos, perpetual secretary of the Académie Française, sagaciously remarked in 1751 on the incongruity of a group which recognized talent as the main index of worth in the midst of a society that was organized on the basis of birth, wealth, or profession. The abandonment of these traditional criteria in favor of one based upon natural abilities was the true mark of the intellectual. Thus merit had to be the principal manifestation for membership in the Republic of Letters.[11] But how was merit to be ascertained?

All of the spokesmen for the Republic of Letters were at pains to insist that public opinion could not be the ultimate judge of talent. To be sure, public recognition played favorably upon the intellectual's vanity and aided his search for patronage. It was even useful for the widespread diffusion of his wisdom. But the only acceptable tribunal of merit had to be a more select group. As D'Alembert said, "it is only for those of the *métier* that is reserved the appreciation of the beauty of a work and the extent of difficulties overcome."[12] Only peers who had intimate experience of the advancement of learning could know enough to be judges of intellectual merit.

As the intellectual recognized his new and self-assigned role, he also evinced a need to differentiate himself from the rest of society. There were a number of patterns of social organization that might have served this function. By the eighteenth century both the trades and liberal professions were organized as corporations to segregate them from other groups in society. Like them, the society of men of letters was distinguished by an occupational role in society. In each case, the peer group was the only body properly equipped to make sound judgments concerning quality of workmanship and to set standards of admission into the company. In practice, however,

11. Duclos, *Considérations*, pp. 231–232.
12. D'Alembert, *Œuvres Philosophiques*, III, 41.

there were a sufficient number of differences that made the intellectual reluctant to align his Republic with the existing corporate organization of the trades or professions.

For one, the Republic's economic position in society was quite different. No true intellectual lived off his talent in the same way as an artisan, selling his wares piecemeal. Unlike the writer, painter, or architect whose livelihood ultimately depended upon his handiwork, he more often banked upon long-range support or an income derived from an occupation not intimately connected with his special talent. To follow the patterns established in the liberal professions was similarly inappropriate, since the lawyer or doctor also earned his living from rendering specific services. In all countries of Western Europe, there was an established tradition of lifetime aristocratic or state patronage which the intellectuals preferred to the restrictions imposed by guild regulations.

An even more basic distinction from the trades and professions turns upon the nature of the intellectual's calling. Genius or talent, being attributes of the individual rather than acquirements, could not be learned in the same way as a trade.[13] The practice in guilds and corporations was to stress training, and in the liberal professions full membership was usually conferred only after the passing of a rigorous set of examinations. Degrees of apprenticeship and study were the operative indicators of status within these communities. No such simple arrangement could prevail among intellectuals whose merit was normally independent of formal training. As Diderot pointed out, genius was born and not made.

Beyond these economic and educational impediments, there were deep-seated psychological motives that kept the man of learning from identifying with either the professional or the craftsman. The typical professional that he knew represented all the values he firmly rejected. Doctors were notoriously impervious to innovation, priests and professors inevitably tied to received authority, and lawyers famed for their skill in distorting truth for the benefit of their clients. Both their pro-

---

13. Dieckmann, *JHI*, II, 151–182; Grappin; Matoré and Greimas; Bry; Tonelli; Fabian, all touch upon this aspect of the changing meaning of the words "genius" and "talent" in the eighteenth century.

fessional organization and function in society perpetuated the very things men of letters were determined to avoid.

The same objections, however, could not be held out against the craftsman, who displayed, as D'Alembert pointed out in the *Encyclopédie*, "the most admirable evidences of the sagacity, the patience, and the resource of the mind" that intellectuals envied.[14] The new seeker of knowledge and the artisan ought to have become brothers-in-arms. Yet the typical intellectual, for all his good will and independence of mind, was unable to turn his back upon the weight of tradition that relegated the manual artisan to a lower echelon of society. Only the exceptional *philosophe*, like Diderot, was able to see beyond the typical but intolerable traditions of duplicity, secrecy, and illiteracy that had prevailed among artisans for centuries. The age-old disdain for *techne* was especially persistent in France.[15]

Notwithstanding all its yearnings, the Republic of Letters could not model itself upon any of the existing groups in society. In spirit, the new intellectual was neither an artisan nor a professional, neither a teacher nor a preacher, neither a political leader nor a mere servant of society, even though he shared many of their occupational concerns and ambitions. He prized judgment by peers, the diffusion of knowledge, and the improvement of mankind. Hence his ambivalent attitude toward the organization of contemporary society, at once aloof because of the nature of his special needs and wistful because of the relevance of his own activity to society.

He reconciled these contradictory tendencies by developing a sense of benevolent superiority that often drifted into arrogance. The intellectual considered himself superior to the craftsman, to the professional, and to the rich and titled, because the object of his concern was more altruistic. He felt superior because he acted out of knowledge rather than habit, because he controlled his emotions and resisted the corrupting influences of immediate and ephemeral gain. He was superior because he was unprejudiced, acting only after reflection, and because he was practiced in the method of arriving at certain

14. D'Alembert, *Preliminary Discourse*, p. 42.
15. Daumas, *Instruments*, pp. 138–140.

understanding. Above all, he was superior because of a devotion to a transcendent ideal of truth.

All the contemporary commentators admitted—and even paraded—this sense of self-confidence and superiority. The scholar Du Marsais said that the *philosophe* "judged and spoke less, but he judges with greater assurance and speaks better."[16] D'Alembert offered an even more pointed opinion when he asserted publicly that

This society [of letters] is really the most useful and noble that a thinking man could wish for. If knowledge tempers the spirit, it also raises it; one of its qualities is even the result of the other, and it must be agreed . . . that not only are [men of letters] superior to others by their lights, but generally less corrupt in their feelings and actions.[17]

In his private papers, he added the more candid if somewhat disingenuous reflection that the intellectual's occasional tone of arrogance, though not politic, was pardonable. Truth, wrote D'Alembert, is irrefutable, and even when presented in a dogmatic guise will prevail. "The essential thing is to be right; the form [of its presentation] is in itself unimportant." And as a final justification for the philosophe's haughtiness he added that "truth is so precious and rare . . . that it is legitimate to pardon those who pronounce it for their enthusiasm."[18]

Sensing his superiority, the intellectual felt both above and responsible for the rest of society. His self-confidence and zeal inevitably led him to assume the air of a conqueror and missionary. He was certain the artisan had to be taught to rationalize his activities, the doctor to understand his diagnosis, the musician to mathematize his art, and the ruler to lead his nation according to natural law. In the metaphoric language of the Enlightenment, the *philosophe* was prepared to lead his fellow man out of the darkness of night with the torch of reason. With his insatiable desire to regenerate and reform all of mankind, the intellectual pictured himself as the true leader of society, placing himself in an elite group within society at large. But

16. Dieckmann, *Philosophe*, p. 40.
17. D'Alembert, *Œuvres Philosophiques*, III, 72–73.
18. D'Alembert, *Œuvres et Correspondances*, pp. 73–74.

what social form would this elitism take in the eighteenth century?

It was a transformed notion of "academy" that ultimately provided the Republic of Letters with a viable institutional framework, at once elitist and community-directed. The term itself was a venerable one, stemming as it did from Plato's Academy. In the Renaissance, the idea of the academy had experienced its first rebirth. Ficino's Florentine Academy had provided a model for scores of Italian circles of learning operating outside the universities. Until the end of the seventeenth century, these groups provided the most important meeting ground for cultural dissenters who objected to traditional modes of thought.[19] So successful was their antischolasticism that, as their original *raison d'être* disappeared, many of these academies slipped into the sterile path of rhetoric or antiquarianism. While surveying the cultural conditions of Italy in 1703, the young Muratori sarcastically commented that "the work of the academies can be reduced to the chase for applause and the attempts to enchant the ears of its patient listeners."[20] The time had come, he announced, to direct them into more productive and useful channels by creating an association of learned Italians, constituting "a single academy or republic of letters, whose object would be to perfect the arts and sciences by demonstration, to correct abuses, and to teach the uses of truth." If they were to respond to the Republic of Letters' deep-seated ambitions, the academies had to take on the task of building the new order in addition to that of destroying the old one.

Two characteristics of the refurbished academy of the late seventeenth century were particularly well suited to the aspirations of the intellectuals. One was that its very form forced its members into a social community that was defined and visible while remaining intellectually open. Whether one analyzes the utopian projects for new academies of a Bacon, a Muratori, a Leibniz, or a Klopstock, the stated objectives of journalists like Bayle, Le Clerc, Masson, or Du Sauzet, or the self-images of

---

19. Pevsner, introduction; Roddier, pp. 45–54; Ben-David, *Minerva*, IV, 30–38.
20. Muratori, pp. 178, 181, for this and the next quotation.

encyclopedia editors, the same preoccupation with open communication emerges. The intellectual not only knew he belonged to a community and that he was morally obligated to share his lights, but he held to the notion that improvement of learning was too large a task for any single individual.

The other and equally important characteristic of the new academy was that it could easily be molded to reflect the intellectuals' sense of cultural superiority. In times past, academicians had formed an elite by virtue of the esoteric nature of their preoccupations. Ironically, it was through their own efforts that the intellectuals menaced their elitist status. By educating increasing numbers, they enlarged the audience capable of following their pursuits, and, in the course of the eighteenth century, there were repeated complaints about the large number of dilettantes and would-be intellectuals who clamored for admission into the hallowed circles of the Republic. J.-H.-S. Formey, the spokesman for the Berlin Academy, made it a special point to single out this problem of *demi-savoir* as one of the academies' major concerns. According to Formey, academicians were, by the very nature of their enterprise and their position, the only legitimate possessors of the key to knowledge. By that very token, they were responsible for stamping out the pretensions of "this legion of half-wits, not even worthy of being called dung-piles . . . who dishonor the Republic of Letters."[21] He added that it was incumbent upon the academies not only to seek new knowledge and propagate it, but also to protect the rest of society from pretense and partial wisdom, lest humanity regress into a state of ignorance. Formey continued his plea with the following observation, pregnant with implications: "The Church guards the sacred trust of religion; the tribunals maintain law; it is up to the Academies to establish sound knowledge, rich in precious fruits, which chases out *demi-savoir* . . ." To carry out this self-assigned task, it was clear that academies had to formulate standards and devise mechanisms to apply them. As the success of the intellectual enterprise made it popular, criteria and means for delimiting

21. Formey (1768), pp. 362–363, 364, for this and the next quotation. This essay was later republished in the *Encyclopédie*, Supplément, I, 93–98, as the article "Académies (avantages des)."

the Republic of Letters became essential. In fact, it was in this direction that the most advanced academies were already moving during the seventeenth century. It was precisely this feature which forced the intellectuals to develop inner techniques for decision-making, thereby placing politics on the doorstep of *scientia*.

The needs of the intellectual community for a new type of organization to carry out its missionary instincts in an elitist fashion coincided in time and direction with the growth in Western Europe of national spirit and new modes of governance. Like any other successful institution, the new type of academy owed its existence and endurance to the manner in which it was able to match its intellectual requirements to the political realities of the age. What is particularly striking is the mutual reinforcement that the Republic of Letters and the emergent state were to provide for each other.

This may be seen in at least three different ways. In the first place, elitism and rationalism, considered essential components of the new intellectual movement, found a strong ally in the centralizing and bureaucratizing tendencies of the absolutist state. The notion of order, control, and power appealed equally to the planners of the state and to the new rulers of the mind, so that the idea of a partnership between them appeared almost automatically.

The precise nature of this partnership varied according to local conditions. In the German- and Italian-speaking areas, where consolidation of political power was slow in coming, the cultural movement was seized upon as a potential tool for national unification. Hence the appropriateness of the pleas of Vico, Muratori, and Leibniz for a total regeneration of the national spirit through new forms of intellectual organization. Wherever the formation of the modern state as a bureaucratic entity lagged, the Republic of Letters was conceived of as the instrument most likely to provide the impetus for a new body politic. Hence also the uncertainty over the proper relationship between politics and knowledge which led to the tortuous juridical theses of Christian Thomasius and his contemporaries over the proper legal status of the Republic of Letters and the more famous pronouncements on academic freedom of Kant and Herder.[22] Together, culture and state were in a process of *Bildung*.

In France, where the crystallization of absolutism was essentially completed by the time the new academies came into being, there could be no debates about the relationship of state and culture. Knowledge was clearly designed to be the ornament and tool of authority, and new institutions, if they were to survive, had to serve the nation rather than form it. In Louis XIV's age, culture was fashioned to reflect his political power and glorify the existence of the nation which he ruled and symbolized.[23] The French Republic of Letters had no other option.

There was even a sense in which the subordination of cultural drives to the absolute state was welcomed. Many of the intellectuals anxious to carve out viable new institutions sought governmental assistance to overcome the resistance of older cultural institutions toward change. The leaders of absolutist France joined the intellectuals in a common struggle against the authority of the University, the Church establishment, and the constituted professions. Indeed, governmental paternalism and royal munificence protected and nurtured the budding new academies during their formative years.

Much of the evidence presented in the previous chapter suggests one instance of the close and mutually agreeable character of the relationship between intellectuals and the Crown. Similar examples for other groups of intellectuals abound. Governmental decrees in 1663 preserved the Academy of Painting and Sculpture from jealous rivals who sought to retain control of the profession through the artisans' guilds.[24] The Academy of Architecture also owed its existence in 1671 to Colbert's commitment to the elevation of the profession of architects by scholarly discussions of the art.[25] Little architectual progress could have been achieved without the backing of the monarchy. Indeed, this academy, like the Academy of Inscriptions and Medals, functioned for a number of years as an

22. Kirschstein, pp. 49ff.; Weil, pp. 84–199.
23. Voltaire, *Lettres Philosophiques*, letter XXIV, and *Siècle de Louis XIV*, ch. XXXI.
24. Montaiglon, I, 211–212; Pevsner, pp. 89–109.
25. Lemonnier, *Procès-verbaux*, I, viii–xi.

extension of governmental authority under Colbert's personal supervision.[26] In retrospect, one can see that Colbert's paternalistic attitude toward the new or refurbished institutions for intellectuals was responsible for creating a unique and notable "academic system." That system, in turn, was constantly displayed as the cultural jewel of Louis XIV's crown, providing his reign with particular distinction and *civilité*.

There was a second way in which the aspirations of the Republic of Letters were easily matched to those of the Crown. Both the government and the intellectuals yearned for the academies to assume judicative as well as leadership roles in French culture. The idea was not only consistent with the intellectuals' self-image, but fitted perfectly into absolutist notions of the exercise of power. In fact, it was such an obvious attribute of the new institutions that the practice long antedated any formal discussion of its merits. The evidence for this was provided in the last chapter, though only in the case of the Academy of Sciences. It was the typical pattern in all the other new cultural institutions as well.

Every Parisian academy performed a judicative task from its very inception. The Académie Française, for example, assumed as its task the preparation of an official dictionary and grammar of the French language. The newly transformed Academy of Painting and Sculpture gathered in special meetings to discuss and establish esthetic norms.[27] Similarly, the new Academy of Architecture's first item of business was a decision on the definition of "good taste" in the arts.[28] Colbert's Academy of Inscriptions and Medals was charged with judging the merits of Quinault and Lully's operas and making pronouncements on the value of Félibien's treatises.[29] In this context, the Academy of Sciences' task as consultant to the Crown appears quite conventional. Like any other organ of the government, the French academies were considered as courts of experts in their particular cultural realm. It was, as we have noted, a notion that coincided with the intellectuals' self-image. The

26. Charles Perrault, *Mémoires*, pp. 34–42; Maury, *Académie des Inscriptions*, pp. 3–19.
27. Félibien; Montaiglon, I, 315–316, 324–326, 342.
28. Lemonnier, *Procès-Verbaux*, I, 3–4.
29. Maury, *Académie des Inscriptions*, pp. 11–12.

partnership between culture and government was once more reinforced by complementary attitudes.

There was yet a third and deeply significant element that converged with the others to make the academic system the natural pattern of organization for higher culture in France. Classicism was the predominant cultural mode of French society in the late seventeenth century, and was articulated and codified at precisely the same time that the academic system was being forged. Indeed, the formalization of that elusive but nonetheless real *goût* or *esprit classique* was, in part, one of the academies' first significant by-products. It was maintained as the prevailing style of French life during the Enlightenment when the Academy of Sciences enjoyed its greatest success.

French classicism expressed itself as a set of rules which defined artistic, ethical, and philosophical norms. The men who fashioned these norms, the Félibiens, Boileaus, Blondels, and Du Hamels, in a sense prepared them under academic, and therefore national, auspices. If it is technically true that learned societies did not officially sanction methodological treatises with the stamp of communal approval, their members often participated directly in the collection, criticism, and even writing of these proper manuals of appreciation. The exact procedure for fixing norms varied from subject to subject, and took on different forms in each academy, from the preparation of a dictionary of the French language to the establishment and teaching of "good taste" in art. Even the Academy of Sciences, which prided itself on refusing to endorse any general system of science, unwittingly set up minimal methodological ground rules for the scientific community. Although no formal manual codifying the rules for scientific acceptability existed, the Academy's preference for or rejection of particular writings was tantamount to the creation of a set of prescripts for the community. Every judgment implied the existence of criteria for excellence. Hence every volume of memoirs and every public comment by spokesmen for the Academy was looked upon as the expression of classical standards to be imitated.[30]

---

30. Through his historical introductions to the *HARS* and in his *éloges*, Fontenelle was able to articulate and transmit these norms to future generations. See Delorme, *Archeion*, XIX, 217–235.

Beyond the setting of norms for subject matter, the academies also expressed the spirit of classicism in the prescription of modes of social behavior and intellectual practice. Classicism implied not only the orderly arrangement of knowledge, but one unadorned by trivialities and pure in its expression. It implied a sobriety brought about by balance, restraint, and discipline that reflected a mastery and control of the passions. All these qualities constituted the "academic style" as much as did precepts taught by fiat or example. Thus the importance given to the staid conduct of academicians at public gatherings had a significance that transcended mere ceremony.

Nothing was so well suited to cultivate this particular "life style" as the patterns of social organization assumed by the new institutions. Without crushing the individual, the academies' communality tempered personal idiosyncrasies and checked eccentricity. In the statutes of each academy, there was an explicit admonition to the members to refrain from the use of vituperative language.[31] The call to moderation involved not merely the prohibition of public anger; it was a positive expression of the sensibilities of a classical age. Moreover, the sublimation of emotions was essential to the ultimate purpose of the academies. Excessive zeal or passionate advocacy ran counter to the essential need for agreement, and could easily have destroyed the communal activity upon which the life of the academies was predicated. To be truly "classical," these institutions had to reflect the sense of order and consensus implied in the spirit of the times.

In no country other than France, and at no time other than Colbert's period of power, did classicism so successfully nourish the growth of the Republic of Letters and, in so doing, fulfill the aspirations of the state. The confluence of these factors provided society with a new set of organizations to regulate the life of the mind and, hopefully, to transform society for the better. The new academic system which had been elaborated in Paris spread with local variations to the rest of the Western world and, for over a century, provided it with a major new institutional framework for intellectual activities. As Fontenelle had said, it was truly an Age of Academies.

31. Aucoc, *Lois*, pp. xxxiv, cviii, cxviii.

So far our discussion has focused on the emergence of the academy as a dominant cultural institution of the late seventeenth century. Most of our attention has been given to exploring the meaning of "Royal Academy," without regard to the phrase "of Science." Were "science" merely the vernacular rendition of the Latin *scientia*, no special problem would be posed. But in fact, there were important differences between *scientia* and *science*.[32] By the end of the Renaissance, *scientia* had come to stand for all secular learning acquired by man's own efforts, as opposed to innate wisdom. Thus, an "academy of *scientia*" could have been the concrete realization of the Republic of Letters' dreams. Perrault's planned General Academy was in fact an academy for the advancement of learning (or of knowledge) that would have encompassed all secular disciplines of the human mind. Such was the conception logically flowing from Descartes's well-known vision of the unity of knowledge. It was also the driving force behind the encyclopedia movement of the mid-eighteenth century.[33]

Instead, the company that Colbert founded was composed of a group of learned specialists who explicitly restricted their studies to mathematics and natural philosophy. In this respect, the Paris Academy of Sciences represents a significant departure from most of the other learned institutions founded in Italy, England, and Germany, as well as the French provincial societies that mushroomed in the eighteenth century.[34] The Paris Academy was from its very beginnings unique in its composition and in the domain of its activity, while remaining typical in its institutional basis.

The explanation for this special situation can be sought on two different planes. On the ideological level, it is generally known that the attempts of philosophers such as Descartes and Leibniz to re-establish a unity in the European mind met with failure. As Cassirer and Gillispie have argued so convincingly, Enlightenment thinkers generally rejected systematic philos-

32. Sas; Ross; Gusdorf, *Sciences Humaines*, I, 9–18.
33. McRae, chs. III and VI.
34. For the French provincial academies' activities, see Delandine, *Couronnes Académiques*; Saunders; Barrière, *Vie Intellectuelle*, pp. 300–304; Roche, "Milieux."

ophy in favor of a more positivistic philosophy of analysis; hence the weakness of a unified concept of culture, and the increasing triumph of pluralism and specialization.[35] The Academy of Sciences' character merely reflected the philosophical trends of the age. But, by itself, this analysis does not offer a sufficient explanation for the differentiation between this Parisian Academy and its sister institutions. To grasp this phenomenon, the local and historical circumstances that surrounded its foundation must be re-examined briefly.

In the previous chapter, we referred to the pleas of Sorbière, Auzout, Huygens, and others for the creation of a society composed of professional scientists. These demands coincided with Colbert's express desire to be surrounded by a group of technical consultants able to provide him quickly with expert advice. His need was not limited to science, but extended to all realms of knowledge potentially useful for the proper administration of the realm. Colbert in fact already had his councilors in history, numismatics, architecture, military affairs, commerce, and finance.[36] The Academy of Sciences was but one expression of the new-found role for the expert within the context of an emerging modern state. Since functional specialization is inherent in the modern bureaucratic state, the group of scientists who formed the Royal Academy constituted an elite as much for the government as for their own community.[37] Hence, like other bureaucrats, the academicians were given a salary tied to their specific function rather than to their literary production.

Since his function matched his aspirations, the scientist welcomed the salary, not only because it allowed him the luxury of being a full-time researcher, but even more because it offered him the special status he sought in French society. A professional bureaucrat could no longer be confused with the cultured polymath so prominent in the earlier period. Nor would he be mistaken for a craftsman or a member of a liberal profession. His position was conveniently linked to his func-

35. Cassirer, ch. I; Gillispie, *Edge of Objectivity*, ch. V.
36. King, pp. 102, 416.
37. Crozier, pp. 210–269, for the French case; Millerson for the English example.

tional role in the state, rather than to the economic fruits of his labor.[38] The existence of an academy of specialists once again reinforced his profoundly elitist values.

Separateness was more pronounced in France than anywhere else in Europe simply because the process of bureaucratization was far more advanced there than in any other nation. Significantly, only the royal academies in Paris, close to the seat of government, experienced this transformation of the *honnête homme* into the budding professional. By contrast, the provincial societies remained for a long time the carriers of a charming but old-fashioned ideal of general culture better suited to amuse the mind and the senses than to harness the specialized wisdom of man to governmental requirements.[39]

Still another set of historical circumstances contributed to the Academy of Sciences' unique situation. It may be recalled that under the prodding of Chapelain and Perrault, Colbert had seriously considered the notion of setting up a General Academy in which he would act as "chief of staff" for French culture. The attempt to realize this conception was probably at the root of the founding of the *Journal des Savants* in 1665 to deal with letters, sciences, the arts, and theology. Its appearance met with such violent opposition, particularly from religious circles, that the *Journal* was temporarily suspended, its editor removed, and its sphere of activities seriously curtailed before it could publish again.[40] Similarly, the plan for a General Academy clashed violently with numerous *corps d'état* firmly entrenched within the state, which bitterly resented the intrusion of new institutions into domains they had previously controlled. Old Regime France was racked with corporate jealousies that could prevent the success of projects having the backing of even as formidable a patron as the Crown.

It was for this reason that Colbert eventually recognized

---

38. This phenomenon is explained as a dysfunctional consequence of bureaucratization by Merton, pp. 195–200.

39. Cochrane, pp. 22–27; Jean Cousin; Trénard; Pichois; Roche, *Annales*, XIX, 896–915. Condorcet's views on the character of provincial academies is discussed by Baker, *RHS*, XX, 257, 274–275. In France, this character persisted despite the formal, legal connections of learned societies with local parlements. Exceptions to the rule are Société of Montpellier and the agricultural societies, discussed in chapter 4 below.

40. Birn, pp. 19–22.

that any plan for a General Academy of *scientia* was utopian. The Sorbonne, the Faculty of Medicine, and the already established Académie Française objected strenuously to any group that would usurp their customary rights.[41] This explains in part the reluctance of the Academy of Sciences to enter into the arena of scientific education and the granting of degrees, as well as its failure to investigate the linguistic problems associated with the creation of new scientific concepts. Theology and politics were likewise excluded from its deliberations, since such matters were already the province of other constituted bodies. On this score, Perrault stated unequivocally in his autobiography that plans for a General Academy were curtailed by the fear of stepping on dangerous grounds.[42] We can appreciate the aptness of Fontenelle's phrase that the Academy of Sciences turned out to be no more than a piece of debris from the original grand design.[43]

Faced with such a complex power structure, Colbert was forced to moderate his desires. Where he was unable to innovate, he tried to gain control of existing institutions and to remodel them. The Académie Française was enticed to become a state institution by offers of prerogatives at court and the privilege of meeting in the King's chambers at the Louvre.[44] Colbert won over the painters and sculptors of the Academy by protecting them from the rival guilds and by offering them financial assistance to refurbish their much criticized teaching practices.[45] He bought the scientists' allegiance by allowing them to set up their own patterns of research and by providing them with the resources they so badly needed.

The creation of an academic system under Colbert's jurisdiction is a testimony to his sense of political realism. But success exacted its price. While appearing to innovate, he was in fact a prisoner of traditions. Even for the Academy of Sciences, which had all the appearances of a new institution, historical precedents were much in evidence. The division between *mathématiciens* and *phisiciens* followed the two distinct

41. Hirschfield, pp. 17–20.
42. Charles Perrault, *Mémoires*, p. 48.
43. *HARS* (1666–1699), I, 7.
44. Académie Française, *Registres*, I, 13.
45. Montaiglon, I, 203–204, 212.

sources which had formed the new science of the seventeenth century, one stemming from the study of the *quadrivium*, the other from newer sources of natural knowledge gleaned from observations.[46] There was no attempt to force the Academy of Sciences' internal organization into a rational scheme that would parallel the structure of the other academies. The formation of each group was treated as a separate problem.

Such an approach was in the end most fortunate, at least for the progress of science. The nature of scientific research is in many respects fundamentally different from literature, which depends upon stylistic elegance; history, which demands bibliographic and linguistic erudition; or the fine arts, which at the time were founded upon the principles of imitation. But the inescapable consequence of Colbert's individual response to France's cultural heritage was to fragment the institutional organization of knowledge. If there was to be a royal academic system at all, it was destined to be a collection of individual academies with only externally similar institutional frameworks.

The emergence of differentiated academic institutions cannot be interpreted solely as a consequence of Colbert's astute accommodation to political realities. It was also an expression of the special needs of science. The French version of the "quarrel of the ancients and moderns," which followed Charles Perrault's reading of his poem "Le Siècle de Louis le Grand" in 1687, brought out into the open something which intimates of scientific circles had already experienced for years. Perrault pointed out that the discovery and elaboration of an understanding of nature is a cumulative and self-corrective enterprise that depends upon the cooperation of generations of individual researchers.[47] Without a pooling of efforts and the existence of central warehouses of knowledge, the continued progress of the moderns was unthinkable. Their superiority over the ancients depended not upon an innate difference in intelligence, but directly upon their ability to profit from the accumulation of knowledge in the past. Modern man could see further only because he stood on the shoulders of giants.

---

46. *HARS* (1666–1699), I, 13.
47. Charles Perrault, *Parallèle*, pp. 42–45, 370–371.

All the communal activities of scientific academies were based upon this premise. In short, what the "quarrel" did was to spell out in detail that certain cultural activities are cumulative in character, while others do not fundamentally partake of historical progress.

The implications of the "quarrel" for the classification of human activities, and thus for the organization of cultural life, were immense. For the first time in history, the fine arts, conceived of as dependent upon the cultivation and refinement of imitative faculties, were explicitly separated from the practical arts and sciences.[48] That separation was already present in embryonic form in Colbert's 1667 statement that the sciences are "taught by precept for man's understanding, and the arts . . . by example to develop man's imagination."[49] The very same distinction was embodied in the differentiated nature and practices of the academies he helped to found. The differentiation was particularly conspicuous in the case of the sciences, since the champions of the moderns always borrowed their historical examples from the annals of science and the practical arts. It seemed a compelling reason for separating scientific activities from other cultural pursuits in France.

Logic and the special historical circumstances outlined above conspired to produce France's distinctive academic system, with the Academy of Sciences' special place therein. With this organization the direction of the nation's cultural life was thenceforth divided among clusters of experts, each regulating its own domain. The system, sanctioned as it was by royal authority, at first was in essential harmony with so many features of contemporary society that an explanation of its prosperity during the Enlightenment in these terms seems like a tautology. It is only because every one of the foundations upon which the academic system seemed firmly based was later seriously questioned that it requires a historical explanation. As long as the values of classicism prevailed, the bureaucratic and absolutist nature of French government continued, and the peculiar needs of the intellectual communities persisted, there could be no serious question about the academies' future.

48. Kristeller, *JHI*, XII, 521–527, XIII, 17–24; Tatarkiewicz.
49. Félibien, preface; Clément, V, 498–499.

Of all the academies, that of sciences seemed to have had the greatest hope for success. None squared better with the varied aspirations of the era. In every respect, it was the embodiment of all those desirable features of an academy we have discussed in this chapter.

The scientific academician appeared as the prototype of the citizen of the eighteenth-century Republic of Letters. As Fontenelle painted him in his popular *éloges*, he possessed all the moral characteristics that would earn him a secure niche as the patron saint of the Enlightenment.[50] Devoted as he was to the pursuit of eternal truth, he could fearlessly abandon all the traditional mores of his times. For the sake of his calling, he was expected to disregard health, wealth, titles, social amenities, and physical comfort. Placing the spiritual life of the mind above all else, he could sacrifice all his being for the secular salvation that the discovery of new knowledge would bring to mankind.

In place of ostentation and vanity, the scientist possessed personal traits that were inherent in the nature of science itself. At various times in the *éloges*, Fontenelle remarked that

His character was that which the sciences form ordinarily in those who make it their sole occupation: seriousness, simplicity, righteousness.
One easily saw that his humility was not a pose but a feeling founded upon science itself.
His character was as simple as his superiority of mind could require. I have already given this same praise to so many persons in this Academy that one would believe the quality to belong rather more to our sciences than our savants.[51]

The enterprise in which the *savant* was engaged forced upon him all the qualities desirable for the building of the new Enlightenment. Skepticism assured him of overthrowing the errors of his predecessors; chaste personal habits and unadorned statements provided him with intellectual balance and lucidity; and independence shielded him from the perils of doctrinaire adherence.

Since the scientist now operated within a society of de-

50. Delorme, *Archeion*, XIX, 225–227; Marsak, *Fontenelle*, p. 43.
51. As quoted by Marsak, *Fontenelle*, p. 43.

manding peers, he was also assured of treading on firm ground in the establishment of true knowledge. By submitting his creation to the collective judgment of his colleagues, he protected everyone—including himself—from the vagaries of his imagination and the subjectivity of individual creation. Composed as it was of specialists, the Academy was the impartial guarantor of the validity of individual discovery. Through its communal procedure of consensus, the new institution was an instrument for separating sound knowledge from conjecture and for certifying the "way of truth." By its publications, it could communicate the new-found truth to others and aid in turning the Enlightenment's dream of progress into a reality.

In sum, the individual scientist was the exemplary eighteenth-century intellectual, and his organization, the Royal Academy of Sciences, the model institution of its time. Beyond this, the enterprise upon which it was embarked was held in the highest esteem during the era. To many, it was the most promising undertaking of the century. Science seemed destined to revolutionize navigation, warfare, agriculture, the arts and crafts, health, and all the varied material amenities of life on earth. It was expected to be of immense service to the state, and above all to civilize mankind by offering it a new secular religion.

Chapter 3

Integration into the
Old Regime

THE SUCCESS of the Academy in the prerevolutionary period was a function of the institution's responsiveness to the internal demands of the scientific community and to the cultural patterns of the era. As its popularity gained strength during the eighteenth century, the Academy gradually outgrew its renown as merely the new jewel of the king's crown or as a club for scientific specialists. Before the century was ended, the Academy and its members had become the principal directors of the entire French scientific enterprise, controlling every aspect of its life, except perhaps the healing arts. In rising to this towering position, the learned society gained its status in a fashion typical of Old Regime political practices and obtained strong backing for its growth from the government. On the eve of the Revolution, the Academy was considered more frequently as a fixture of Old Regime society than as the symbol and agent for the advancement of science.

While this rise to pre-eminence cannot be chronicled in its minutest detail, a number of key developments need to be examined, especially because they were later to be the source of many of the difficulties academicians encountered during the Revolution. Three interrelated lines of development in particular illustrate the manner in which this relatively new institution, while carrying out its scientific mission, also became an integral part of the culture of the Old Regime. The first is the progressive growth and institutionalization of the Academy's power. This is closely related to—though distinct from—the visible signs of the Academy's growing prestige during the eighteenth century. A third line to be followed is the Academy's internal structural evolution, which paralleled contemporary cultural patterns of social organization, thereby reinforcing its ties to the Old Regime.

The Academy's growing control over scientific endeavor manifested itself on all levels of activity, ranging from acts legally sanctioned by the Crown to the *de facto* domination of the pettiest transactions in the scientific community. From the description of this evolution it will become clear that the Republic of Letters—at least in the sphere of science—was becoming a highly aristocratic commonwealth, more closely fitting Montesquieu's conception of a republic than any notion of democratic republicanism. That distinction was to become of considerable significance as the Revolution progressed and the very existence of the Academy was repeatedly threatened.

Initially, the 1699 regulations, emanating from the king and confirmed by letters-patent registered by the Paris Parlement in 1713, gave the Academy considerable responsibility and status. In them, the society was defined as a royal institution, to be administered directly by one of the Crown's ministers.[1] The king reserved the right to approve the election of new members, to appoint officers of the society, to guarantee funds for its experiments and for annual pensions for its older members, as well as for the redemption of tokens (*jetons*) given out as additional favors for attendance at each meeting. In return, the Academy was expected to keep in constant communication with the scientific community in France and abroad, to stay abreast of all scientific news, to read and discuss important publications, to repeat significant experiments carried on elsewhere, and to criticize the works written by its own members. The mandate was already ample. It was further augmented by the written confirmation of two specific powers which were to become the cornerstone of the Academy's con-

1. Aucoc, *Lois*, p. lxxxiv.

siderable authority over science and technology. One has already been cited in passing, the mandate to inspect machines sent by the Crown for the granting of a royal *privilège*, the Old Regime form for patent rights. The other concerned a matter more vital to the membership, the exclusive privilege given the Academy to have works of its members printed without obtaining the approval of the royal board of censors.

The assignment of printing rights—traditionally under the jurisdiction of the Paris Faculty of Theology—to a group fully and directly under royal control was directly in line with the Crown's efforts to wrest power away from the University.[2] As the book trade expanded and with it the need for censors possessing specialized talents other than theology, the Crown had gradually increased the number of censors, from four in 1653 to about eighty a century later.[3] Each new appointee was paid out of the royal coffers and was held responsible to a "Director of the Book Trade," who was himself a Crown official. The censors, moreover, were each assigned a specialty consistent with his training. Was anybody better suited to control scientific publications than the new Academy, composed as it was of proven experts in science?

The move was of considerable importance in assuring the society an outlet for its writings. For the individual academician, it greatly simplified administrative procedure. Any work read before the company and approved by the membership was automatically invested with this legal sanction, without further ado. Arrangements for the actual printing and sale of the works varied, depending upon the desires of the author and the commercial practices of the printer or book dealer.[4] Whatever the special circumstances, the common practice with each book approved by the Academy was to cede printing privileges to the publisher chosen by the author.

Similar practices were followed for collective writings.

2. Olivier-Martin, pp. 55–60.
3. Pottinger, ch. IV; Grosclaude, pp. 452–461; Estivals, *Statistique*, part I, ch. I. Lists of these censors, many of whom were academicians, are given in the *Almanach Royal* beginning with the year 1742.
4. D'Alembert, for instance, used Antoine Boudet as printer and the David family as distributors, whereas Bossut chose C. A. Jombert. Bailly preferred the De Bure family.

At periodic intervals beginning in 1699, contracts were signed between the Academy and a printer known as the "imprimeur et libraire de l'Académie Royale des Sciences."[5] According to one such contract of 1713, the privilege to print and distribute the annual *Histoire et Mémoires* was transferred by the Academy to Claude Rigaud, director of the Imprimerie Royale, who agreed to take on this task in return for the option of first refusal on other collective writings taken from the Academy's annals.[6] In 1721, the right to print winning prize essays was transferred first to Jombert and later to Martin, Coignard, and Guérin. In the 1730's, similar cessions were made by the Academy for the reprinting of the *Histoire et Mémoires* for 1666 to 1710 as well as for the publication of the *Machines et Inventions Approuvées par l'Académie.*[7]

At first the privilege was employed principally for academicians. Their writings appeared either as part of the "Mémoires" section of the *Histoire et Mémoires*, or, when the works were too bulky, as separate tomes bearing the familiar statement "approuvé par l'Académie Royale des Sciences." The findings of nonacademicians, when they were reported at all, were paraphrased and inserted in the section entitled "Histoire," drawn up by Fontenelle. For the first decades of the century, this practice was apparently dictated by the mediocre quality of the observations submitted for discussion by nonmembers. In 1750, the Academy's secretary, Fouchy, looking back upon early practices, condescendingly noted:

We cannot hide the fact that, especially in its early days, the Academy more often praised the good intention of authors than the excellence of their work. Ideas often inherently sound were badly expounded, or mixed in amidst a large quantity of irrelevant matters.[8]

5. I have found royal edicts pertaining to the Academy's printer and bookseller signed on the following dates: 6 April 1699, 13 Feb. 1704, 29 June 1717, 21 Jan. 1734, and 19 March 1750. Among the favorite printers for collective publications were Léonard, Mabre-Cramoisy, J. Boudot, and Anisson.
6. Lalande, "Collection," pp. 62–68.
7. The text and dates of these decisions are generally found on the unpaginated leaf containing the "Privilège du Roy" bound in the first volume of the publications. See also Lalande, "Collection," p. 77.
8. *Mémoires de Mathématique*, I, ii–iii.

A noticeable change was brought about in 1720 by the opening of prize competitions which generally drew excellent contributions from the nonacademic scientific community. It was thereupon decided to capitalize upon the ambiguous language of the printing privilege to have the winning prize essay published without further censorship.[9] A precedent for the publication of scientific writings by outsiders was thus set. It was further reinforced in 1729 by the decision permitting another nonacademician, the engineer Jean-Gaffin Gallon, to edit all descriptions of machines examined and approved by the company. Seven important volumes appeared between 1735 and 1777, including the works of numerous technicians, most of whom were completely unaffiliated with the learned society. The Academy was beginning to extend its realm of operations beyond that originally envisaged by the government.

One more printing enterprise, begun in 1750, firmly fixed the practice of permitting the Academy to endorse the scientific productions of nonmembers. In that year the *Mémoires de Mathématique et de Physique Présentés à l'Académie Royale des Sciences, par Divers Sçavans, et Lus dans Ses Assemblées,* commonly known as the "Mémoires des savants étrangers," was initiated. In this series, with the prestigious impress of the Academy upon its covers, the select works of "outsiders," including aspiring academicians, would find an outlet and wide distribution. In effect, the Academy had created a new scientific journal, setting itself up as an editorial board. Its prerogatives were thereby expanded and its controlling power generally increased.

By stretching the original printing privilege to include nonacademicians, the Academy had considerably extended its jurisdiction, in both a legal and extralegal sense. But the process reflects less any conscious hunger for power than the demands success itself had imposed upon the Academy. In turn, of course, the extension of its domain of activity insured future success. To recognize the mechanics of this circular phenome-

9. According to Lalande, "Collection," p. 75, this decision was made by the Academy on 5 Feb. 1721. A slight change in the wording of the *privilège* between 1717 and 1734 seems to have legitimized the Academy's innovation.

non is to grasp the ultimate source of the society's spiraling prosperity within the scientific community. Once the Academy had become an established public institution with delegated royal powers it was *ipso facto* endowed with a magnetic influence. Every aspiring scientist hoped that his work would achieve wide public recognition through the Academy's approval and that he would thereby gain entrance into the select company. The Academy, in turn, by setting high standards and banking on its rising prestige, could impose professional norms upon the rest of the community and assure itself of recruiting the best talent in the nation. The society was thus able to improve its own standards, gain further prestige, and become an even more attractive institution. It was an automatic, self-feeding growth mechanism that seemed limitless once it had been initiated.

There were other means by which the Academy gradually gained control of much of the publishing apparatus of French science. An astronomical almanac, the *Connaissance des Temps*, had been appearing annually since 1679. Although it had included a number of academicians among its editors, it was operated by the printer as an autonomous commercial venture until 19 January 1702, when its printing privilege was legally ceded to the Academy.[10] Thereafter, this convenient ephemeris employed by every astronomer, ship captain, and surveyor carried on its title page the phrase "par les ordres de l'Académie." It became the learned society's customary task to appoint its editor, so that when the publication enlarged its scope in 1766 by including short astronomical articles, the Academy had still another scientific journal under its jurisdiction.[11]

The monthly *Journal des Savants* fell under the Academy's dominion in a slightly different set of circumstances.[12] Before

---

10. For details, see Condorcet, *Eloges*, pp. 159–160; Lalande, *Bibliographie Astronomique*, pp. 341–344; Delambre, *Astronomie Moderne*, II, 683–685, and *Astronomie au Dix-Huitième*, pp. 752–755.
11. Lalande, "Collection," p. 71. In a letter of 29 Jan. 1774 from the minister in charge of the Academy (the Duc de Lavrillière) to its director (Macquer), the Academy is specifically asked to name a replacement for the retiring editor, Lalande (BN, MS. fr. 12305, fol. 471). The Academy generally allotted 800 livres from its funds to pay the editor (JT, pp. 91–94).
12. BN, MS. fr. 22225, fols. 101–107; Birn, pp. 25–29.

1699, the periodical had often reported events taking place in the society and was its principal—though not official—outlet for public announcements. During the seventeeth century, control of the *Journal* was in the hands of a single editor responsible to the chancellor. For many years, the editor had been the academician Jean Gallois. In 1702, as part of the efforts of Chancellor Pontchartrain and the Abbé Bignon to coordinate the cultural life of the country, control of the periodical was transferred to a group of paid editors, each a specialist in a particular scholarly domain. Fontenelle was selected as the official book reviewer for the sciences and made directly responsible to Bignon, who held both the printing privilege and the title of editor-in-chief. Considering that, for decades, Bignon was the minister of state responsible for the administration of academies, that he had been named president or vice-president of the scientific society for over thirty years, and that Fontenelle was concurrently the Academy's secretary, it is hardly extravagant to call the *Journal* a quasi-official publication of the Academy. Whether it was a requirement of law or simply carried out as a practice, the editorial board was thenceforth composed of academicians. Through them as individuals, the learned society was able to exercise considerable control, far beyond the technical limits of its power as an institution. Few incidents reveal this better than the society's reproof of the academician and astronomer Lalande, who had dared to use the pages of the *Journal des Savants* to criticize the Academy's handling of some editorial changes in the *Connaissance des Temps*.[13] On that occasion, the editor-in-chief of the *Journal* was asked to "take care that in the future nothing concerning the Academy or academicians be inserted without prior consultation."[14] Despite this polite administrative language, it was transparently clear that the learned society was piqued by its temporary loss of control over the *Journal*.

The Academy was able to make its authority felt among printers in countless other ways as well. For example, the daily *Journal de Paris*, under the editorship of Antoine-Alexis Cadet de Vaux, often opened its pages to academicians such as Lalande

13. *Journal des Savants*, May 1767, pp. 335–338.
14. Bertrand, p. 162.

and Lavoisier who would publicly explain and justify decisions made by the learned society.[15] In the choice of articles printed and books reviewed, and in decisions to temper criticisms directed against the Academy, Cadet de Vaux relied heavily upon the counsel of his brother, the chemist and academician Cadet de Gassicourt. Such personal connections, not unusual in the small world of literati in Paris, established ties that were mutually useful to the society and the newspaper. Because the *Journal* depended upon the Academy's good will to supply it with important information, it was possible for the Academy to bring considerable influence to bear upon it. When informal persuasive powers failed to curb journalistic excess and the issue was deemed sufficiently significant, the Academy could also resort to strong-arm techniques. A case in point is its successful efforts to restrain the Jesuit *Journal de Trévoux* from continuing to publish Father Louis Castel's bitter attacks against the society and the academician Réaumur.[16] The Academy's simple threat to block the renewal of the *Journal*'s printing privilege was sufficient to bring Castel and the periodical into line.

It must not be concluded on the basis of these random examples that the Academy was consciously pursuing a secret plan for the domination of the press and other scientific activities in the nation. To assume this would imply the formulation of a far-sighted "imperialistic" policy, carried out consistently for a period stretching over a century and in the face of total changes of the Academy's personnel. No such blueprint existed. In practice, the society reacted on an ad hoc basis, asserting and extending its powers as the situation permitted. But behind all its individual acts was the ever-present but unconscious conviction that, as the elite body of French science, it was best suited to determine what is proper for the development of science. Notwithstanding its innocence in these matters, appearances clearly point to the Academy's strengthening of authoritarian rule over the scientific life of Old Regime France.

A similar pattern of haphazard but effective accumulation of power can be detected in focusing on the Academy's activi-

15. Duveen and Hahn, *Isis*, LI.
16. Saurin; Schier, p. 18; Pappas, pp. 18-21.

ties in the realm of technology. Because of the peculiar administrative state of patent arrangements in France, this growth is
not visible through an examination of legal documents alone.
The basic laws governing the granting of a royal *privilège* or
"brevet d'invention" for a new or improved invention remained
essentially unchanged until a full-fledged patent system was
established by the National Assembly in January 1791. Yet,
during the eighteenth century, practices for administering the
laws evolved in such fashion that the Academy or its individual
members were placed in a central position of control.[17]

Three types of patent arrangements existed before the
Revolution. The least frequently used was the outright sale of
a new device, or "secret," to the administration. It was more
customary for the inventor to seek either a financial reward
(an "encouragement") or the temporary grant of a monopoly
to manufacture and market his invention. This last option took
the form of a *privilège* granted by letters-patent. It will be remembered that the initial role of the Academy, as understood
by Colbert and Huygens, had been to provide technical advice
on the merits of proposals whenever these were too complex
to be judged by the ministry. By the time this function was
incorporated into the Academy's 1699 regulations, the practice
was to funnel all technological projects through the society,
regardless of their nature.

In practice, the Academy judged more than merely technological issues. The minister's advisors found it impossible
to separate technical from legal and economic considerations.
Their job as consultants also included determination of an invention's novelty to prevent the Crown from issuing a right of
monopoly where one already existed, and to decide if the project was technically and commercially feasible. Functionally,
the formulation of official advice involved the investigation of
priority claims, the determination of the application of scientific
principles to ends contemplated by the inventor, and a judgment of the project's utility. The academicians were thus

17. The history of this development is pieced together from Bondois; Isoré;
McCloy, ch. XII; Gillispie, *Isis*, XLVIII, 154–158; and Parker, "French Administrators," pp. 87–101.

forced to develop a knowledge of legal intricacies and marketing conditions in addition to their expected expertise in scientific matters.

To carry out its task, the Academy made certain demands of the inventor. One was a written description of the device accompanied by a drawing and scale model to be placed in the hands of the learned society.[18] If approved, the model had to be deposited in the Academy's collections for furture reference, and the drawing was prepared for publication after expiration of the *privilège*. Both of these requirements produced serious tensions between the artisan and his judges, arising from the different goals pursued by each. The former, though anxious to obtain a favorable report, was also intent upon maintaining the ownership of his invention in order to capitalize on it. Traditionally, inventions were jealously guarded as commercial secrets of immense potential value.[19] The disclosure of technical details was considered a baring of secrets, to be sold for cash or offered only in return for capital investment. It was therefore difficult for an inventor to bring himself to release his secret to a select group of academicians without the assurance of a favorable report in advance. Even more troubling was the realization that, by discussing his invention before a public institution, he was in effect relinquishing ownership.[20] The embryonic idea of the modern patent system, by which a limited monopoly is granted in exchange for placing the invention in the public domain, was hard for the artisan to comprehend, let alone accept.

For the Academy, on the contrary, the system coincided with its deepest convictions about the advancement of science. Like the contributors to the *Encyclopédie* who shared their beliefs, the scientists were convinced that the key to progress in the practical arts was rationalization of the trades and the adoption by them of the true "scientific method." It was only by following tested scientific practices of objectivity and publicity that the artisan could be extricated from blind traditional-

18. Aucoc, *Lois*, p. lxxix.
19. Lalande, *Art du Tanneur*, preface.
20. Réaumur, p. 9; Parker, "French Administrators," p. 95.

ism and made to realize how fruitful the perfection of his craft
might be for the entire community.[21] It was for this very reason
that, as early as 1675, Colbert had instructed the Academy to
begin a description of the mechanical arts. Throughout the
eighteenth century, the members of the Academy continued to
collect and publish information about the crafts, eventually
forming an impressive collection of twenty-seven folio volumes, the *Description des Arts et Métiers*.[22] By the imposition
of their own professional standards on the artisans, the scientists
aimed at fashioning a revolution in the technological realm comparable to the one they were accomplishing in science.

This rather confident elitist aim had already been expressed
in a memoir written around 1720 on the "utility the Academy
could be for the kingdom if it received the help it needs."[23]
The author, who is assumed to be the famous metallurgist and
entomologist Réaumur, urged that the "arts would profit from
the principles of scientists" only if academicians were assigned
tasks related to practical concerns. By establishing proper, rational rules of conduct, making suggestions for the perfection
of the trades, and urging short cuts for industrial practices, great
progress could be achieved. Réaumur proposed that academicians should be appointed inspectors of manufacturing and
of civil engineering, and be allowed to sit on the Bureau de
Commerce, which set national economic policy. In short, in
his proposal the Academy was to be given technocratic direction of all enterprises remotely concerned with science, in
order to insure their eventual progress. Within the limits set by
the complexity of Old Regime practices, this was in fact the
manner by which academicians came to dominate technological
activities in France.

Given the unpremeditated character of this development,
it is difficult to ascertain the stages by which academicians became the controllers of technology. The direction of the evolution is nonetheless clearly marked. Once the practice of re-

21. Lalande, *Art de Faire le Papier*, pp. iii–iv; Rémond, pp. 20–22; Gillispie,
Isis, XLVIII, 398–407; Hahn, *Studies on Voltaire*; Parker, "French Administrators," pp. 90–91.
22. *HARS* (1666–1699), I, 199–200; Lalande, *Art du Tanneur*, preface; Cole
and Watts; Birembaut, *AIHS*, XIX, 67–68; Proust, *Diderot*, pp. 182–191.
23. Maindron, *Académie*, pp. 103–110.

ferring all requests for patents to the Academy was established, there was gradual extension to cover a large number of other technical problems encountered by the central administration. As early as 1725, the Bureau de Commerce was requesting that the Academy carry out experiments to verify claims made by petitioners.[24] By the 1730's, individual academicians, receiving salaries as consultants or inspectors, were being asked to provide their expert services for the French bureaucracy. During the rest of the century, Dufay, Hellot, Macquer, Berthollet, Montigny, Desmarest, Vaucanson, Le Roy, Vandermonde, and others often rendered the determining judgment in administrative decisions, either as members of a committee of the Academy or as individuals.[25] The distinguishing feature seems not to have been their particular position in the government hierarchy but their professional skills and the government's faith in their impartiality. It was precisely their membership in the Academy of Sciences that symbolized and guaranteed such qualities.

Beginning with Réaumur, members of the Academy were appointed to important administrative positions because of their prominence as academicians. Dufay, Hellot, Macquer, and Berthollet were placed in key jobs in the cloth industry, where they could apply their interest in the chemistry of dyes; Réaumur, Hellot, Guettard, Macquer, and Darcet were associated with the improvement of the ceramic industry; Réaumur, Hellot, Morand, Duhamel, Dietrich, and others played similar roles in mining and metallurgy.[26] Indeed, wherever one turned the bureaucracy had its academician. Some random examples will indicate how their decisions could affect the industrial life of France in general and the lives of petitioners to the Crown in particular.

In 1780, an academic committee composed of Macquer, Cadet, and Baumé concluded its report of a project for setting

24. Parker, "French Administrators," p. 93.
25. See Bonnassieux and Lelong, pp xxvii–xxviii, and Bacquié, for general assessments. For specific examples involving academicians, see AN, F^{12} 992–994; Paris, Bibliothèque Mazarine, MS. 3596; BN, MS. fr. 12305–12306; Birembaut and Thuillier; and Doyon and Liaigre, part II.
26. Guerlac, *Chymia*, V, 77–98; Birembaut, *Enseignement et Diffusion*, pp. 381–385, 403–405.

up an alum factory at Javel by saying "no advantage can result from it for commerce unless it is set up far from Paris."[27] As a result, the proponents of the plan were not immediately granted an exclusive monopoly. In another instance, the bitter arguments between Holker and Demachy over a vitriol factory were arbitrated by the Academy, with Macquer as principal judge.[28] In still another dispute, between the Genevan Pierre de Rivaz and the master clockmakers of Paris, the Academy intervened on more than three separate occasions, producing lengthy reports by D'Onsenbray, Camus, La Condamine, Deparcieux, Duhamel du Monceau, Bouguer, and Hellot.[29] Taken as typical cases, these reveal the extent of the Academy's involvement and its decisive role in matters going far beyond its original mandate.

As might be expected, the Academy was called in as an expert witness in litigations involving technical matters. On one such occasion, the report of the academic committee accurately summed up its historical position in these words:

At its origins the Academy was principally set up to examine new inventions forwarded to them on the king's behalf. Insensibly, it has become a voluntary tribunal to which individuals directly refer for a judgment, and it has considered it an honor to respond to the trust placed in it by the public. Since then, the work it has undertaken for the *Description des Arts et Métiers* has considerably augmented its jurisdiction. The Parlement and the Lieutenant-General of the Police have frequently consulted it on matters relating to the arts or mechanics. We see that it [the Academy] has never sidestepped an opportunity to be useful.[30]

On the eve of the Revolution, the Academy had truly become an adjunct of royal authority. As a consultant and arbiter in the realm of technology, it was the final court of appeals.

One further instance of the Academy's growing control of the technological world should be cited. It concerns the role assumed by the Academy in the more scientifically minded

27. *Correspondance des Propriétaires.*
28. Letter of Demachy to Macquer, undated, but written after 1772: BN, MS. fr. 12305, fol. 261.
29. Le Roy; Rivaz.
30. AdS, Reg 28 May 1788, fols. 134r and v.

craft organizations of instrument makers.[31] As a result of the joint activity of the academicians Cassini IV and Bailly, an association of "engineers in optical, mathematical, and physical instruments and other tools useful for science" was organized by the Crown in 1787. The major rationale for creating this new "corps" was to enable instrument makers to circumvent the rules of traditional artisan corporations that prevented them from working simultaneously in metal, wood, and glass without obtaining a permit from each corporation. In the machinations that led to the creation of this new group, the Academy was not involved as a body. But according to the letters patent approved by the Paris Parlement on 19 May 1787, the Academy was chosen as the body which was to screen applications for membership in this select new company. Since the Academy considered a knowledge of mathematics, physics, and mechanics to be a necessary prerequisite for membership, the group remained small and exclusive.[32] Each member took special pride in having received the approbation of the Academy of Sciences.

It was precisely this exclusiveness that added to the Academy's prestige. Petitioners who had received the Academy's approval tended to parade their success in public and to attempt to capitalize on the learned society's name. The phrase "approuvé par l'Académie Royale des Sciences" carried much the same weight in France as does "by special appointment to Her Majesty the Queen" in England. Since the phrase could serve as an advertisement for a commercial product, we may reasonably assume that manufacturers often submitted their products to the Academy with this sole purpose in mind. No matter how insignificant the matter appeared, the Academy always handed down pronouncements garbed in the traditional language of officialdom.[33] The result was sometimes ludicrous, turning

---

31. Cassini, pp. 86–94, 217–225; Bailly, *Recueil*, pp. 120–125; Devic, pp. 105–109; Daumas, *AIHS*; Chapin, *French Historical Studies*, V, 382–383.
32. AdS, Reg 13 June 1789, fols. 162v–163r, and 8 July 1789, fol. 190r.
33. An exception to this rule is cited in Lalande, "Collection," p. 121. According to a vote of the Academy on 2 Dec. 1769, Malincot's new wigs were refused a hearing. See Doyon and Liaigre, p. 445, for a different outcome of the Malecot [*sic*] affair.

learned academicians into unwitting sponsors of such varied products as "Sieur Vicq's inalterable and always fluid Argonne ink," a new facial cosmetic, and a shoeshine wax called the "cire coquette."[34]

Artisans were not the only group to benefit from the Academy's prestige. The title of academician carried with it a remarkable variety of personal prerogatives, guaranteed by either law or usage. Thus, in 1719, every member of the learned society was granted the right of *committimus*, which allowed him to transfer any personal litigation to the Paris courts of law.[35] It is most likely that academicians were also exempt from military service and from the "guets et gardes."[36] In 1786, through the efforts of the Duc de La Rochefoucauld, Calonne granted the scientists an exemption from the 10 per cent withholding tax imposed on all royal pensions.[37] Such privileges were not taken lightly, especially since they were customarily reserved for the highest ranks of society.

On a different level, the title of academician gave scientists a considerable advantage in securing remunerative employment. A number of positions requiring technical skills were reserved for them. Positions in the royal manufacture of tapestries and as inspector of ceramics were usually held by academicians. Many of the teaching positions at the Collège Royal were entrusted to academicians; and after the letters patent of 1772, academicians were explicitly given preference over other candidates.[38] A similar situation prevailed with the Chair of Docimasy created at the Mint in 1778. It was set up for the academician Balthazar-Georges Sage and was thereafter reserved for fellow academicians.[39] The holder of the Chair of Hydrodynamics at the Academy of Architecture was also

34. *Nouvelle et Seule Manufacture*; Duveen and Klickstein, *Bibliography*, p. 40; Mercier, XI, 261.
35. Maindron, *Académie*, pp. 43–45.
36. This was a privilege also granted to members of the Académie Française and the Academy of Painting and Sculpture. See Aucoc, *Lois*, pp. xxxiii, xliv, cxx.
37. Messier comments on this in the margins of his copy of *Connaissance des Temps* for 1786, pp. 120 and 236–237, deposited in the Harvard University Library. See also AN, F¹⁷ 1021B, dossier 1, for the ministerial correspondence.
38. "Documents sur le Collège de France," p. 410.
39. Lalande, "Collection," p. 84, cites a letter of Necker dated 18 Nov. 1778 to this effect. See also Birembaut, *Enseignement et Diffusion*, p. 388.

required by law to be a member of the Academy of Sciences, and the same held true of all examiners of mathematics for the numerous military and naval schools throughout the land.[40] Academic domination at its extreme was exemplified in the school of mathematics at Reims where the Academy was requested to select a professor, who was then forced to submit his curriculum to the inspection of the Academy.[41]

It may be assumed that there were other ways by which the Academy strengthened its commanding position and created an image of itself as a major *corps d'état* of the land. In Old Regime France, power and prestige mutually supported each other to the point where they were often inseparable. Both were underscored by public ceremonial events, the symbolic meaning of which could hardly escape the public eye.[42] The Academy, no less than other *corps d'état*, developed its share of luster and pomp.

One occasion for display was the semiannual public meeting held at the Louvre. Like the sovereign law courts and the Collège Royal, the Academy took particular care to arrange stately *rentrées publiques* following its official vacations at Easter-time and during the fall.[43] On these occasions, specially selected academicians would read carefully prepared papers, and the secretary would eulogize the memory of deceased members and present new academicians to the public. Accounts of these gatherings always stressed the majesty of the event, and the newspapers generally reported them as they would other significant social affairs.[44] Some special events at the Academy were also publicized with great enthusiasm, particularly the visit of important heads of state. Such occasions were evidence of the high distinction bestowed upon the Academy. One meticulous academician, Lalande, carefully listed these visits, which, in addition to the King of France, the Dauphin,

40. On the Chair of Hydrodynamics, see a decision of the Conseil du Roi, 2 Feb. 1777, AN, O^{1}1931^8, pièce 3. For the military schools, Hahn, *Enseignement et Diffusion*, p. 519.
41. Lalande, "Collection," pp. 79–80, 84–90; Féry.
42. Ford, pp. 55–58.
43. Aucoc, *Lois*, p. xc. For descriptions of these public exercises, see Trembley, pp. 103–117, and Halem, pp. 341–342.
44. *Journal Gratuit*, pp. 36–37, 72; Lalande, "Collection," p. 50. Comments about the *rentrées* appeared regularly in the *Mercure*.

and members of the French royal family, included Peter the Great in 1717, the King of Denmark in 1768, the King of Sweden in 1770, Emperor Joseph II in 1777, Czar Paul I in 1782, Prince Henry in 1784, and even the son of an African king in 1785.[45]

Another symbolic event that enhanced the Academy's prestige was the privilege, conferred upon it in 1774, of presenting newly elected members to the king at Versailles, together with the annual gift of the *Histoire et Mémoires*.[46] On such occasions the various academies competed for royal attention. One privilege that created some stir was the right to genuflect before the monarch at court, granted to the Académie Française but denied to the Academy of Sciences.[47] That record of such details has been preserved is an indication of the importance everyone attached to what now seem trivial matters. Some consolation for this rejection was afforded when the Academy of Sciences achieved the minor triumph of receiving tickets to attend the coronation of Louis XVI at Reims.[48] Another ceremony carefully observed was the *Te Deum* sung in the nearby Church of the Oratory whenever an important member of the royal family was born or married and the special masses celebrated when one was ill or died; the members of the Academy attended these ceremonies in large numbers with their colleagues of the Academy of Inscriptions and Medals, with whom they also shared a special mass on the feast day of St. Louis.[49]

Although these rituals bore little direct relationship to the scientific work of the Academy, their importance is more significant than appears on the surface. Each of these details was a part of the image the public held of the learned society. Ac-

45. Lalande, "Collection," pp. 11, 27. These visits are usually cited in the "Histoire" section of the *HARS*. Special speeches prepared for these occasions may be found in the collected works of Fontenelle, D'Alembert, and Condorcet. See also Maury, *Académie des Sciences*, pp. 179–182.
46. Lalande, "Collection," p. 8; AdS, Lav 890.
47. Letter of Courtanvaux to Macquer dated 10 March 1774 in BN, MS. fr. 12305, fol. 222.
48. Lalande, "Collection," p. 8.
49. Letter of Father Bertier to Father Louis Cotte, undated but written before 1783, in which Bertier grants the privilege of saying mass for the Academy: in Laon, Bibliothèque Municipale, MS. 16C A6. See also Lalande, "Collection," pp. 82–84.

ademicians were the high priests of the scientific and technical world, and also, as a group, formed a recognized and respected public establishment, decked out with all the standard Old Regime trappings. On the eve of the Revolution, the Academy of Sciences stood not only for the promotion of science, but also symbolized, as much as any other public institution, the royal administrative apparatus. It fully merited every word of its title, the Royal Academy of Sciences.

If the Academy's scientific role and its public stance were often confused, it was in part the group's own doing. As has been suggested, one of the sources of its success was the ability to satisfy the scientific demands of the age while functioning within the framework of contemporary society. In many instances these two demands reinforced each other. The over-all organization of the scientific community along elitist lines, for example, was readily accepted by a society that cherished corporative privileges and respected proper authority. In Old Regime society, the Academy was simply taken as another of the powerful, legally constituted, and specialized *corps d'état*, with its appointed place in the contemporary *Who's Who*, the *Almanach Royal*. But the full extent to which the learned society was integrating itself into French society while at the same time answering the needs of its profession cannot be perceived without penetrating into the institution's innermost procedures.

Under the direction of the Abbé Bignon, a number of structural innovations were incorporated into the 1699 regulations, ostensibly to rejuvenate the Academy. Through these regulations, its size was increased, attendance and productivity were encouraged, and the continuing support of the Crown was guaranteed. All three of these measures clearly worked to the institution's benefit. Yet most of the articles of the regulations deal with the society's internal organization, spelling out in great detail its structure, the election, tenure, and duties of its officers, the procedures for the appointment of new members and their voting rights, and the precise information about the location, frequency, and length of meetings. So meticulously framed is this document that one might even expect a seating chart to accompany the regulations. Bignon did in fact prepare one, and gave specific instructions about the seating at the Academy's first meeting under the new regulations.[50]

Such extreme concern with minute detail might be written off as the perennial occupational disease of "constitution makers." Every framer of institutional bylaws has a tendency to imitate legal style. But to dismiss the matter out of hand would be to ignore the importance of Bignon's regulations, for they were an attempt to legitimize a body by formalizing its practices. With a set of acceptable bylaws, the Academy assured itself of a semilegal existence independent of changes of personnel or of interest on the part of the government. By setting down institutional practices in writing and having them published, Bignon was effectively taking the Academy out of the personal domain of the Crown and making it a public institution. He was conferring upon it a second, more enduring lease on life, or, as Fontenelle put it, fashioning a *renouvellement.*

Bignon also introduced new features which, like many other clauses in his regulations, reveal an unusual concern with an orderliness and hierarchy within the Academy that had generally been absent in the seventeenth century. These features stemmed from his apparent desire to introduce into the Academy the notions of etiquette, status, rank, and propriety so prevalent in Old Regime society. It was as if he yearned to turn it into an institution fully integrated with that society. We may assume that, in the process, much of the intimacy and freedom that had existed when the society was smaller and less structured gave way before this carefully conceived plan regulating every aspect of the Academy's life.

Most novel was the creation of a complex personnel structure where none had existed before. Two basically different but concurrent ways of organizing membership were adopted, one relating principally to scientific interests, the other tied to a combination of occupational affiliation, residence, and seniority. Because the latter defined hierarchical status within the society, we will refer to it as the vertical organization. In many obvious

---

50. Aucoc, *Lois,* p. xc; *HARS* (1699), p. 14. For an account of the seating arrangements later in the century, see Lalande's diagram in "Collection," p. 23, and Birembaut, *RHS,* X, 152–154.

ways, it is Bignon's effort to make the Academy a microcosm of French society.

Within this vertical structure academicians were conveniently divided into three groups. One of the new and special classes was composed of members known as the *honoraires*. Usually selected from the high clergy, nobility, or persons holding important government positions, they were representative of the highest type of amateurs or patrons of science, who had been excluded from membership in 1666. With a few notable exceptions—including the Abbé Bignon himself—honorary members rarely attended working meetings of the society. They were more often in evidence at the *rentrées publiques*, when an important dignitary paid an announced visit, or when the selection of a new academician was scheduled to take place. Normally, their official functions in society kept them from playing an active role in the institution's proceedings, unless they became officers of the Academy. Every year, the Crown selected from among them a president and vice-president, who formed the official link between the society and the Crown.[51] The total number of *honoraires* was always small, increasing from ten according to the 1699 statutes to twelve beginning in 1716.

Another group, the nonresident academicians, likewise played a minor role in the institution because of their inability to participate regularly in meetings. Two major categories may be distinguished among them, the most numerous being the *correspondants*.[52] Each of these corresponding members resided outside Paris, in France or abroad, and communicated with the society through one of the regular members. No limit to the number of correspondents "attached" to a regular academician was set, but in practice they rarely numbered as a group more than one hundred, and only a small percentage was regularly engaged in significant scientific correspondence. The other and more select category of nonresident academicians was known as *associés*. In the eighteenth century, eight

51. A list of officers to 1770 is in Rozier, I, xix–xxii, and a more complete one is in Lalande, "Collection," pp. 101–102.
52. Lists of correspondents appeared annually in the *Connaissance des Temps* and the *Almanach Royal*.

positions were traditionally reserved for foreigners and became known as the class of *associés étrangers*. In 1716, still another class, the *associés régnicoles* (more commonly *associés libres*), was created for prominent French scientists unable to attend meetings regularly but not sufficiently high on the social ladder to be named honorary members. By 1765, their ranks had risen to a statutory dozen, which also included members of religious orders, scientifically inclined officers of the army or navy, or prominent doctors or surgeons. Among all the nonresident members, the *associés libres* were the most active, always including several members who participated in academic life as much as any regular academician.[53]

There remains the most powerful and significant group of academicians, numbering slightly more than fifty, which composed the core of the institution during the eighteenth century. These were the "professionals," the working academicians who were in constant attendance at the biweekly meetings, sitting on all the individual ad hoc committees appointed to expedite the society's daily business, and generally monopolizing the pages of the *Histoire et Mémoires*. For them, the Academy was the principal focus of activity, even though they might concurrently be teaching science or directing a technical governmental enterprise. These other functions were so closely linked with academic status that, in the end, all the scientific activities of the individual, whether carried out at the Louvre, in the Jardin du Roi, or at the Sèvres porcelain manufacture, stood for the same thing. What mattered was professional involvement with science, symbolized as it was by membership in the Academy.

Within this last group, there were hierarchical distinctions not unlike those that prevailed in contemporary guilds. Most of them were based upon seniority within each scientific discipline. The highest level that could be attained within the Academy was the rank of perpetual secretary, a post of truly commanding influence. It was held by only four persons during the century, two of whom have left a permanent mark on French

53. Chapin, *RHS*, XVIII, 10–13. Clerics like Pingré were in the class of associates because of the regulation prohibiting them from filling a seat in the hierarchy leading to a pension. See Aucoc, *Lois*, pp. lxxxvi, xciii.

culture: Fontenelle and Condorcet. The office of treasurer, also held for life and thus called "perpetual," was of minor significance, since the finances of the society were controlled by a standing committee of the entire body. None of the holders of the office—the two Couplets, Buffon, Tillet, and Lavoisier—is particularly remembered for his tenure. Both the secretary and the treasurer, as well as the annual director and vice-director, were elected from among the subgroup of "professional" academicians known as *pensionnaires*. As the title suggests, their position entitled them to a stipend from the government, which averaged around 2,000 livres a year.[54] Below this privileged class stood the *associés* and, until 1785, a third and still lower class, known first as *élèves* then as *adjoints*, also existed. It was customary to move up the ladder step by step from the lowest rung, so that advancement within the organization depended upon the death or retirement of senior scientists. The numbers in each academic class were prescribed and limited, and competition for a position even on the lowest rung of the ladder was fierce.

Each class of academicians had special duties and privileges, the most important of which belonged to the pensioners. Election to a post as officer of the society and membership on standing committees was generally reserved for these elders.[55] There were three such committees, dealing with financial, secretarial, and printing matters, on which the secretary and treasurer as well as the annual directors sat as members *ex officio*. An even more important feature of the 1699 regulations was that voting eligibility was dependent upon an academician's place in the vertical hierarchy. When the Academy dealt with questions relating to science, the honorary members, pensioners, and associates had the franchise, but never students or adjuncts. On matters pertaining to the administration of the

54. At the beginning of the eighteenth century, stipends varied greatly. On the eve of the Revolution, the three pensioners received 3,000, 1,800, and 1,200 livres a year, depending upon their seniority. For details see JT; Lalande, "Collection," pp. 7, 41–42; letter of Malesherbes to Fouchy dated 18 Nov. 1775 in BN, MS. n.a.f. 5151, p. 32; and Maindron, *Académie*, chs. VI and VII.

55. According to Lalande, "Collection," p. 43, one associate was made a member of the treasury committee from 1725 to 1748. This is the only exception I have found.

Academy—including the critical election of new members—
voting was restricted to the *honoraires* and *pensionnaires*. Since
the former were at times absent, this meant that most of the
internal affairs of the Academy were in general decreed by a
group of some twenty elders. A more elitist or authoritarian
system compatible with scientific expertise could hardly have
been conceived.[56]

The division of academicians into hierarchical "classes"
according to social position and seniority was a significant reflection of the habits of French society. Whether or not Bignon
consciously introduced these innovations to mirror societal
norms, he did manage to incorporate the elitist notions held by
the academicians themselves into a rational system consistent
with Old Regime values. Although talent remained the major
criterion for election and entrance to the Academy, it was completely disregarded once the scientist was admitted into the
fold. Within the organization, status was made dependent upon
factors essentially irrelevant to science. It proved to be a workable and even beneficial arrangement for the Academy as long
as societal and scientific values did not conflict.

Because of the special position the Academy held in French
society and the particular relationship it had with the government, a great number of extrascientific factors did in fact in-
trude into the institution's election practices. According to the
statutes, the king made appointments to membership in the
Academy, relying upon the recommendation of the learned
society. In practice, the Crown did not always respect the
Academy's suggestions or refrain from putting pressure upon
academicians before elections took place. As might well be
expected, governmental influence in appointments was strong-
est with regard to honorary membership and the *associés libres*,
and considerably weaker in the case of working academicians.

A large number of elections to honorary membership were
made with little consideration for the scientific merits of the
candidates. All the ministers of state who, by their office, were
placed in charge of the academies were automatically elected
when a vacancy occurred. Bignon, Breteuil, the two Amelots,
Maurepas, and Malesherbes are cases in point. It may also

56. Duclos noted this point in "Honoraire."

be assumed that dignitaries such as Louvois, Law, Fleury, De Castries, and Loménie de Brienne were elected either by pressure from outside or simply out of a sense of diplomatic courtesy. The same is true for a number of foreign associates, including the Princes Jablonowski and Löwenstein and Czar Peter the Great. According to the well-informed academician Pierre-Charles Le Monnier, the Crown also considered the appointment of *associés libres* within its prerogative. In a letter to a candidate for one of these positions, he pointed out that

When d'Argenson created them he said clearly he considered these positions of *associés libres* to be at his disposal, leaving us the others to fill as we choose, from which it follows that if these places are given out as they used to be and if in fact the Court owns them, you should not lose time in making proper arrangements there.[57]

This situation helps to explain the selection of some academicians whose scientific credentials are difficult to establish.

But the Crown also interfered with the Academy's choice of regular membership, in which scientific competence was of crucial concern. Among such instances, there are a few in which the Academy did not choose to stand on its presumed freedom, such as in the elections of two now famous scientists, Lavoisier and Lamarck. In 1768, Lavoisier was the Academy's first choice for a vacant seat, but the ministers overrode the election and appointed Antoine-Gabriel Jars in his stead.[58] In 1779, probably through ministerial protection, Lamarck was picked over the Academy's stated preference, Descemet.[59] To understand the Academy's reluctance to protest this infringement of its prerogatives, it must be realized that the votes in the Academy had been close, and that the Crown's final choice was actually the Academy's second choice. But when interference was more blatant or less justifiable, the Academy did not hesitate to react. Two instances can be cited—the Bordenave affair in 1774 and the Fontanieu affair in 1778—in which the Crown pressed its desire to have questionable candidates ap-

57. Letter of Le Monnier to Bory dated 25 Sept. 1761 in BN, MS. fr. 6349, fol. 41.
58. Lalande, "Collection," p. 119; Sage, *Opuscules*, p. 235; Guerlac, *Chymia*, V, 97–98.
59. Lalande, "Collection," pp. 9, 119; Cassini, "Mes Annales," p. 17 n.

pointed to academic positions.[60] So strong were the Academy's feelings on this matter that on both occasions delegations protesting the minister's action were immediately dispatched to Versailles where they placed their complaints at his feet. Macquer's explanation of the Academy's unusual action, written in a private letter, clearly indicates the sentiments academicians held about the impropriety of governmental influence in certain crucial areas of the society's activities. He took great pride in the fact that the "elite" of science is chosen on the basis of talent

and without regard to their estate (provided there is nothing dishonorable or too lowly therein); I have never understood the Academy to be a political body in which there must be a proportional representation according to professions . . . I have suffered on certain occasions from seeing conniving mediocrity win out over talent unsupported by patronage . . . I have often feared . . . that if the liberty of elections were hampered, this dangerous mediocrity would prevail and ultimately bring about a shameful ruin of the Academy . . .[61]

As director for that year, Macquer used the same strong language in presenting the Academy's grievances to the Duc de Lavrillière at Versailles. In both cases, an administrative solution accomplished the ends demanded by the Academy without publicly countermanding the minister's authority. Both new academicians were honorably retired or, to use contemporary phraseology, "veteranized."[62]

The academicians accepted the values of Old Regime society, but also recognized there had to be limits to the intrusion of extrascientific considerations in their institution. As Le Monnier had suggested, appointments to regular positions had to remain within the purview of the learned society. Without this measure of internal freedom, the scientific mission and ultimately the renown of the Academy would be seriously com-

60. For details on Fontanieu, see AdS, Reg 14 and 18 March 1778, fols. 92v–93v. On Bordenave, see AdS, Reg 16 and 19 March 1774, fols. 95v–97v; AdS, dossiers Bordenave and Vicq d'Azyr; and BN, MS. fr. 9134, fols. 99–101.
61. Draft of letter from Macquer to Mesnard de Chouzy dated 24 March 1774: BN, MS. fr. 9134, fol. 133r and v.
62. A special category of *vétérans* (retired members) was created to allow younger and more active academicians to reach the position of *pensionnaire*. See Lalande, "Collection," pp. 47, 123.

promised. If it were to maintain its standing as leader of the scientific community and consultant to the Crown, it was essential that the Academy maintain a degree of autonomy from society at large. For its continued success, the learned society vitally needed to forge a harmonious relationship between those aspects of its organization that mirrored the general culture and those essential for the proper furtherance of science.

A balance between these two tendencies was indeed discovered in the early decades of the eighteenth century. In a way, the discovery of this golden mean constituted the very essence of the Academy's institutional life. As this delicate balance was slowly fixed into habit, practices, and laws, the Academy's future became inevitably tied to the continued stability of the demands made upon it by both science and society. Gradually and imperceptibly, success turned the Academy into an association that had a vested interest in the permanence of its environment. Though dedicated to progress, the Academy was in fact becoming an essentially conservative institution. The great challenge to its existence came from the progress of science and the development of society, which were constantly undermining this desired stability. Its survival therefore depended greatly upon the strength of these forces of change and the institution's ability to cope with new challenges.

Chapter 4

The Growth of Science

SO FAR, our discussion of the Academy's establishment and prosperity has concentrated on the evolution of the institution, with little reference to the changing nature of its environment. In fact, both the scientific community and French society were undergoing important transformations throughout the century, which, in their political phase, climaxed in the dramatic events of the 1790's. As will be seen in the next chapter, the new demands placed upon the Academy by French society even before the eruption of the Revolution were a threat to its institutionalized practices. They were all the more dangerous because the Academy, like many institutions flushed with success, was losing the flexibility that had characterized its early years. The increasingly entrenched and self-satisfied academicians were not easily moved to accept or even to recognize societal change. When they were disposed at all to consider the effects of evolving conditions upon their institution, it was to the growth of science itself that they naturally gave their attention. In their eyes, the Academy's principal function was, after all, to promote the advancement of science. Our first concern will therefore be with the evolution of science itself, particularly with its pronounced growth.

During the eighteenth century, the phenomenon characteristic of all levels of scientific activity was expansion. There was growth in the size of the profession, in the public which it addressed, in the quantity of its publications, in the scope of its activities, in the precision of instruments demanded by experimentalists, and in the range of scientific application. We easily recognize the phenomenon to be a particular instance of the general pattern of scientific development in post-Renaissance Western civilization, but we are without any close understanding of its process in eighteenth-century France.[1] As will become evident, growth was neither uniform in all disciplines nor important simply in quantitative terms. It was accompanied and often fostered by marked shifts of interest from one domain to another, and the patterns of growth differed depending upon the special needs of the particular discipline. Our concern with the complex interrelationships of these changes arises principally from the realization that they created immense strains within an institution that had been organized and shaped to meet a different set of circumstances at an earlier time.

It is impossible to estimate the size of the French scientific community. The very phrase "scientific community" is an anachronism in the Old Regime because there was no category corresponding to the term. In the statistical records that exist, government officials never bothered to separate men of science from men of letters, quite understandable if we recall the intimate connection between the Republic of Science and the Republic of Letters.[2] The term "savant," when employed at all, retained its older meaning of man of learning rather than specialist in the study of nature. Perhaps the only functional sign of membership in the professional community was admission to the Academy of Sciences. But since its numbers were fixed by law to about fifty-six working scientists, membership statistics provide no direct evidence of the expansion of science.

The difficulty of measuring the growth of the scientific community must not prevent us from considering it a significant phenomenon. Every observer of the panorama of activities presented by eighteenth-century science, including the more reflective contemporaries, has noted an expansion discernible

1. Some important suggestions are to be found scattered in Mornet, *Sciences de la Nature*; *L'Encyclopédie et le Progrès*; Morazé; Roger, *Sciences de la Vie*; and *Enseignement et Diffusion*.
2. The distribution of occupations for tax purposes in 1695 does not include a specific category for scientists, and the situation remained unchanged a century later. See *Correspondance des Contrôleurs Généraux*, I, 565–574; Tourneux, "Un Projet"; and AN, O¹611, O¹666–686.

on two mutually encouraging levels. For one, scientific information was being diffused among the literate classes of France
at a rapid pace. Second, the success of scientific enterprise was
continually providing more scientists with a livelihood. Science
was thus becoming at once a more popular and a more professional discipline.

In no other period of French history have scientific principles and inventions held such a predominant share of public
attention.[3] From as early as the physics demonstrations of the
Cartesian Rohault to the very eve of the Revolution, when all
minds were abruptly turned to political events, scientific or
pseudoscientific matters were much in vogue. Evidence for this
abounds. Every intellectual of the Enlightenment made at least
a stab at scientific research, often initiating his career with an
essay or treatise related to the "new philosophy."[4] Locke, Voltaire, Montesquieu, Hume, Holbach, Diderot, Rousseau, Kant,
and dozens of lesser minds could boast of familiarity with
scientific problems that often constituted an integral part of
their more philosophical literary productions. The greatest
publishing venture of the century, the *Encyclopédie*, or the
"Dictionary of Sciences, Arts and Crafts," was, to a large extent, an expression of the current enthusiasm for scientific
information. Such information was among the favorite topics
of conversation in the threadbare reading rooms of provincial
towns, the sophisticated *salons* of Paris, and even at court.[5]
Periodicals carried announcements of public lectures and demonstrations alongside traditional political news. Science was the
true passion of the century at all literate levels of society, in
every urban center of France, and even among the progressively
minded gentlemen-farmers.

Although science was generally popular throughout the
century, there were special patterns to this popularity. In the
early years of the Academy's existence, there was a considerable vogue for anatomy, reflected in the statistics for anatomical
publications during the last quarter of the seventeenth century.

3. P. Lacroix; McCloy.
4. P. Smith, pp. 34–40; Guerlac, *Daedalus.*
5. Delorme, "Académie et Salons"; Vernière; Mornet, *Origines Intellectuelles,*
chs. III and IV of each part; Barrière, *Vie Intellectuelle,* pp. 309–360.

For that period, there were twice as many titles as in the preceding and subsequent twenty-five-year spans.[6] According to
Fontenelle, it was the academician Du Verney who was most
responsible for the fad of collecting and displaying dried
anatomical parts that momentarily captured the fancy of aristocratic circles. Anatomy was also made a popular subject of
discussion by the spectacular discoveries of "worms" (thought
to be the source of generation), the availability of the micro-
scope, and the handsome plates illustrating anatomical dissec-
tions which the Academy repeatedly published. In the 1720's,
the fad for anatomy—and especially for grotesque embryonic
monsters—gave way before an interest in insects, conveniently
summarized in Réaumur's six-volume masterpiece produced
from 1734 to 1742. His *Mémoires pour Servir à l'Histoire des
Insectes* was favorably reviewed in the *Journal Historique de
Verdun* by a writer who was quick to note the source of its
widespread success. Entomology, he pointed out, gave direct
and immediate pleasure even to those whose understanding of
science was superficial.[7] The ease with which natural history
was comprehensible to the layman lies at the root of the mania
for collecting that caught the public. Any person of means
could become a scientific amateur overnight by purchasing a
"cabinet," and displaying its wondrous contents to his circle
of friends. Indeed, random collecting was the virtuoso's favorite sport.

The vogue for natural history began much earlier than
Mornet's pioneering study would have us believe.[8] In addition
to anatomy and entomology, botany was also popular, particularly at the Jardin du Roi where Tournefort, Vaillant, and the
Jussieu dynasty lectured. The crowd of listeners, at first largely
confined to pharmacists seeking information on medicinal
herbs, was so swelled by curious enthusiasts that a larger amphi-
theater was eventually required. By the 1740's, another related
vogue, this time for conchology and lithology, took hold of
the French public. It became sufficiently significant for the

6. Roger, *Sciences de la Vie*, p. 182 n. 92; Doyon and Liaigre, p. 126 n. 74;
Fontenelle, "Eloge de M. Du Verney," *HARS* (1730), p. 124.
7. Quoted in Torlais, *Réaumur*, p. 124.
8. Mornet, *Sciences de la Nature*.

creation of a regular commercial market for the sale of dozens of important *cabinets* of natural history at auction. In an exhaustive study of these miniature private museums of the pre-Revolutionary period, Laissus has identified over two hundred names of *cabinet* owners in Paris alone.[9] Some were famed scientists such as Adanson, Duhamel du Monceau, Bernard de Jussieu, Lamarck, Réaumur, and Romé de l'Isle. But, for the most part, the list is made up of dozens of amateur collectors from the ranks of the nobility, government administration, the military establishment, the clergy, the medical professions, architecture—and even the Duc d'Orléans's *chef de cuisine.*

The fad for natural history is evidenced in many ways. From midcentury on, dozens of dictionaries, guides, and almanacs of natural history—which included horticulture—made their appearance.[10] Periodical journals were also founded, and numerous private lecture courses dotted the capital. Not only was there a lively book trade for specialized treatises on natural history, but particularly for a modern and more secular version of cosmological treatises dealing with the wonders of nature. Mornet, in his study of private libraries, noted two titles that were particularly popular.[11] One was the Abbé Pluche's *Spectacle de la Nature*, which turned the lessons of nature to moralistic ends. The other, more popular and universally known, was Buffon's great *Histoire Naturelle*, adorned with magnificent engravings and the author's elegant style.[12] Popular also were the journals edited by Gautier d'Agoty, Toussaint, the Abbé Rozier, Buc'hoz, and the Abbé Grosier, who devoted a large number of pages in their publications to natural history. From 1752 on, the number of such journals increased, as did the number of pages in each devoted to natural history, presumably in response to the subscribers' interests.[13] The number of

9. Laissus, *Enseignement et Diffusion*, pp. 659–670. See also Lamy; and Townsend, I, 13–30.
10. Hérissant, pp. 17–40; Grand-Carteret. Lettsom's book on natural-history collecting, translated into French in 1775, was quite popular, as had been Turgot's manual of 1758.
11. Mornet, "Les Enseignements des Bibliothèques Privées." This study has recently been complemented by Furet.
12. An important discussion of this book by Roger is given in his critical edition, pp. cxiv–cxlix.
13. Kronick, pp. 87–106.

private courses in natural history multiplied noticeably after midcentury, so that lecturers such as Valmont de Bomare and Fourcroy could both earn a livelihood from their performances and develop a substantial public following.[14]

In the provinces, patterns of growth can also be perceived to have followed the Parisian lead, with a lag of about a decade. The popularity of natural history and experimental science in general was almost assured, since provincial circles were principally populated by amateurs whose interest was more often the contemplation of nature, or the application of scientific knowledge than its progress.[15] In Lyon, Bordeaux, and Rouen, for example, learned societies and educational institutions seldom dealt with the mathematical sciences, or even with chemistry, which demanded a considerable body of substantive knowledge and apparatus that amateurs would not normally possess. In the few cities where important institutions of science existed—such as Montpellier—the balance between amateur and professional science was more like that of Paris.[16]

Another of the popular forms of scientific activity was the field trip. Botanizing, for example, became a respectable and relatively simple endeavor that attracted the enthusiast and professional alike, providing each with a different kind of satisfaction.[17] For the former, it was a sport that filled his time usefully without demanding endless preparation, and for the latter it was an important means for identifying local flora. Others found vicarious pleasure in the growing body of travel literature, which could be at once charming and informative as well as surrounded by the prestige of science. So avid were the readers of travel accounts that they were often unable to distinguish between real and imaginary voyages.[18] Small won-

---

14. The comments of Raymond-Latour, I, 146–208, and Smeaton, *Fourcroy*, pp. 6–18, corroborate my detailed studies of scientific lectures announced in the *Journal de Paris* (1777–1789).
15. Mornet, *Origines Intellectuelles*, pp. 145–152, 298–300; Roche, "Milieux," pp. 157–176; Proust, *L'Encyclopédisme*, pp. 59–70.
16. For Lyon, Bordeaux, Rouen, and Montpellier, see respectively Ruplinger, and Trénard; Barrière, *Académie de Bordeaux*; Martin; and Dulieu. Other important recent studies of provincial academies are by Roche (Châlons-sur-Marne), Tisserand (Dijon), Jean Cousin (Besançon), and Torlais, *RHS* (La Rochelle).
17. Rey; Bonnet.
18. Atkinson; Adams.

der, since the authors of descriptions of fictional expeditions have always consciously imitated the style and format of real travelers, and ever embellished their already exotic stories.

The eighteenth century is rich in significant scientific expeditions led by Frenchmen, particularly those sponsored by or associated with the clergy and the Company of the Indies.[19] In the late seventeenth century, even though the scientific aim of the voyages was often subordinated to religious or economic purposes, every expedition was an occasion for the further collection of data. All travelers, and especially the Jesuit Fathers, regularly offered their services to the advancement of science. Whatever his motive, the traveler generally brought back exciting reports. Indeed, the accounts of his visits to foreign lands often launched his career and reputation. Many academicians, such as Tournefort, Adanson, Maupertuis, Bouguer, La Condamine, Bougainville, and Rochon, were barely known prior to their experience in the field. The names of other travelers, such as Plumier, Commerson, Sonnerat, Poivre, and Dombey, also came to be known in polite circles of French society. The fame they derived from this vogue for travel accounts resulted from a curious combination of fascination for the exotic, the craving for new knowledge, and a latent sense of colonialism.

Accounts of voyages were not solely concerned with natural history. Many of them provided data for formulating and testing highly speculative theories of physical anthropology or ethnography, then hardly recognized as subjects worthy of scientific investigation.[20] The contributions these voyages made to geography and cartography were much more meaningful and just as easily understandable to the layman as the description of an exotic animal or the drawing of a luxuriant plant. Above all, the popularity of voyage literature rested upon its promise to resolve major scientific debates. In this respect, two kinds of expeditions were particularly savored by the public at large as well as by the professional community of scientists— one geodesic, the other astronomical.[21] In 1735, two missions

19. Fournier; Huard and Wong; Steele, pp. 12–25; Broc, pp. 137–142; Faivre.
20. Stocking, pp. 138–141.
21. Chapin, "Expeditions"; Woolf; T. B. Jones; Faivre.

were sent out, to Peru and Lapland, to measure the length of a degree of the meridian to settle the dispute over the shape of the earth provoked by Newtonian theories. Twice in the 1760's, full-scale scientific expeditions were sent out to observe transits of Venus in the hope of obtaining accurate dimensions for the solar system. Both received considerable publicity, which further enhanced the image of science in the public eye.

Fads in science were not confined to disciplines still in a fact-gathering stage of development. The imagination of the educated populace was equally receptive to the scientific demonstrations of itinerant lecturers, to inventors of new mechanical devices, and to discoverers of panaceas for man's persistent ailments. The "enlightened" century had its share of the superstitious who unwittingly welcomed magic and quackery in the guise of science. Particularly captivating were the imponderable fluids whose strange powers were only then being discovered and exploited. The discussions about electric, igneous, and magnetic substances left their mark upon the French public and the scientific community as well.

In the 1750's, the respected academician Nollet brought experimental physics into immediate popularity with his demonstrations of static electricity, which were imitated with even greater success by his followers, Brisson, Sigaud de la Fond, Deparcieux, Lefèvre-Gineau, and Charles.[22] To keep abreast of the new fads, the true adept of science now had to enlarge his scientific *cabinet* to include instruments of experimental physics. By the end of the century, the *cabinet de physique* was fetching a handsome price at public auctions, often surpassing that of natural-history collections. Even the popular fairs on the edges of Paris boasted demonstrators of physics armed with all the latest spectacular paraphernalia.[23] One of the more famous of these, Comus, dispensed electric-shock treatments as medical therapy while expounding on the virtues of the Leyden

22. Daumas, *Cabinets de Physique*; Torlais, *Nollet* and "La Physique Expérimentale." It is possible that this tradition stems from Rohault and Polinière. See Corson, p. 404.
23. *Journal de Paris*, 26 and 27 March 1780; Campardon, I, 214–215; Torlais, "Un Prestidigitateur Célèbre."

jar. So successful was his showmanship that he managed to win the coveted title of "physicien du Roi."

If electric matters were the talk of the town in the 1760's, they were soon replaced by that of aeriform fluids discovered and immediately popularized in the next decade. Priestley's and Lavoisier's revelations were more fully appreciated by their contemporaries for the vast possibilities of application that they promised than for the theoretical revolution they eventually forced upon chemistry.[24] Within a short time after the news of Priestley's discoveries was announced in Paris, Bucquet had introduced them into his lecture demonstrations and, by 1777, every course dealing with natural history, chemistry, or physics was duty-bound to make a prominent place for the new "airs."[25] Sigaud de la Fond, his nephew Rouland, Fourcroy, and probably others as well developed special sets of lectures on "pneumato-chemistry" which drew sizable crowds and enhanced the sale of copies of their updated textbooks. For the charlatans and system-builders, these new substances, often released in the laboratory by the action of the "igneous fluid," were especially appealing, since they provided further "evidence" for the existence of a universal fluid pervading and regulating the world, including both the animal and human economy.[26]

The most spectacular application of these aeriform fluids was provided by the Montgolfier brothers, who initiated the era of ballooning in 1783.[27] Public imagination was instantly captivated by the remarkable feat made possible by the new discoveries of science, and the capital and court were flushed with excitement over the new flying engines. A new world, long hoped for in fiction, had finally been conquered in reality. Not only were the newspapers full of descriptions of successful ascents, in which every phrase uttered by the intrepid airborne adventurers was repeated, but the theme swept through other media so that playwrights, poets, song writers, and engravers

24. For a contemporary discussion, see Poissonnier and Bertholon.
25. Bucquet; Sigaud de la Fond, pp. vii–viii, xiii–xviii; and Smeaton, *Fourcroy*, p. 7.
26. Examples of these views may be found in abundance in the writings of Para du Phanjas, Marivetz, Court de Gébelin, Ledru (dit Comus), Miollan, and Bienvenu. See also Ritterbush, pp. 15–56.
27. Rouland, pp. 310–372; Croÿ, IV, ch. XXXI; Dolffus.

were kept busy turning out marketable mementos of the new era. Through it all, science gained ever greater attention.

It was precisely at this juncture that debates over another "scientific" question reached a climax in Parisian circles. In 1784, reports of two separate investigations of mesmerism were made public, which generally condemned the rationale behind the cures effected by Mesmer and his French disciple Deslon. An imposing roster of scientists from the Academy, including Lavoisier, Bailly, and the visiting Benjamin Franklin, announced that their experiments had failed to reveal the existence of "animal magnetism," by which Mesmer explained his therapeutic results.[28] His fate as a scientist was sealed, and his theories discredited. But enthusiasm for Mesmer continued, fanned in part by the pamphleteering which followed the publication of the official reports. Mesmerism had been a favorite topic of conversation since Mesmer arrived in Paris in 1778, and his clientele remained more impressed by the success of his cures than the soundness of his principles. In particular, he had obtained the support of women who savored the unorthodox tactile methods of treatment, which were always carried out by male doctors behind closed doors.

Whether or not they were reputable from a scientific or moral viewpoint, these fads added to the burgeoning interest in science. On the eve of the Revolution scientific enterprise was no longer confined to a small band of virtuosi as it had been a century earlier. Its right to existence as an accepted part of French society firmly established by practice, the probability that this activity would be wiped out even by a major political upheaval was remote. Science had irreversibly become an important part of French culture during the Enlightenment—and this represented a major accomplishment in which the Academy and its members played their part.

Diffusion of science was also accompanied by the strengthening and broadening of the professional's position. Although the often superficial infatuation with science occasionally attracted and encouraged the charlatan, it also provided much needed financial opportunities for the accomplished scientist.

28. Delaunay, *Monde Médical*, pp. 331–355; E. Schneider; E. B. Smith, pp. 484–493; Duveen and Klickstein, *AS*; Darnton.

As has already been suggested, the book trade significantly reflected the popularity of science. Many a scientist earned part of his livelihood by writing for the expanding market in popular texts. Others, more proficient in the oratorical arts, took to public lecturing or sought sinecures in the growing number of teaching and consulting positions opened up by the expansion of scientific activity.[29]

The writing of popular expositions on cosmology was not confined to England and Holland, where excellent models existed. In France, Fontenelle and Voltaire paved the way for a high-level popularization which was emulated throughout the century by Newtonian stalwarts such as Buffon, Le Monnier, the Marquise Du Châtelet, Maupertuis, and the Abbé Sigorgne. Some limited their activities to the translation of the writings of Newton, Keill, Hales, 'sGravesande, or Musschenbroek, while others struck out on their own.[30] Among astronomers, the most popular writings came from the pens of La Caille and Lalande, followed by more elementary treatises written by Bion, Dicquemare, and Mentelle. In experimental physics, the names of La Caille, Nollet, Brisson, Sigaud de la Fond, and Paulian predominated, while in chemistry the translations of Boerhaave, the editions of Lémery's lectures, and the writings of Macquer and Fourcroy held a primary position. In addition to the popular works of the Abbé Pluche and Buffon in natural history, there were translations of Nieuwentyt and Derham, which also served to shape popular concepts of nature.

Popular expositions of current science went through such a large number of editions that it is safe to assume, even without meaningful statistics, that they played a significant role in providing scientists with a means of support. Another profitable type of income must also have originated from the cataloguing, fashioning, and selling of *cabinets* of instruments and natural specimens. But even more central to the changing life of science were the positions opened for men of science in the instructional institutions that flowered throughout the Enlightenment. Of all the consequences of the growth of science, none had a

29. Duveen and Hahn, *Isis*, XLVIII, 416–418.
30. Lalande, *Bibliographie Astronomique*, pp. 394ff.; Brunet, *Physiciens Hollandais*, pp. 101–153; Rochedieu.

greater and more lasting impact upon the professional. The role of the scientist in education set a pattern that was to be of considerable importance in transforming scientific enterprise during the French Revolution.

Several scientific subjects had been an integral part of the medieval curriculum of the faculties of arts and medicine at the University of Paris. Mathematics and astronomy held a central place in the quadrivium, and anatomy, physiology, and botany were taught as sciences auxiliary to the arts of healing. But as the character and importance of these subjects changed, so did the development of scientific instruction. Its evolution during the course of the century, however, was by no means uniform.

Mathematics was given an increasing importance and broader meaning not only in traditional institutions of learning, but, above all, in the technical and military schools that began to multiply and flourish at midcentury. At the Collège Royal, for instance, chairs in Greek and Latin philosophy were turned over to the mathematical sciences in 1769.[31] Four years later, a chair in Syriac was transformed into one for mechanics. In naval and military schools, mathematics—always considered an essential part of the curriculum—was enlarged to include the physical sciences.[32] With every curricular reform, the number of teachers was increased. Certain textbooks, such as those of the professional scientists Camus, Bossut, and Bézout, became standards against which the mathematical and physical knowledge of students was tested. The caliber of examiners insured that such knowledge would be of very high quality. For the "mixed" mathematical sciences, such as hydrography, fortification, and mechanics, special courses were set up, even in schools of drawing and architecture.[33] There were, moreover, a number of new teaching positions created during the Old Regime which enhanced the relative importance given to physical sciences in education. Nollet's Chair of Experimental Physics at the Collège de Navarre and Bossut's Chair of Hydrodynamics are only two of the better-known examples within the

---

31. Archives de la Seine, MS. 2AZ 2DD3; "Documents sur le Collège de France," p. 406.
32. See the articles by Russo, Hahn, and Taton in *Enseignement et Diffusion.*
33. Birembaut, *Enseignement et Diffusion*, pp. 452–453.

official institutions of the capital.[34] In the private sphere, in both Paris and the provinces, large numbers of scientists offered courses that became quite popular. Among them, the most conspicuous and renowned were given at various institutions called "Musée" or "Lycée."[35]

A somewhat different evolution occurred for those sciences originally associated with healing. While they often continued to be taught in conjunction with medicine, surgery, or pharmacy, the character of the lectures was transformed. A less utilitarian approach was adopted, and the subjects were increasingly considered as sciences in their own right.[36] As the number of professors increased, there was a corresponding specialization among them, so that by the end of the century medical students customarily followed lectures given by instructors of "auxiliary" sciences who had little or no experience—and probably little concern—with the practice of medicine. At the Jardin du Roi the ascendancy of the chemists Macquer and Fourcroy over Dr. Bourdelin, of the botanist Desfontaines over Dr. Le Monnier, and of the anatomist Vicq d'Azyr over the surgeon Ferrein is typical of the increasing value placed on the sciences as subjects worthy of study for their own sake.[37] The new professors were identified more and more with their specialties and correspondingly less and less with the medical profession.

The shifting popularity of other sciences was also reflected in the creation of new types of courses offered in Paris and elsewhere. A combination of the popular ferment over natural history and the recognition of potential commercial and national value in the study of the earth sciences was responsible for the initiation of lectures in mineralogy, which began seriously with Daubenton in 1745 and culminated in the creation of the Royal Mining School in 1783.[38] At an earlier date, the need for surveyors and cartographers had been partially satis-

34. Lacoarret and Ter-Menassian, pp. 146–150; Hahn, "The Chair of Hydrodynamics."
35. Taton, *RHS*, XII, 130–138.
36. Huard, *Enseignement et Diffusion*, pp. 184, 210–212.
37. Crestois; Contant; Laissus, *Enseignement et Diffusion*, pp. 308–314.
38. Birembaut and De Dainville in *Enseignement et Diffusion*, pp. 367, 484–489.

fied by the creation of a school of civil engineering into which specialized training in trigonometry and geography was incorporated. The same need for instruction in "practical" geometry and map-making was also responsible for the private courses on these subjects taught by Dupont, Lucotte, Maclot, Robert de Vaugondy, and Perrard.[39] Regardless of the type of education or its sponsor, such developments reflected the general awakening of society to the new needs and interests in science.

Sufficiently varied and detailed evidence has now been adduced to allow us to proceed to an examination of the effect these changes in the scientific life of the Enlightenment had upon the Academy. Growth posed at least two interrelated challenges to the institution, which was operating on practices that had been crystallized at a time when the scientific community was perhaps only one-tenth as large as it had become by 1789. A long-range problem was connected with the inevitable specialization that a numerical increase forced upon scientists. As the pace of progress in the individual sciences quickened, the nature of professional arguments became increasingly technical and thereby restricted to a more specialized audience. But the more immediate consequence of growth was the increasing competition for seats in the Academy and the concurrent need to distinguish between the professional and the amateur. Both of these concerns manifested themselves clearly with reference to elections in the Academy.

The average age for admission into the Academy slowly increased during the century. In the decade of the 1730's, the age of new academicians averaged a little above 28. By the 1780's, the figure had risen to about 39.[40] At the same time, the number of unsuccessful candidates for positions also increased.

39. Announcements of these courses are given in *Journal de Paris* on 28 July and 10 Nov. 1780, 6 Jan. 1783, and 2 Dec. 1786.
40. The progression by decades is 28.2, 31.4, 34.2, 35.2, and 38.8, counting only working academicians. In the last decade, no scientist under 30 was elected, contrasting with the election of Clairaut in 1731, when he was 18. His election required a special dispensation by the minister, as indicated in Brunet, *Clairaut*, p. 18. For a comparison with ages of academicians in the 1840's, see Tudesq, p. 457.

Elections of chemists will serve to typify the Academy's growing pains. In 1776, there were eight candidates for a seat as
adjoint chimiste; for another vacancy in 1772, there were nine;
for one in 1778, seventeen names were cited.[41] As a consequence, a number of competent chemists who earlier in the
century would have been assured academic status were never
elected, while others had to wait years before entering the hallowed company. An example is Darcet, who had taught chemistry at the Collège Royal for a decade before he became an
academician. His admission came in 1784 when he was 59,
and only through a clever subterfuge. He was named *associé
chimiste surnuméraire*, that is, above the quota allotted for
chemistry. This can be contrasted with the young Dufay's rise
in the chemistry section from *adjoint* in 1723 to *pensionnaire*
eight years later. Dufay's speedy promotion was helped by the
absence of serious competitors.

The supernumerary category was the Academy's initial
response to the pressure of numbers, although it was meant to
be a temporary measure. By 1785, five such extra academicians
had been squeezed into an organization meant to accommodate
only fifty-six, and their presence seriously distorted the institution's personnel structure.[42] Pressure was particularly strong
in the areas of science that had been disregarded at the beginning of the century when the Academy's vertical divisions were
first established. Until 1785, the society had maintained a separation of the working academicians into six disciplines: three
each for the "mathematical" (exact) sciences and the "physi-

41. 1766: Cadet, Baumé, Lavoisier, Jars, Demachy, Sage, Valmont de Bomare,
and Monnet (see AdS, Reg 1766, fol. 142r); 1772: Baumé, Bucquet, Veillard,
Laborie, Mitouard, Demachy, and Romé de l'Isle solicited the position, in
addition to H. M. Rouelle and Darcet, who were considered but did not
seek the nomination (see Lalande, "Collection," p. 119); 1778: Bucquet,
Cornette, Mitouard, Demachy, Veillard, Grignon, Bosc, Quatremère d'Isjonval, and Fourcroy solicited; H. M. Rouelle, Darcet, Fontanieu, Bayen, Monnet, Laborie, Parmentier, and Berthollet did not, but were considered (see
Lalande, "Collection," p. 129).
42. The five included Le Gentil, Bailly, Desfontaines, and Brisson, in addition
to Darcet. The practice of appointing supernumeraries was also a procedure
used by the court to introduce into the Academy's ranks scientists who had
political support but insufficient scientific recognition. Among those who
were forcibly introduced as chemists into the Academy this way were
Fontanieu and Angiviller (see p. 81 above and AdS, Reg 5 Sept. 1772, fol.
329r).

cal" (natural) sciences. In the former category were geometry (mathematics), astronomy, and mechanics; in the latter, anatomy, chemistry, and botany. Although, in theory, promotions were allowed only within each discipline, in practice a number of advantageous transfers were made, generally to advance a scientist toward pension status rather than as a reflection of changes in his professional interests.[43] As long as disciplinary commitments and the pressure of numbers were weak, such adjustments were tolerated. But as the number of candidates multiplied and scientific specialization was accentuated, specialty affiliation took on greater significance. In the Academy, the word "classe" which had generally been used to designate hierarchical divisions ("classe des pensionnaires") was increasingly used to separate academicians according to disciplinary interests ("classe de chimie"). Opposition to transfers gained strength within the academic ranks, and, together with an incipient revolt against the franchise restrictions and the pressures of supernumeraries, it drove the society into considering a reform of its structure.[44] It was Lavoisier, as director for the year 1785, who drafted, engineered, and pushed through a complete personnel overhaul for the learned society.

The reorganization of 1785 is an example of the Academy's answer to contemporary pressures and a tribute to Lavoisier's mastery in administrative juggling.[45] By one new set of regulations, he succeeded in doing away with all supernumeraries without jeopardizing any academician's seniority. He also increased the number of academicians possessing the franchise and realigned the existing disciplinary divisions to match contemporary research activities. This was accomplished by abolishing the *adjoint* class and by creating two new disciplinary sections: experimental physics (as a new exact science) and natural history and mineralogy (as a new section among the

---

43. A major exception to the rule is Buffon, who switched from mechanics to botany in 1739 (Hanks, p. 134).
44. A promise to refrain from appointing supernumeraries had been made by Amelot in 1778, but was in fact not kept (see AdS, Reg 18 March 1778, fol. 93r and v). On the question of franchise reform, see Hahn, *RHS*, XVIII, 17.
45. For the texts connected with this reorganization, see Lavoisier, *Œuvres*, IV, 555–593, and the forthcoming section of Lavoisier, *Correspondance*, which the editor René Fric has generously allowed me to read in manuscript.

natural sciences). At the same time, Lavoisier had the chemistry section redesignated "chemistry and metallurgy" and the botany section "botany and agriculture."[46]

Arranging for a more representative distribution of disciplines within the membership was a sensible and direct response to the changing needs of scientific activity. But whereas the Academy recognized these structural imbalances, it refused to react in a similar fashion to the increasing size of the community. Lavoisier proposed a transformation of the old personnel structure consisting of seven academicians for each of six disciplines (totaling forty-two) into a new one with six academicians for each of eight disciplines (forty-eight).[47] But the increase of six positions was all but absorbed by the five supernumeraries who had to be incorporated into the new organization.[48] None of the academicians was deluded by the apparent increase in numbers. In fact, the possibility of using the occasion to boost membership was considered by the society and firmly rejected. The matter is of some importance in reconstructing the Academy's attitude in the face of external pressures.

In a private meeting of senior academicians, the issue of increasing membership had been raised and Lavoisier spoke directly against it. He gave two reasons for his position.[49] One was that increments would increase the confusion already reigning in meetings. The Academy would thereby be "converted from the most respectable of European learned societies into an academic club." Instead of a working association, it would become a debating society, if not an oversized *salon*. His second point, kept until the end for dramatic effect, was clearly

---

46. At an earlier date, Macquer had proposed renaming the botany section "natural history" (see BN, MS. fr. 9134, fols. 44–45).
47. Only working academicians affiliated with a specific discipline were affected by the major structural changes he proposed. The *associés libres* (twelve members) and the two perpetual officers remained untouched by the reforms.
48. After the reorganization was legally sanctioned on 23 April 1785, there were in fact three new academicians elected (Fourcroy, Charles, and Broussonet). These openings resulted from three vacancies in the Academy before the reform (one in astronomy and two in anatomy), the six "new" positions minus the five supernumeraries and Gua de Malves, who was reinstated as a regular academician from his post as *vétéran*.
49. Lavoisier, *Œuvres*, IV, 567–568.

the more important. Lavoisier argued that an increase in size was tantamount to a lowering of the standards of admission. The dilution of quality would inevitably lead to a debasement of the title of academician. He held up the specter of mediocrity and, echoing Formey, the fear of "demi-savoir, more dangerous even than ignorance, and accompanied by quackery and intrigue."[50] Basically, his stand was a reiteration of the elitist tradition of maintaining a high sense of professionalism within the Academy.

The argument was especially important and effective in 1785, closely following as it did upon the public exposure of Mesmer's charlatanry by a committee of experts. It was particularly apposite because the general enthusiasm for science had effectively blurred the line between professional and amateur that had been consciously traced at the time of the Academy's foundation. Like his fellow academicians, who supported him unanimously, Lavoisier envisaged the task of his group to be the maintenance of standards, even at the expense of creating dissatisfied candidates. Indeed, the difficulty of entering the inner sanctum might well serve as an effective stimulus for would-be entrants to make them excel in and multiply their accomplishments.

Lavoisier could have used still another argument, although he was too politic to broach a subject offensive to his colleagues' sense of pride. Even though the Academy was the principal scientific organization of its time, it was not the only body that could offer satisfaction to the aspiring scientist. Since 1699, a number of other learned societies or recognizable groups had been formed, precisely in response to the growing numbers and the increasing specialization of the scientific community. By the 1780's, there were several other public scientific societies in Paris, as well as a number of other institutions vying for the allegiance of professional scientists. Whether or not they constituted a threat to the Academy's primacy, they clearly provided alternatives for scientists unable to find a vacancy in the senior society.

Three of the new institutions corresponded to the profes-

50. Lavoisier, *Œuvres*, IV, 569. Condorcet was led to similar views (see BI, MS. 876, fols. 95–96; BN, MS. n.a.f. 23639, fols. 293–296).

sions directing the healing arts: medicine, surgery, and pharmacy. They reflected the general tendency of the Enlightenment to accelerate the progress of all arts by making them more scientific. It was the intention of the promoters of these new institutions to provide the practices of medicine, surgery, and pharmacy with a rational basis founded upon communally verified observations rather than upon dogma, hearsay, or blind custom. This scientific attitude, together with the benefits derived from the study of physics, anatomy, physiology, chemistry, and botany, inevitably led to the foundation of new learned societies, aping more or less consciously the organization of the Academy of Sciences. It was only logical that science's characteristic form of social organization should be adopted to bring about the desired transformation of the arts and crafts.[51]

Proposals for the creation of an academy of medicine date back to 1718, but at first its creation was successfully checked by the jealousy of the firmly entrenched Paris Faculté de Médecine, which fought to retain its monopoly over the profession.[52] The earliest successful effort to create a modern institution of the healing arts came from the surgeons, whose major antagonist was the less powerful barbers' corporation. In 1731, following a minor victory in their struggle for emancipation from their craft origins, the surgeons formed an academy which rapidly took its place alongside the Royal Academy of Sciences as a major government-sponsored society.[53] The Académie Royale de Chirurgie was given similar statutes, which placed it under the jurisdiction of the same minister, was granted similar printing privileges, and was urged to follow the practices of the senior society. Fontenelle willingly offered his registers, eulogies, and annual historical essays as models for the surgeons to adopt. Not only were they embraced without question but, shortly thereafter, the Academy of Surgery began to publish its series of *Histoire et Mémoires* and prize-winning essays, following in the footsteps of its more

51. Hahn, *Studies on Voltaire*, XXV, 836.
52. Paris, Faculté de Médecine, MS. 578; Paris, Académie de Médecine, MS. box 114; Delaunay, *Monde Médical*, pp. 309–310.
53. Dubois, *Mémoires*, XVI, i–lxxvi; Delaunay, *Monde Médical*, pp. 175–180, 190–198; Boisseau; Huard, *Académie Royale de Chirurgie*; Doyon and Liaigre, pp. 120–123, 137–140.

venerable sister institution. Within the profession of surgeons, the title of academician quickly took on the same connotation it had for scientists.

A Société Royale de Médecine was not set up until 1777 and, although it acquired the same trappings of a learned society as had the Academy of Surgery, it came to possess somewhat more extensive functions.[54] It grew out of the government's need to obtain information and advice on controlling the epidemics then raging throughout the country. A "committee of correspondence with provincial doctors" was set up in 1776 under the directorship of the king's "first doctor." Probably because its members were so well acquainted with academic practices, the committee developed rapidly into a proto-learned society, electing its own members, offering prizes, and holding well-publicized *séances publiques*. Hence, when it finally took on the title of Royal Society of Medicine after having won its power struggle against the Faculty of Medicine, there could be little cause for surprise. What differentiated this group from the Academy of Surgery was its role as technical consultant to the Crown, and its accumulation of jurisdiction over the medical community. The examination of secret remedies, cosmetics, and mineral springs, and even the publication of a newspaper, fell under its control, so that, within a short time, the Royal Society was placed in the formidable position of principal judge in all matters relating to the profession.[55] Only the awarding of medical degrees and the control of teaching remained in the hands of the Faculty. In its rapid emergence as the dominant institution of progressively minded doctors, the Society of Medicine paralleled the Academy's role within the scientific community.

For pharmacy, the corresponding group was the Collège de Pharmacie, founded in 1777 as an extension of the corporate interests of the guild of apothecary-druggists.[56] Its major official function was to direct the teaching and to control the

---

54. Corlieu, pp. 199–228; Foucault, pp. 26–31; Smeaton, *AS*, XII, 228–244.
55. These views are expressed in almost identical terms in a letter of Retz to Vicq d'Azyr, dated 31 March 1784, in Paris, Académie de Médecine, MS. box 167. See also Genty, *Progrès Médical. Supplément*, XI, 33–37, XIV, 17–21; and Desgenettes, II, 170.
56. Prevet; Cazé.

examinations of aspiring pharmacists in Paris. But long before the Collège was actually transformed into a full-fledged "Société" in 1796, it had already shown signs of becoming the learned society for the pharmacological sciences. Local government officials took it as the proper agency through which to settle technical issues, inspect apothecary shops, and offer candidates for appointment as censors. Like other academies, the Collège began to hold annual public meetings in which papers of a technical nature were read, prize-winners announced, and eulogies delivered. These proceedings found their way into print, either through articles in the *Journal de Médecine, Chirurgie et Pharmacie* or in the Collège's own *Almanach*. On the eve of the Revolution, pharmacy, like medicine and surgery, had forged its own specialized institution which symbolized the advancement of the art, just as the Academy of Sciences stood for the communal improvement of the sciences.

Outside the clearly defined specialties associated with a scientific profession, there was also a movement toward the multiplication of specialized institutions. In a manner quite similar to that of the creation of the Society of Medicine, an association of agronomists coalesced into a learned society.[57] The original stimulus came from the joint desire of the Marquis de Turbilly and the ministers Trudaine de Montigny and Bertin to form committees of specialists to advise local intendants on measures for improving agriculture. The Parisian group of private councilors who first met in 1761 was quickly transformed into one of the many local societies of agriculture set up during Bertin's tenure as comptroller-general. Soon thereafter, a volume of proceedings appeared and the organization began to elect members and to deliberate on agricultural matters like any other academic body. By 1785, the group had taken on a national character and assumed the role of an effective governmental consultant and promoter of its specialty of rural economy. Like the other learned societies, the Société Royale d'Agriculture had taken its place in the growing academic family.

In addition to these societies, there were other groups that

57. Justin; Passy; Smeaton, *AS*, XII, 267–277; Bourde, pp. 1109–1121, 1193–1203, 1302–1313.

crystallized within the scientific community and began to take on some of the functions once uniquely assumed by the Academy of Sciences. Outside Paris, a Société Royale des Sciences for medicine had been founded in Montpellier as early as 1706.[58] Its mission was so clearly related to the Academy's avowed promotion of science that a formal affiliation between the Montpellier and Paris groups was established. Efforts to transform animal medicine into a respectable science through the creation of special schools at Lyon (1761) and Alfort (1766) turned these institutions into natural centers for the promotion of veterinary medicine.[59] The desire to improve naval affairs led to the creation of an Académie Royale de Marine at Brest in 1752.[60] Some of the military schools—particularly the Ecole du Génie Militaire at Mézières—were in part turned into specialized research centers for the advancement of military science. In every situation requiring social organization, the attitudes of rationalism, communality, and publicity displayed by learned societies in Paris were embraced without question. The Academy of Sciences, as the most successful organ for the promotion of science, was the natural model to which the entire community looked for guidance. So promising was the academic model that even visionaries like the Abbé St. Pierre imagined that the creation of a new academy for every subject would prove to be the panacea for society's ills.[61]

The proliferation of all these groups in France is further evidence of the growth of science and of the prestige held by the Academy. In theory, it should also have provided alternatives for scientists unable to find a place in the Academy of Sciences. A professional could begin to satisfy his ambitions and gain status in his community by seeking these intermediary positions that would partially reward his efforts. Indeed, by the end of the Old Regime, such specialized institutions were becoming the "waiting rooms" for entrance to the Academy. Fourcroy, for example, was an active member of the Society of Medicine and Sabatier was a leading figure in the Academy of Surgery for many years before election to the Academy

58. Dulieu, *RHS*, XI, 250–262.
59. Hours; Huard, *Enseignement et Diffusion*, pp. 206–209.
60. Doneaud du Plan; Charliat.
61. Siégler-Pascal, pp. 90–101, 257–259; Perkins.

of Sciences itself. The new learned societies were also answering other needs of the community. They provided further outlets for the publication of scientific research and offered an opportunity for a fuller critical exchange of views among specialists than the Academy, by its very nature, could provide.

Appearances notwithstanding, the presence of these lower societies did not fully satisfy the increasing pressures for status among younger professionals. Since the number of positions was limited by statute, only at the moment of foundation, when entirely new positions were available, was some relief afforded. Membership in learned societies was generally awarded for life, and few scientists relinquished their positions in less prestigious societies when they had finally entered the ranks of the Academy of Sciences. There was a tendency to accumulate titles without making room for younger professionals. As a consequence, the membership in all of these Parisian societies increasingly overlapped and the old tensions—which new institutions might have alleviated—returned.

Among amateurs, a parallel multiplication of associations occurred, although for very different reasons. Amateurs banded together for the mutual exchange of information and the sharing of pleasure rather than for the rigorous advancement of the discipline or for purposes of acting as government consultants. Hence, when they organized themselves, they generally lacked government affiliations and the formality of academic organizations.[62] Membership, which was not limited in number, was not a function of scientific competence, but generally of wealth, social status, enthusiasm, notoriety, or even religious sentiment. These groups met informally in *salons*, at provincial academies, around popular lecturers, or even in Masonic lodges. Because they were often short-lived, unaffiliated, voluntary associations, they have left very little trace on the pages of history. Only a few are still known, although they once existed in great abundance.

A special Freemason group, later known as the Société des Neuf Sœurs, was organized in Paris in 1769 to discuss scientific matters.[63] Another group of scientists and artisans brought to-

62. Rose; Roddier, p. 54.
63. Amiable, "Les Origines Maçonniques," and *Loge Maçonnique*; Hans.

gether by Pahin de la Blancherie in 1778 attracted sizable crowds and led to the publication of a scientific "gossip column," the *Nouvelles de la République des Lettres et des Arts*.[64] In the 1780's, a bevy of new institutions for amateurs appeared on the Parisian scene: Court de Gébelin's Musée; its offshoots, the Musée de Paris on rue Dauphine and the Musée de Monsieur on rue St. Avoye; another independent Musée National et Etranger on rue Mazarine; the two Lycées; a Société and Institution Polymatique; a Société Encyclopédique; and a Société Philanthropique.[65] Many of these groups devoted their attention primarily to the further diffusion of scientific knowledge through paying lectures and ephemeral publications. So widespread and persistent were advertisements for these amateur groups that one somewhat bewildered journalist remarked that "if we do not become learned, it will not be for lack of opportunities. It seems as if the word is out to make us absorb science through all our pores, whether we wish to or not."[66]

Because of their amateur character, these groups failed to provide the institutional outlets necessary to satisfy the serious professional. Though he might occasionally lend his name as a sponsor, or even teach for a fee, the scientist never confused voluntary associations with the official learned societies that served as the principal vehicles for raising the level of learning and providing expert advice to the administration. Privately sponsored institutions always retained a well-intentioned but dilettantish character, whereas government-affiliated societies encouraged a more professional spirit. So constant was the link between professionals and "official" organizations that, at least in Paris, the habit became a tradition. Every institution aspiring to respectability sought government affiliation and support and, conversely, every promising professional group was absorbed by the Crown or the city. The process was clearly in line with Colbert's original conception of a bureaucratically organized culture.

Thus the needs of a larger professional scientific com-

---

64. Bellier de Chavignerie; Rabbe; Tourneux, *Mémoires de la Société de l'Histoire de Paris*, XXIX, 37–50; Bergman, pp. 289–291.
65. *Discours sur les Découvertes*, p. 3n; Dejob, pp. 123–152; Smeaton, *AS*, XI, 257–267; Scheler, I; Birembaut, *RHS*, XI, 267–273.
66. *Affiches*, 28 Nov. 1781, p. 192.

munity could not adequately be satisfied by the organizational pattern of the Old Regime. The Academy of Sciences had a limited membership, which it was not willing to increase; the other professional groups were filled with academicians; and the popular private societies lacked the prestige and power to be proper substitutes. Necessarily, the age at which scientists reached the highest academic status rose, the standards of admission tightened, and the distinction between the professional and the amateur became an even more sensitive subject than it was earlier. The primary consequence of this set of circumstances was the reinforcement of the already strong elitist character of the Academy. In relation to the total population of scientists, academicians of the late eighteenth century were rarer, older, and more distinguished than their counterparts a century earlier. The benefits of their position in terms of influence, prestige, and the ability to earn a living had risen considerably. It was only normal that their vested interest in the existing institutions was also on the rise.

Once this mutually reinforcing set of circumstances is perceived, it is easier to understand the few but significant instances in which the Academy as a body objected to, or even prevented, the creation or development of independent new scientific institutions in France. Its motive was not mere jealousy or even insecurity, although such emotions surely played their part. In each case, not only was the Academy's prestige, power, or even control over the scientific community at stake, but also, as will become apparent, the entire tradition of elitism and government affiliation. We know about only three such cases, but despite our fragmentary knowledge, it is important to examine them as symptomatic of the strength wielded by the Academy and the fears its action engendered in the elimination of budding new scientific organizations.

Two of the cases involve the creation and disappearance of organizations bent upon accelerating industrial progress in France. The earliest was an obscure group known successively as the Société Académique des Beaux-Arts and the Société des Arts. Its origins go back to an unidentified group of artists who met periodically at the Galleries of the Louvre before the

Regent's death in 1723.[67] The famous English clockmaker Henry Sully, and his confessor, the Curé de St. Sulpice, revived the group in 1728. They enlisted the support and protection of the Comte de Clermont, who provided them with a meeting hall at the Petit Luxembourg and money to distribute as prizes.[68] We know almost nothing about the group's meetings, except that its principal concern was to "perfect the arts" by the close union of theory and practice.[69] To that end, the membership always included both scientists and artisans, many of whom later achieved respectable positions in their chosen professions. An early version of the regulations—possibly never enforced —states that the Société's focus was to be "on geography, navigation, mechanics, and civil and military architecture, but without neglecting any of the other arts, be they useful or simply pleasurable."[70] The aspirations of the Société des Arts included the encouragement of inventions, their description and publication, and the offer of members' considered judgment on any matter relating to technology—a program that came close to the one followed by the Academy of Sciences.

It was precisely this imitation of the Academy in function, institutional form, and public posture that gave rise to a conflict with the older and better established institution.[71] To many academicians the existence of the Société must have seemed redundant and unnecessary. Others interpreted the Société's desire to describe machines and trades and to publish accounts of inventions as a threat to academic authority. The academicians met the challenge by electing to their own ranks the Société's most learned and active members (La Condamine, Clairaut, and Fouchy) and by blocking the Société's elevation to the status of a government-sponsored institution.[72] The Academy, however, paid a price for the elimination of its po-

67. BN, MS. fr. 22225, fol. 7; Sully, pp. 407–408.
68. *Mercure*, Dec. 1728, p. 2893; *Mémoires de Trévoux*, Feb. 1733, pp. 357–359.
69. D'Alembert, *Œuvres Philosophiques*, XI, 414–415.
70. BN, MS. fr. 22225, fol. 1; *Reglement de la Société des Arts*, pp. 6–9.
71. Bertrand, pp. 95–97; Hahn, *Studies on Voltaire*, XXV, 834. Contemporaries often referred to it as another Academy, as for example the geographer Delisle's sister in her correspondence, in Paris, Assemblée Nationale, MS. 1508, fols. 49 and 64. I owe this reference to Françoise Weil.
72. Remond de Sainte-Albine, pp. iii–v, 41–43, 60–67; Jules Cousin, I, 108–110.

tential competitor. Its reputation of benevolent paternalism was seriously blemished by persistent and unflattering rumors, which circulated throughout the rest of the century.[73] The Academy was also prodded into incorporating certain of the Société's projects, foremost among them the printing of descriptions of new inventions.[74] The engineer Gallon, who eventually edited the *Machines et Inventions Approuvées par l'Académie*, was himself a member of the Société des Arts. It is no coincidence that the first announcement of that very publication by the Academy was made in December 1733, just at the time that the Société was being forced out of existence.[75]

Without more information, it is difficult to be certain of all the Academy's motives in throttling the Société. Clearly, its novel structural organization, which called for the inclusion of unlettered artisans making a living from their trades as well as of devotees of the fine arts, was at odds with prevailing conceptions of learned societies in Paris. It countered the trend followed by the founders of the academic system. Moreover, from the very beginning there was no guarantee that high standards would be set by the Société; and after the Academy had plucked off its promising members, there was even less assurance. Under these circumstances, overt royal approval was unthinkable. The Société's inability to enter the ranks of government-affiliated institutions was the kiss of death.

All attempts to revive the Société met with similar difficulties. In the 1750's, the clockmakers Julien Le Roy, Thiout, and Berthoud tried to establish an Académie d'Horlogerie out of the remnants of the Société but failed to win necessary support from the ministry.[76] Some twenty years later, a group of wealthy and enlightened Parisians organized a Société Libre d'Emulation to "encourage discoveries and perfect the arts and crafts."[77] Its leader was the physiocratic disciple the Abbé Baudeau, who took as his principal model the Society of Arts in London. Founded in 1776 at the pinnacle of the Anglophile movement, the Société Libre had an auspicious start with the

73. Pidansat de Mairobert, VI, 187.
74. Lepaute, p. xxiii.
75. *Mémoires de Trévoux,* Dec. 1733, pp. 2196–2198.
76. Berthoud, I, xlvi.
77. Birembaut, *RHS,* IX, 150–161, and *AIHS,* XIX, 80–81.

financial backing of some three hundred public-spirited members. Turgot, Necker, the ducs de Chaulnes and Charost, the Marquis de Puységur, the farmer-general Paulze, Dupont de Nemours, and wealthy academicians such as La Rochefoucauld d'Enville, Lavoisier, Condorcet, and Cassini de Thury all subscribed.[78] Meetings were announced and held, prizes offered and distributed, and, for three years, the group flourished. But, by 1780, interest had already lagged, and its activity seems to have ceased shortly thereafter. A number of factors extraneous to our study contributed to its demise, not the least of which was the failing mental health of its organizer, the Abbé Baudeau.[79] But even without this circumstance, the seeds of its own failure were already sown at its foundation. Following the British model, the association was deliberately private, attaching a great deal of importance to its title of Société Libre.[80] The stigma of amateurism associated in France with a voluntary association made its continued existence precarious.

For the Société Libre d'Emulation, we have even less direct evidence than for the Société des Arts to indicate the manner in which it was prevented from developing. It is likely that the powerful hand of the Academy was working behind the scenes. As the voluntary association began to consider candidates for the award of prizes, it was increasingly forced to encroach upon the Academy's prerogatives as arbiter. Its judgments were necessarily suspect, for membership in the association was open to any wealthy subscriber, regardless of his scientific competence. Hence the Academy, seeking to protect standards, prescribed that "The Société d'Emulation must not concern itself with theory or science, but only utility."[81] While the latter was proper for amateurs, science had to remain the domain of professionals. It was shortly after this announcement that the Société's growth began to lose momentum.

There are further indications that the form of the institution adversely affected its chance of survival. A third serious

78. AN, T 160[16] and 160[19]; *Journal de Paris*, 29 June 1779.
79. Letter of Dupont de Nemours dated 19 Nov. 1780; in AN, T 160[19], no. 492.
80. "Projet de Règlement pour une Association qui se Proposerait de Contribuer au Progrès des Arts," AN, T 160[22], no. 6.
81. *Journal de Paris*, 13 May 1780; *Gazette Littéraire de l'Europe*, July 1777, pp. 54–55.

attempt after the Revolution to found an institution linking science and technology met with complete success. This time the Société d'Encouragement pour l'Industrie Nationale (1801), with aims quite similar to those of the group that had failed twenty years earlier, had the government's fullest backing and was considered as an "official" organization.[82] It was precisely this governmental sanction that made the difference. Even under the Old Regime, most government-affiliated institutions managed to survive while private associations like the Société Libre d'Emulation were floundering. The cases of the Société Royale d'Agriculture and the Académie Royale de Marine, each with more restricted but similar utilitarian aims, bears out the distinction.

The third case of the disappearance of a private scientific institution is equally revealing of the mentality that guided the Academy's actions before the Revolution. The Academy's alleged motive for scuttling the Société Linnéenne de Paris was the denial of a new scientific doctrine that ran contrary to the academicians' cherished beliefs. The group in question was formed by an enthusiastic circle of young naturalists enamored of the Linnaean system of classification, already popular in northern Europe and England and at Montpellier. Their meetings began in 1788 shortly after a visit from the English disciple of Linnaeus, James Edward Smith.[83] They continued to meet regularly throughout the year of foundation and into 1789 as well, only to stop abruptly.[84] We know, through letters exchanged between the French botanists Broussonet and L'Héritier de Brutelle and Smith, and later by the comments of another Linnaean enthusiast, Millin de Grandmaison, that senior botanists at the Academy were personally responsible for the disbanding of the burgeoning association.[85]

The trick they employed was to force academicians and aspirants to make a choice between the Academy and the Société Linnéenne. Thus the academician Broussonet, who had been one of the originators of the private association, rarely

82. Tresse, *RHS*, V, 246–256.
83. The same Smith founded the Linnean Society of London in 1788.
84. Paris, Bibliothèque Mazarine, MS. 4441.
85. The correspondence is part of the Smith papers at the Linnean Society of London, and is described in Dawson, pp. 19, 66–67. See also Millin, p. 30.

attended its gatherings after May 1788. His colleague André
Thouin contributed little to the group, even though he had
also been enthusiastic at its inception. The attendance and participation of L'Héritier de Brutelle, who was not at first a
member of the Academy, was also sporadic. The troubles he
suffered at the time of his candidature for a seat in the Academy
early in 1790 were related to the suspicion that he had not completely broken his ties with the Linnaean circle.[86] Apparently,
threats of exclusion from established institutions and the pressure placed on its members to refrain from participation were
sufficient to bring about the Société's demise.

Given these indications, one might well advance the position that this was an obvious and serious case of scientific politics. Did not the three most powerful botanists of the Academy
have a vested interest in preventing the spread of Linnaean
classification? Adanson, Antoine-Laurent de Jussieu, and Lamarck had each published his own principles of systematic
botany, and all were at odds with Linnaean notions.[87] Jussieu
in particular was finishing his monumental *Genera Plantarum*
just as the Société Linnéenne was emerging under his very eyes.
We know that he and Lamarck sullied L'Héritier's scientific
reputation when the latter sought admission to the Academy,
a move which all knowledgeable observers interpreted as the
product of professional disagreements. Clearly the antagonism
was based upon doctrinal differences among naturalists.

But to conclude from this, as did Millin, that the Academy
was supporting an official doctrine of botany to counter Linnaean intentions would be to misread the evidence. There was
in fact no agreement among the senior academicians about
which of their systems was correct. Jussieu rejected Adanson's
philosophy, and the latter resented the false accusation that he
had plagiarized from Jussieu's uncle, Bernard de Jussieu. Lamarck had developed a different set of taxonomic principles,
which flatly ignored the alleged existence of natural classifica-
tions that Adanson championed. There could not conceivably
have been a single official academic doctrine. The only con-

86. Stafleu, "L'Héritier," p. xxxi.
87. Pittsburgh, Hunt Botanical Library, MS. AD 298; J. E. Smith, I, 121–125;
Cain, pp. 204–207; Stafleu, and Sneath in *Adanson*, pp. 241–244, 490.

sensus possible was to disagree, and to allow a variety of views to persist simultaneously. To permit such philosophical pluralism was indeed the Academy's basic credo.

The academicians' fundamental objection to the Linnaeans was not their principles per se, but their commitment to a single dogma. None of the botanists was ever persecuted for his preference for Linnaean doctrine, which had been introduced into the Academy by Guettard decades earlier. Broussonet and Thouin, for instance, were asked only to refrain from participating in partisan activities or at least to help to liberalize the naturalists' philosophical spectrum. From the Academy's point of view, such a precept was meant to protect freedom of opinion, not stifle any single opinion.[88] Sectarian societies were to be avoided at all costs because they violated the sacred principle of open-mindedness which the Academy was bound by tradition to champion.

As might be expected, the Academy's pluralistic philosophy prevailed and the Société Linnéenne fell into a state of torpor. When it was successfully revived in 1791, the new group of naturalists eschewed the alleged dogmatism or narrowness of the defunct circle. They now adopted the name Société d'Histoire Naturelle and broadened both their activities and membership to avoid further trouble.

In the elimination of the Société Linnéenne, there were also familiar hints of the Academy's long-standing disapproval of voluntary associations as a viable means for advancing science. Membership in the group was secured by enthusiasm for and adherence to a taxonomic method, without any reference to scientific competence. Two of its members, Lerminier and Pierre-Rémi-François Willemet, had barely finished their secondary schooling.[89] Others worked in the questionable tradition of indiscriminate collecting, which considerably limited the value of their findings. Louis Bosc, for example, had been as concerned with the moral lessons culled from the study of natural history as with its delineation as a science.[90] As an intimate friend of Madame Roland, he was at the time more

88. Cuvier, *Recueil*, III, 89.
89. *Biographie Universelle*, XXIV, 248, XLIV, 632.
90. Rey, pp. 248–253.

properly classifiable as an incurable romantic than as an aspiring naturalist. Millin de Grandmaison was a less than competent naturalist, who wisely chose to abandon this pastime to become a leading numismatist.[91] With such members, there was little hope that significant work could emerge from the voluntary association.

The Academy's reaction to the growing needs of the scientific community was primarily governed by what had become established academic traditions. These included most prominently the maintenance of high standards and doctrinal freedom in matters of science. It was important for the Academy to pursue this policy even if it turned itself into an enemy of many of the new scientific circles and got the reputation of being an unsympathetic tyrant. Its "sacred mission" was to protect the advancement of science at all costs. If this meant assuming the role of an unpopular but benevolent despot, the Academy was willing to suffer the consequences.

Considering the fact that the Academy was organized and functioning according to the same corporate and authoritarian patterns generally followed in Old Regime society, the risk seemed extremely small. To disapprove of its action was tantamount to criticizing the fabric of society and disowning the institution altogether. But this was precisely what took place. A growing chorus of critics outside the Academy—and even some within—began to question the very traditions that had brought the Academy to the pinnacle of its power. One by one, gradually, all the bases upon which these practices were founded were being questioned and, even before the Revolution broke out, the institution was under severe attack.

91. Auguis.

Chapter 5

Changing Demands of Society

*I*T *IS* evident that the academicians were well aware of the challenges posed by the growth of science. Their decisions to reinforce elitist traditions, to give preference to government-affiliated institutions, and to alter the academic structure were self-conscious responses to changing scientific conditions. Yet one cannot discern in their actions the same awareness of the forces of change impinging upon the Academy from outside their own community. With their gaze fixed on the progress of their profession, most scientists failed to consider how changes in eighteenth-century society might bring into question the very premises upon which their institution was based. Nevertheless, in at least three fundamental and interrelated ways the position of the Academy was slowly being undermined. Utilitarian demands placed upon the royal institution multiplied at such a pace that the delicate balance between its research and consultative functions was jeopardized. A growing egalitarian sentiment fostered dissatisfaction with the hierarchical organization and elitist principles cherished by the Academy. Romanticism challenged the assumptions of the classical age, including the very rationale behind both the academic system and the kind of collective activity in which the academicians were engaged. Rousseau and his disciples even raised doubts about the desirability of scientific progress itself.

Without amounting to an overt repudiation of the Academy of Sciences as it had been established in the age of Louis XIV, these movements laid the foundations for serious attacks upon the institutions which were to emerge triumphant during the Revolutionary period.

As we have seen, the utilitarian object of the Academy was recognized from its very inception as an essential part of its function. It was a theme reiterated throughout the century by all spokesmen of science from Fontenelle to Condorcet.[1] Everyone referred to the possibilities held out by science for the improvement of man's well-being and in the service of the Crown.[2] But the relative importance of these activities changed considerably in the decades preceding the French Revolution, and in direct response to the stepped-up pace of economic development in France.

Urban centers developed quickly, paralleling the rise in the population of France from around twenty-one million at the beginning of the century to over twenty-five million on the eve of the Revolution.[3] Simultaneous with this surge came the beginnings of an agricultural and industrial revolution which, while failing to develop as impressively as in England, placed considerable pressure on the Academy. Technological innovations requiring academic attention were numerous on both sides of the Channel, regardless of the difference in production figures.[4] In addition, during the second half of the century, the government actively sought to improve communications within the country by setting up technical schools and to contribute to the nation's prosperity by subsidizing nascent efforts of economic and agricultural regeneration.[5] Since the Academy held within its ranks an unparalleled reservoir of talent, it was natural that many of its members were drafted to plan or implement the progressive policies of the administration. Thus the Academy, as the government's principal

1. Typical in this regard are the comments of Fouchy, *Eloges*, p. 417.
2. "Réflexions sur l'Utilité dont l'Académie des Sciences Pourroit Etre au Royaume," in Maindron, *Académie*, pp. 103–110.
3. L. Henry, p. 455.
4. Landes; Crouzet.
5. Artz, pp. 74–86.

consultant on technological affairs, was bound to feel the repercussions of the marked quickening of the tempo of French
life.

In one sense, this set of circumstances proved a distinct
boon to the Academy. Its renown and value to French society
could only grow as its function in the kingdom became more
essential. That fact was underlined by the constant contrasts
made in public between the different academies.[6] It was normal,
even in the perennial attacks leveled against the venerable
Académie Française, to praise its scientific sister for its utility.
Voltaire sarcastically referred to the older body as an "academy
of vain words," whereas he expressed more admiration for the
scientific organization, which he liked to portray as an "academy of useful things."[7] He was echoed by the popular writer
Louis-Sébastien Mercier, who was so fond of the distinction
that he proposed to scrap the other and "frivolous" institutions
altogether: "As for the académies française et de belles-lettres,
they are declining and will collapse by themselves; their futile
and petty tastes, their little intrigues, their pedantic ways, their
uselessness have already brought disdain upon them. . . ."[8] On
the other hand, he insisted that "The Academy of Sciences
deserves our respect and tribute because it has amassed discoveries, . . . and is the only academy in France whose name
one dares to utter abroad. . . ."

An ideology of utilitarianism had its merits for popular
consumption, especially since it would lend support to the
Academy's other activities. In an increasingly materialistic age,
it was politic to emphasize the tangible contributions made by
the scientific organization, even if research motivated by scholarly considerations alone were temporarily ignored. Though
such a public policy might well bring short-range dividends,
it was fraught with danger if overdone. By repeating the utilitarian argument too often, the academicians were liable to
convince the public that their function rested solely on their
role as technological consultants. Should this role suddenly be

6. Lesuire, p. 63.
7. Voltaire, *Correspondance*, II, 423.
8. Mercier, X, 189, V, 140.

assumed by another agency, the professed rationale for the Academy's existence would disappear. Such dangers were not perceived, for many of the academicians were beginning to find their time increasingly devoted to consultative tasks.[9] It was as if they were losing their identity as men of learning to become civil servants, a role for which they were psychologically unprepared.

As indicated earlier, technological projects usually came to the Academy from two sources. One was the private individual who, for the sake of personal fame or financial gain, needed the company's formal and public stamp of approval. The other was the government, which was in constant need of professional judgments to make intelligent administrative decisions. Until reliable statistics on the number of cases handled by the Academy in either category are compiled, we can only assume that the work load of the Academy was measurably increased after midcentury. This assumption seems to be borne out by converging evidence from the archives of the Bureau de Commerce, the minutes of the Academy's meetings, and McCloy's conclusions about the number of successful inventions made in France during the eighteenth century.[10]

There was more than merely a quantitative difference during the course of the century. The types of projects submitted shifted noticeably from those associated with navigation and cartography to problems requiring chemical, metallurgical, and mineralogical knowledge and skill. Even the academicians' medical understanding was tapped in ways that would have been considered unusual earlier in the Enlightenment.[11] Projects for surgical and mechanical devices continued to pour into the Academy in great volume, accompanied increasingly by working models of high quality. Another prominent change was the diminishing concern for automata, planetaria, and musical instruments, in the face of increasing interest in self-winding watches, water pumps, and threshing machines, bearing out the

9. Good examples of this trend may be found in the careers of Hellot, Montigny, and Lavoisier. See Birembaut and Thuillier; Paris, Bibliothèque Mazarine, MS. 3596; McKie, *Lavoisier*, chs. VI, XVI–XIX; Vergnaud, pp. 124–127.
10. McCloy, pp. 192–194; Bourde, p. 114.
11. McCloy, ch. XI.

lawyer Servan's comment that "what characterizes this century most is the love of the useful and the disdain for the merely curious. . . ."[12]

The growth of cities and its attendant problems was also a major preoccupation of the Academy, especially after 1750. In 1765, for example, the society offered a prize of 2,000 livres for the best essay submitted on the improvement of street lighting.[13] Lavoisier, who as a young man had competed in this contest, was later called upon to examine at least a half-dozen projects on the same topic.[14] By far the largest number of urban affairs brought to the Academy's attention were related to hydraulic phenomena. The problem of water—a resource essential for human life, the major source of power in pre-industrial society, and a dependable vehicle for transportation—took up much of the Academy's time. Once-popular plans for elaborate fountain works for royal palaces were gradually replaced by schemes for water purification, storage, and distribution, and by suggestions for new fire-fighting equipment and improved water mills.[15] Major proposals for bringing more water to Paris—in particular those of Deparcieux and Perronet for diverting the waters of the Yvette or the Bièvre, and of Brullée, Fer de la Nouerre, and Trouville for rerouting the Seine—were repeatedly discussed and judged by academicians from the 1760's until the early years of the Revolution.[16]

Serious problems of public health also engaged the learned society. The new understanding of gases inaugurated by the age of pneumatic chemistry led to a series of projects intended to make the job of emptying cesspools safer.[17] In 1789, a major report urging the removal of slaughterhouses from the center of Paris because of their toxic emanations was prepared by a "blue-ribbon" committee of academicians which included Daubenton, Tillet, Laplace, Lavoisier, and Bailly.[18] It followed

12. Servan, pp. 16–17. For a typical list of projects submitted to the Academy from 1746 to 1779, see Doyon and Liaigre, pp. 443–449.
13. Maréchal, pp. 4–5; Boutteville, p. 21.
14. Lavoisier, *Œuvres*, III, 1–105; Storrs, *AS*, XXII, 251–275.
15. Bernal, p. 295; McCloy, pp. 115–117.
16. Perronet, pp. 545–557; Bouchary; Duveen and Klickstein, *Bibliography*, pp. 15–18.
17. Duveen and Klickstein, *Bibliography*, pp. 246–248, 349–350.
18. E. B. Smith, ch. VIII; Duveen and Klickstein, *Bulletin of the History of Medicine*.

closely upon a series of lengthier and more detailed reports drafted by Bailly in the name of another ad hoc committee on the condition of Paris hospitals. This committee had devoted hours to the inspection of local hospitals, dispatched Tenon and Coulomb across the Channel to investigate English practices, and presented its conclusions, backed with statistical evidence which irrefutably demonstrated the shocking state of hygiene in the capital.[19] The hospital report won for the Academy the admiration of the humanitarian public and provided the basis for a complete reform of hospital administration in Paris. A similar report on the state of prisons by Tillet, Tenon, and Lavoisier in 1780 had sparked a discussion about prison reforms.[20] All these reformist preoccupations were closely linked to urban growth, which pushed the population of Paris during the eighteenth century from 500,000 to 670,000, and to a developing humanitarian consciousness.[21]

Pressures for academic participation in civic affairs grew in step with the burgeoning of the state and the decentralization of power after the Regency. Whereas under Louis XIV most official communications to the academies were channeled through one minister in charge of academies, during the reigns of Louis XV and XVI requests for advice came from all ministers and local authorities as well. In 1787, for example, the Minister of War, Ségur, requested that the Academy form a special committee to visit and report on the metallurgical plants of Wendel at Le Creusot.[22] For this purpose, the government provided Lavoisier, Fourcroy, Monge, Berthollet, Vandermonde, and Périer with an expense-paid eleven-day trip. Another academician, Coulomb, was dispatched to examine the feasibility of a plan for a comprehensive canal system put forward by the Estates of Brittany. The Estates of Normandy demanded a thorough examination of adulterated cider which had allegedly caused several deaths in Rouen. An intendant in Poitou pleaded for assistance in fighting insects which were

19. Mac-Auliffe.
20. Duveen and Klickstein, *Bibliography*, pp. 56–57; Duveen, *Supplement*, pp. 8–9. See also BN, MS. n.a.f. 22742–22751, 11358.
21. Mols, II, 30–31, 164, 512–514.
22. Aubry, pp. 59–60, for Le Creusot; Bertrand, pp. 170–171, and Billingsley, pp. 99–110, for details in the following sentences.

about to ruin the local crops. Others, such as the police chief of Paris, Lenoir, made repeated demands upon the Academy to assist in affairs connected with the administration of the capital.[23] Even courts of law requested the opinions of the learned society. All these demands were flattering, and the Academy proudly noted that "in no instance has the body ever sought to shirk the opportunity to make itself useful."[24]

The reform spirit which swept over a large part of the intellectual elite during the thirty years before the outbreak of the Revolution brought the learned society into even closer contact with the state. Enlightened officials such as Trudaine de Montigny, Bertin, Malesherbes, Sartine, De Castries, and Turgot were constantly in touch with the society and its members. Turgot and his associates even envisaged a more central role for the Academy as a national regulative agency to replace functions performed by corporate or provincial groups. In 1776, his close friend Condorcet, who was assistant secretary of the Academy for several years, circulated a plan to strengthen the society's national position by linking it to provincial academies throughout the country.[25] Condorcet's scheme—though primarily aimed at making local societies more useful to scientific progress—would also have coordinated the activities of French savants everywhere under the leadership of a corresponding secretary in the capital. According to one version of Condorcet's plan, more grandiose than any contemplated by Colbert, all academies of the realm would come under the control and aegis of a single minister of the Crown. Had this dream been realized it would have provided a ready-made instrument for extending the power of the central administration. We know that the idea in this form was shelved following unfavorable reactions by provincial academicians.[26] But though it failed for the Academy of Sciences, it was appropriated with great success in the refurbished Royal Societies of Agriculture and the new Royal Society of Medicine. Correspondence net-

23. Orléans, Bibliothèque Municipale, MS. 1422.
24. AdS, Reg 28 May 1788, fol. 134v.
25. BI, MS. 870, fols. 161–173; Boston, Public Library, MS. Ch I.5., 34–36; Bouillier, "Divers Projets"; Baker, *RHS*, XX, 255–267, 274–280.
26. Baker, *RHS*, XX, 267–274.

works with provincial groups gave these societies the means for gathering important data and offered a convenient administrative channel for effective consultation. In this manner, the scientist found himself more closely tied to the government than ever before.

Increasing demands made upon the Academy were manifest in three other concrete ways. Pressures of growth forced the Academy to seek more space, money, and personnel to fulfill its responsibilities. Above all, space was necessary for the storage, use, and display of the collections of machines and specimens which had been accumulating over the years at an ever-increasing rate. As early as 1745, an inventory of the Academy's belongings covered 145 pages.[27] To these treasures were added D'Onsenbray's gift of instruments in 1753, the machines and papers willed by Réaumur, and the Duke of Orléans' natural-history *cabinet* in 1786, as well as countless new models of machines deposited with the learned society. Many of the items remained uncatalogued and were indiscriminately piled up in locked cases. Some relief was in sight in 1785 when the Academy secured additional space in the Louvre.[28] By 1789, machines were being catalogued and transferred into new, enlarged quarters. But the equally pressing need for space for the establishment of a chemical laboratory temporarily checked the society's efforts to display its collections properly. A new effort at expansion was made in 1792.[29] Ironically, the Academy was just beginning to solve its space problems when it was disbanded.

Money for experimentation was also a fundamental necessity for the society. The 12,000 livres annual grant awarded to Réaumur in 1721 for meeting the costs of preparing the *Description des Arts et Métiers* and promised for the Academy upon his death did not materialize for the society until 1778; by then, the sum proved less than adequate.[30] Minor expenses for testing were mounting and had to be taken out of funds destined for the advancement of "pure" science. Lavoisier re-

27. Maindron, *Académie*, p. 122; Lalande, "Collection," p. 8.
28. AdS, Reg 18–28 Nov. 1789, fols. 224v–228r.
29. AdS, Reg 25 and 29 Feb. 1792, pp. 84–85.
30. Maindron, *Académie*, pp. 111–112; AdS, Lav 864.

minded the finance minister, Cahier de Gerville, that "the Academy has always followed the practice of dipping into its own allocations for expenses of all kinds, even those made at the behest of the government which hold greater interest for the administration than for the progress of science."[31] When larger amounts were required, often involving trips or the purchase of expensive equipment, special provisions for funding had to be made. After 1785, for example, the *Connaissance des Temps*, recognized by the Minister of Navy as "of an indispensable utility for the navy," was subsidized by a special annual grant of 1,600 livres.[32]

More time and personnel were also needed to meet the exigencies created by increased consultation. So gradual was the process, however, that few academicians complained of its effects. Only Lavoisier who, as a farmer-general, was conditioned to observe academic practices from a detached bureaucratic viewpoint, was conscious of its consequences for the advancement of science. In 1787, he convened a group of young and energetic nonacademic scientists to assist in the publication of the Academy's much-heralded *Description des Arts et Métiers*. His address welcoming them reveals the predicament faced by the learned society:

Notwithstanding the zeal of members of the Academy, most of them are so burdened with work, either related or unrelated to science, that it is not possible to hope that this enterprise [*Description*] will proceed with rapidity . . . that they [academicians] were forced to solicit exterior assistance and that young men destined for a scientific career could become worthy of the government and the Academy in no better way than by devoting part of their time to this type of work.[33]

Lavoisier was one of the few academicians who sought remedies to alleviate the pressures to which he and his colleagues were subjected. But devices such as this gentle "coercive co-option," or the addition of collaborators for his own scientific work (Laplace, Meusnier, Seguin), were only temporary solutions.

31. Letter of Lavoisier to Cahier de Gerville, draft dated 29 Dec. 1791, in Clermont-Ferrand, Archives de Chabrol, M. A copy of the letter is in JT, p. 8.
32. Copy of letter of De Castries to Breteuil, dated 17 June 1785, in JT, p. 5.
33. AdS, Lav 503; Lavoisier, *Œuvres*, VI, 67.

It was the possibility of working in Lavoisier's laboratory more than a charitable desire to relieve him of pressures that attracted others to his assistance.[34]

For all their scientific acuity, the academicians held a myopic view of their own affairs and displayed little sense of perspective about their predicament. Their tendency was to react to a particular stress with an ad hoc response rather than to consider the problem in its broader context. Only when the secretary was forced to speak in public for the Academy were these problems explicitly discussed. Even in such circumstances, the self-contradictory statements uttered by intelligent secretaries like Condorcet reflect the confusion which existed in their minds about the proper role of the Academy in the face of the changing demands of society.

Condorcet considered that there was a continuous spectrum of activities in which a scientist could properly be engaged, ranging from internally motivated research to consultation demanded by outside agencies. But he assessed these functions differently and in contradictory ways. In his 1783 eulogy of Duhamel du Monceau, he could maintain that

His career, useful, glorious, and peaceful, is one of the happiest the annals of science can display. It will mark an era in these annals, because his name is linked to this intellectual revolution which has directed the sciences toward public utility and to which no one has contributed more than Duhamel. No doubt this revolution will be a lasting one. The idea of the general good of mankind will be the guide of savants in their research. . . .[35]

Yet, a year later, he asserted that Macquer

knew that learned societies, embracing as they do the whole range of sciences, must be concerned with the progress of science. It is only in academies that research which does not have an immediate application that is striking to the curiosity of the public can be appreciated . . . If, seduced by a vision of immediate utility, the learned societies devoted themselves exclusively to practical research, the march of science would be retarded to the detriment of this same utility for which it would be unwisely sacrificed.[36]

34. Duveen and Hahn, *RHS*.
35. *HARS* (1782), p. 154.
36. *HARS* (1784), p. 28.

And in a private letter to Turgot, Condorcet could say sarcastically that his colleague Borda "is what is referred to as a
good academician, because he speaks up in meetings and looks
for ways of wasting his time writing pamphlets, examining machines, etc."[37]

Condorcet's conflicting analyses of the value to be attached
to consultative work is symptomatic of his inability to resolve
the dilemma presented by the changing conditions in society.
He recognized the dangers that crass utilitarianism posed for
the advancement of science; it would dry up its creative fount.
At the same time, he could also hail the very utilitarian revolution that was sapping scientists' time and energies as a harbinger
of a better world. The only way to reconcile these two apprais-
als was to claim—and without having much tangible evidence—
that they are mutually reinforcing, and to cherish the hope that
some balance of enthusiasm for pure science and for government consultation would be maintained in the academies. In the
face of the irreversible growth of government, urbanization,
and technical demands, such a hope was illusory. The equilibrium established early in the century between the demands
science and society placed upon the Academy was now impossible to preserve. With the advantage of hindsight we see that,
to serve both ends adequately, an alternate form of scientific organization or a drastic revision of the Academy was necessary.
For the men of the Old Regime who did not recognize the
problem, it was unlikely that anything but minor changes in the
Academy's structure and function would be contemplated.

A number of ideological transformations in the Enlightenment also directly impinged upon the Academy's position in
Old Regime society. Foremost among them was the growing
egalitarian sentiment founded upon the belief in a set of natural
laws superior to tradition. Coupled with this devotion to equality were libertarian ideals, imported from England, which mil-
itated against government intervention. While preparing the
ground for the French Revolution, these movements were also
bringing both of the Academy's cherished traditions of elitism
and adjudication into question.

37. C. Henry, *Correspondance Condorcet-Turgot*, p. 215.

Before the Revolution, this process may be observed on two quite distinguishable levels. Within the ranks of the Academy, the spirit of libertarian equality manifested itself in efforts to reform the internal organization of the society. A group of academic reformers used the Academy as a stage upon which to display their more general criticism of inequalities in Old Regime society that seemed particularly unacceptable in a true Republic of Letters. On another level, a few nonacademic enthusiasts, inspired largely by English examples, proclaimed that the quasi-corporate character of the Academy was detrimental to progress. In their eyes, the free marketplace of scientific and technological ideas was violated by the existence of government-affiliated institutions which, through their exercise of political power, assumed nefarious regulatory functions. These critics likened the Academy to the privileged corporations which were currently stifling free trade and economic prosperity. The reservations of critics within and without the Academy were clearly stimulated by the changing political atmosphere of the later decades of the eighteenth century. We will examine the reflections of the constructive critics first.

It was in the politically pregnant *Encyclopédie* that the first important public criticism of the Academy of Sciences' internal structure was made. The secretary of the Académie Française, Duclos, opened the attack in the article "Honoraire," written in part to exhibit the superiority of his institution over others.[38] His Academy, founded long before Colbert's creation, had stubbornly defended its practice of observing complete equality among its forty members, despite sporadic attempts to introduce distinctions based upon birth or wealth. The 1754 election to the literary society of the Comte de Clermont, a prince of the blood, and the insistence that he be treated like any other academician had spectacularly confirmed this tradition in the face of seemingly overwhelming odds.[39] To resist such pressures was to demonstrate the *de facto* autonomy enjoyed by the Académie Française and the tenacity with which it clung to its cherished traditions.

38. *Encyclopédie*, VIII, 291, col. 2. See also the article "Eleve," *Encyclopédie*, V, 506, col. 2–507, col. 1.
39. D'Alembert, *Œuvres Philosophiques*, XI, 405–425; Meister, pp. 106–107.

But Duclos also recognized the uniqueness of his Academy's practices. Many learned societies had created a class of privileged *honoraires*, who were accorded precedence in rank without responsibility of attendance or participation. Duclos implied that distinctions based upon birth or wealth were a violation of the code of ethics of men of letters, and potentially detrimental to their intellectual independence. D'Alembert, in his "Essai sur les Gens de Lettres," echoed Duclos' sentiments by declaring that talent was the only acceptable yardstick for measuring merit within the world of intellectuals. To accept any other criterion would be to corrupt the true meaning of a Republic of Letters which D'Alembert insisted is guaranteed by its "democratic form."[40] In an unguarded moment, he even went so far as to indict the academies in which "the spirit of despotism reigns" as a consequence of the intrusion of social considerations. His choice of words, setting democracy and republic against despotism, reveals how easy it was to transport political terminology into the sanctuary of science. In remarks made by the other encyclopedists Jaucourt and Condorcet, one senses the same readiness to insert current political concerns in reflections made about the constitution of learned societies.[41]

Under pre-Revolutionary conditions, however, an open attack against honorary status was unthinkable, since the custom had been legalized in written regulations and had behind it the weight of tradition. Having accepted the class of *honoraires* when its statutes were promulgated in 1699, the Academy of Sciences was in no position to opt for a radical innovation. There was some concern that the election of honorary members unworthy of the Academy would lower its standards, but this could not be expressed openly.

The only move envisioned by academic reformers was to dilute the power of honorary members by merging them with the class of *associés libres* or by extending the suffrage. Under the original statutes, the *honoraires* were given a vote in all academic affairs including elections, whereas the *associés* enjoyed the franchise only on matters concerning science. This

40. D'Alembert, *Œuvres Philosophiques*, III, 95–97.
41. Jaucourt. Condorcet's views are expressed in BI, MS. 885, fols. 103–107, and APS, MS. 506 C 75. See also Bouillier, *Séances et Travaux*, CXV, 638–642.

arrangement gave the least qualified academicians a dispropor-
tionately strong voice in elections and internal matters concerning the society, a curious state of affairs considering that
few deigned to attend meetings. By this merger of *honoraires*
with *associés libres*, who were both better qualified and more
diligent, the voting power of honorary members could be attenuated without attacking them frontally. Three such plans
have come down to us: one in 1769 by D'Alembert, another
around 1774 by Macquer, and a third by Bory in 1788.[42] It is
true that none came to fruition before the Revolution. Each
one, however, is a testimony to the willingness of certain academicians to consider reforming their society in a manner not
unlike the programs of contemporary political reformers.

The correspondence between the philosophical premises
of enlightened society and the reform movement within the
Academy is nowhere so evident as in the second area of change
proposed by academicians themselves. It was by all odds the
most important area of reform, since it struck at the inequity
existing among the regular academicians who carried the brunt
of the society's work load.

Regulations passed in 1699 placed every working member
of the Academy in a particular class and subdiscipline of science. As was mentioned earlier, the class system (*pensionnaire,
associé,* and *élève* or, after 1716, *adjoint*) was generally based
upon seniority calculated from the date of initial election to the
learned society. Membership in each class carried with it certain
privileges denied to those in lower ranks. Thus the *pensionnaire*
—but not the *associé, élève,* or *adjoint*—received a regular stipend as well as tokens for attendance (*jetons*) and voted in
elections of new members and on the affairs of the society. Attendance sheets which academicians signed upon arrival at
meetings were therefore carefully divided into columns for
pensionnaires and others.[43] Even seating was arranged according to rank and discipline, so that the *élèves* would be placed at
the furthest remove from the center of activity—and on uncom-
fortable benches. There were other protocols observed by

42. D'Alembert, *Œuvres et Correspondances,* pp. 47–48. Macquer's proposals
are in BN, MS. fr. 12305, fols. 65 and 71; Bory's are discussed by Chapin, *RHS.*
43. AdS, feuilles de présence.

academicians that underscored the hierarchical distinctions separating the working academicians into a set of classes carrying
different privileges and obligations.[44] A senior member of the
Academy was in effect an academic aristocrat, enjoying title
and status confirmed by financial security and political power.

This arrangement was first successfully challenged in 1716,
when a new regulation of the Crown substituted the class of
adjoint for *élève* and widened the franchise system within the
learned society.[45] The new *adjoint* was allowed to vote on matters related to science and was permitted to sit around the central meeting table with those of higher rank, provided absentees
left chairs empty. The *associé* was now granted the privilege of
participating in the nomination of new *adjoints* in his own subdiscipline, though not the right to vote in the election. To
justify these changes, it was pointed out that the term "student"
was no longer appropriate for entering academicians of a certain age. D'Alembert remarked that the term was "offensive in
itself and apt to dissuade subjects worthy of the Academy."[46]
Because this reform was offered in the guise of a scheme to remove a barrier to the development of the Academy, it met no
serious objection.

A completely different reaction set in when a new step
toward the erosion of class distinctions was proposed in 1759.
The academician Chevalier D'Arcy, whose love of libertarian
ideals had induced him to emigrate from his Irish homeland to
France, presented his reform program in terms of sweeping
ideological considerations. He began his impassioned speech at
the Academy by proclaiming that "equality has always been a
precious possession of all thinking beings."[47] He went on to
move that the *adjoint* class be eliminated and its members incorporated into the rank of *associés*, that the *associé* be given
the same voting rights as the *honoraire* and *pensionnaire*. Thus,
with the exception of monetary rewards, D'Arcy was in effect
demanding complete equality within the Academy. It was a

44. Lalande, "Collection," p. 30; Maindron, *Académie*, ch. IV; Sieffert; Birembaut, *RHS*, X, 150–154.
45. Aucoc, *Lois*, pp. xciii–xcv.
46. D'Alembert, *Œuvres et Correspondances*, p. 37. There is a similar appraisal
in Lavoisier, *Œuvres*, IV, 558.
47. AdS, dossier 14 March 1759.

radical proposal in which political egalitarianism was appropriated to an academic setting. After D'Arcy's death, Condorcet
emphasized that very point in his eulogy. He wrote that D'Arcy

looked upon it [the Academy] as his country, and as a Republic
in which he yearned to be known as a respected citizen—but no
more than a citizen. The rights, and above all the liberty of the
Academy, seemed to be his primary concern, and the slightest
affront to these sacred objects was enough to excite his wrath.
He felt that liberty was worth even more than peace—a republican
position it is dangerous to press, and which it would be easy to
abuse by applying it to the regime of a learned society where, unlike political associations in which the privileges of citizens are
at stake, the principal object must be the progress of science.[48]

D'Arcy's motion was envisioned by academicians in precisely
those political terms, as a criticism of the *pensionnaire*'s prerogatives. It was bound to fail, if only because the class under
attack controlled the majority of votes needed to effect any
reform in the Academy.

In the ensuing debate, the *pensionnaire* Duhamel du Monceau rose to oppose D'Arcy's motion and expressed the sentiments shared by the majority of voting members.[49] The essence
of his argument was that the proposed change would only create turbulence where serenity and wisdom should prevail. He
reminded his colleagues that *pensionnaires*, by virtue of their
long years of service in the Academy, would have a greater attachment to the institution and would thus be more liable to
vote in the true interests of their corporation than younger and
less committed academicians. Moreover, Duhamel du Monceau
pointed out, the class system was designed to stimulate ambition
and zeal. "The more prerogatives there are in this last degree,
the more academicians will strive to reach it," he added. His
views carried the day, to the disappointment of the seven *adjoints*, six *associés*, and three *pensionnaires* who had originally
sponsored D'Arcy's proposal.[50] It was politically naive of these

48. *HARS* (1779), p. 67; C. Henry, *Correspondance Condorcet-Turgot*, pp.
221, 301.
49. AdS, dossier 14 March 1759. The speech was read to the Academy on 5
May 1759.
50. The proposal failed by a vote of 11 to 2 (AdS, dossier 14 March and 5
May 1759, and Reg 5 May 1759, fols. 423-424). The two *pensionnaires* who
voted for the proposal were D'Alembert and Le Monnier.

"underprivileged" academicians to expect satisfaction from a frontal attack.

A decade later, in 1769, D'Alembert made another attempt to attenuate the class distinctions by advancing a milder version of D'Arcy's proposal.[51] He asked for the fusion of the *adjoint* and *associé* classes, but would restrict their new voting privileges to elections in their own subdisciplines. His case was pleaded with less fervor and idealism, and included references to other academies in which a greater measure of equality reigned. D'Alembert particularly singled out for comparison the sister Academy of Inscriptions and Belles-Lettres, which had done away with the class of *élèves* in 1716.[52] He appealed to his senior colleagues to remember their younger days, when they too suffered from inferiority in status even though their professional qualifications were the equal of their seniors. In concluding, D'Alembert explicitly disclaimed any notion that he was moving toward reform out of a spirit of innovation for its own sake. Showing greater political acumen than his predecessor, he finished by insisting that his sole interest rested in the improvement of the lot of most academicians "without doing injustice to others."

Notwithstanding these disclaimers, the 1769 reform bill was also voted down, on much the same grounds. This time Nollet was the spokesman for the *status quo*, carrying with him a majority of the voting members of the Academy and eventually the decisive voice of the king, expressed through the Comte de St. Florentin.[53] D'Alembert was not without some support, notably from the influential and liberal *honoraire* Malesherbes.[54] It would seem, however, that D'Alembert's co-authorship of the *Encyclopédie* and his close association with Voltaire placed him firmly in the camp of enlightened reforming liberals, despite all his protestations. The academic aristocrats were not ready for change.

51. C. Henry, *Correspondance Condorcet-Turgot*, pp. 35–50.
52. D'Alembert, *Œuvres et Correspondances*, pp. 38–39. He made the same point in the article "Eleve" in *Encyclopédie*, V, 506, col. 2–507, col. 1.
53. Nollet's speech is in BI, MS. 881, fols. 43–52. Guettard had also prepared a rebuttal which was probably not read. See BMHN, MS. 2193, fols. 201–202. For the final outcome, see AdS, Reg 31 Jan. 1770, fols. 14v–15r.
54. BI, MS. 876, fols. 91r–93v. See also the letter of Malesherbes to D'Alembert in C. Henry, *Bullettino di Bibliografia*, XVIII, 568–570.

The elimination of the *adjoint* class was not achieved until sixteen years later, when Lavoisier guided his reform bill through the Academy with considerably less trouble than his predecessors had. It will be recalled that the major idea behind the reorganization of 1785 was to increase the number of sub-disciplines without substantially increasing the membership. In the process of achieving this "legerdemain," Lavoisier managed to do away with the lowest class of academicians without centering his arguments on the merits of this move. The principal appeal of his reform plan was its addition of two new disciplines and a resultant redistribution of members, bringing almost every academician advancement toward the powerful status of *pensionnaire.* The lure of receiving pensions earlier and the backing of influential *honoraires* such as the Duc de La Roche-foucauld d'Enville and the Duc d'Ayen, who had a voice at court, assured the success of Lavoisier's reform bill.[55]

This reduction of internal inequities within the Academy, wrested from it only after several decades of unsuccessful attempts, did not put an end to its internal problems. Although the old *adjoints* had gained the right to vote on scientific matters by virtue of being incorporated into the class of *associés*, attempts to give them a vote in elections failed completely. The majority of *pensionnaires* continued to maintain that their exclusive right to vote was an essential privilege which must not be shared.[56]

Other inequities within the Academy were decried even when they occasionally proved of personal advantage to reform-minded academicians. There was, for instance, great dissatisfaction with the arbitrary interference of the Crown through its ministers, which often resulted in the naming of supernumerary academicians. We saw earlier how strongly the society reacted in the cases of Bordenave and Fontanieu. To prevent a recurrence of this unpleasant event an article outlawing the appointment of supernumeraries was appended to the 1785 reform bill and became law.[57] It was generally felt that such discretionary actions infringed upon the society's sacred

55. Lavoisier, *Œuvres*, IV, 555–598, and the forthcoming *Correspondance* for 1785.
56. Lavoisier, *Œuvres*, IV, 566–571.
57. Aucoc, *Lois*, p. ciii.

independence. Other instances of ministerial intrusion were also noted, the most blatant of which was Condorcet's appointment to the position of assistant secretary against the wishes of a large number of academicians.[58] This *affaire Condorcet* was a flagrant case of pressure politics by the very liberal faction that deplored its use. The joint efforts of Malesherbes, D'Alembert, and Bossut, working through Turgot and the unsuspecting Fouchy, managed to win Condorcet the position, blocking the Buffon-Bailly faction that also had aspirations of taking over the helm of the Academy of Sciences. Clearly it was the sort of high-level intrigue that has always characterized such elections. The unexpected result of the *affaire Condorcet* was to muster additional support for change from among the most conservative of academicians. The instinct to rationalize institutions through reform, however inconsistent and unsuccessful actual efforts might have been, was decidedly gaining adherents in the Academy. Nonetheless, on the eve of the Revolution, the academicians were far from having removed all of the internal inequities of the 1699 regulations.

Throughout all the discussions about internal reform, a tacit assumption was shared by all the academicians. Disputes about the proper relationship of classes and disciplines could rage on, but no one considered questioning the elitist nature of the institution as a whole. Wherever the egalitarian spirit predominated, it still applied only within the membership of this closed Republic of Letters. Academicians firmly believed the title they had earned upon election set them off from the rest of society by providing them with an inviolable certificate of scientific merit which placed them above the common man. Whatever their individual differences as academicians, they looked upon the institution as a corporate body possessing the legal right and moral duty to assume an authoritarian posture in the world of science.

It was precisely this assumption that came under severe fire during the last decades of the century, further preparing the ground for the Academy's demise during the Revolution. In retrospect, we are able to understand how criticism of the ac-

58. Lalande, "Collection," p. 8; E. B. Smith, pp. 448–449; Baker, *RHS*, XX, 233–255.

ademic form, with its elitist presuppositions and authoritarian posture, could emerge from the anticlassicist movement of the late eighteenth century. But, for the contemporary academician, such attitudes were so completely foreign to the established traditions that he could not garsp them well enough even to refute them logically. His characteristic reaction was one of cavalier dismissal.

Criticism of the academic institution by nonacademicians developed on several levels. The most obvious originated with those who were rebuffed by the Academy.[59] It was an expression of the would-be scientists' frustration and anger, which repeated failures at times turned into paranoic frenzy. Most of the time, however, they covered their wounded pride by expressing personal injury in terms of the general ideologies prevalent in France during the eighteenth century. The two ideologies that most conveniently suited their purposes were an anticorporate bias and a romantic version of the notion of genius.

Two currents of thought fed the growing anticorporate feelings. One was indigenous, arising in free-trade advocates who had taken their lessons from Quesnay's physiocratic school. Grievances against the guild system, which impeded the natural flow of the economy, were easily transposed into an attack against the concept of the *corps d'état* in general.[60] Arguments supporting economic and political liberalism were modulated into sentiments favoring cultural liberalism, which rejected the interposition of self-propagating groups like the Jesuits, the universities, and even the guild-like academies. The movement also found support from administrators such as Bertin, Malesherbes, Turgot, and Dupont de Nemours who, in favoring a centralized cultural flow moving directly from the state to the individual, also resented the presence of intermediary bodies.[61] Although this spirit of liberal reform was temporarily checked by Turgot's downfall in 1776, it remained a vital force both in Paris and in the provincial towns.

From across the Channel came the other source of inspira-

59. This view was advanced most forcefully by Gillispie, "The *Encyclopédie*," and "Science in the French Revolution."
60. Olivier-Martin; Palmer, I, 86–99.
61. Rose; Grosclaude, pp. 449–462; Manuel, p. 25.

tion adopted to combat the French cultural tradition of strongly entrenched *corps d'état*. French travelers who visited England were always struck by the vitality and success of British cultural life, even though it operated under a totally different regime.[62] Leaving aside the moribund universities at Oxford and Cambridge, the English pattern was viewed by the French as an institutionally open society organized around common-interest groups rather than hierarchic corporate structures. Such successful scientific and technological institutions as the Royal Society, the Society of Arts, and the Lunar Society in Birmingham were living examples of associations that made no pretense of acting as supreme arbiters for the scientific community. A free and competitive scientific press also underscored the special state of organization of English science, in which the rigid distinctions between amateurs and professionals considered essential by the French seemed not to exist.[63] This foreign example, when coupled with the success of privately sponsored local enterprises such as the *Encyclopédie*, the Lycées, and Musées, offered tangible evidence that the academic system erected in Colbert's time was not the only possibility.

Viewed against this background, the emergence of Rousseau as the principal writer to whom rebuffed critics turned when searching for ideological justification comes as no surprise. His prize-winning essay, the *Discourse on the Sciences and the Arts* (1750), was well known as an attack upon the corrupting influence of corporatism on morality and virtue. It was particularly significant as the link that tied cultural liberalism to the new romantic concept of genius. When joined, they provided a potent arsenal from which the critics could assail the concept of academism in France.

Rousseau's *Discourse* was perhaps the most provocative and widely read public attack against the very foundation on which the whole academic system rested. It was there that he overtly challenged the central beliefs shared by the entire Republic of Letters, namely that public enlightenment through

62. Brissot, *Licée de Londres*, prospectus, pp. 1–11; Faujas de Saint-Fond, I, 59–62, 99–105.
63. Daumas, *Instruments*, pp. 123–143; Taylor, pp. 6–80.

the widespread use of human reason was the most fruitful of human activities. Rousseau asserted that, on the contrary, products of erudition and learning which "civilize" mankind also sap him of his native honesty and vitiate his stoic virtues.[64] He managed to attack knowledge on precisely those grounds that all leading academicians adopted without question: that discourse and literacy lead men to sobriety and Pyrrhonism. The urbanity of the "polished" man, Rousseau maintained, corrupted "the sublime science of simple souls," and the habits of suspending judgment paralyzed his instinct to act according to the dictates of moral wisdom.[65] Clearly, the academies encouraged all the faults Jean-Jacques perceived in his own times. This explains his covert attack against the very type of institutions from which he was soliciting a prize,

those famed societies simultaneously responsible for the dangerous trust of human knowledge and the sacred trust of morals—trusts which these societies protect by the attention they give both to maintaining within themselves the total purity of their trusts, and to requiring such purity of the members they admit.[66]

The irony of his position vis-à-vis these academies is made manifest only when it is pointed out that Rousseau himself had been deemed "impure" by the Academy of Sciences, which had rejected his novel but inept system of musical notation nine years earlier.[67]

In carrying his criticism of the learned society one step further, Jean-Jacques exhibited a curious kind of masochistic logic. While on the one hand adopting a characteristic Pauline pose that learning by itself is vain and empty, he also leveled his biting remarks at the indiscriminate bestowal of titles to men of learning who did not really deserve to be praised. The award of prizes by academies for wit and learning rather than for profound virtues manifested in action reinforced the blindness of

64. Pire, pp. 173–198.
65. Rousseau, *Œuvres Complètes*, III, 30, and comments by Masters in Rousseau, *First and Second Discourses*, p. 67 n. 11.
66. Rousseau, *Œuvres Complètes*, III, 26. The editor Bouchardy takes this passage at face value (p. 1254) despite the contrary opinion of contemporaries like Brissot, *De la Vérité*, p. 132.
67. Guéhenno, pp. 159–162; Rousseau, *Œuvres Complètes*, III, 21.

society toward essential moral values. "One no longer asks if a man is upright, but rather if he is talented."[68] Rousseau also pointed out that the institutionalization of such rewards by civilized nations encourages a "crowd of obscure writers and idle men of letters who uselessly consume the substance of the state."[69] It is the very success of this Republic of Letters that Rousseau deplores, for it leads an increasing number of men into a state of paralysis by allowing them to bask in luxury. Conversely, it discourages the potentially useful citizen, the moral man.

The wise man does not chase after riches, but he is not insensitive to glory, and when he sees it so poorly distributed, his virtue, which a little emulation would have animated and made useful for society, languishes and dies out in misery and oblivion.

In the long run, this is what must everywhere be the result of the preference given to pleasing talents rather than useful ones, and which experience since the revival of the sciences and arts has only too well confirmed. We have physicists, geometers, chemists, astronomers, poets, musicians, painters; we no longer have citizens. . . .[70]

When replying to the criticism his essay provoked, Rousseau was careful to disassociate his real views from those falsely imputed to him. He was not so naive as to believe in the possibility of man's return to the purity of nativism. Nor was he criticizing the pursuit of knowledge in an absolute sense. It was the relative importance society attached to that pursuit that irked him profoundly. The system of communal judgment and rewards imposed by institutions like the academies encouraged false wisdom while debasing true virtue. All the meanest attitudes Rousseau had discovered among the polished and urbane members of the Republic of Letters—flattery, jealousy, hate, deception—were bred by the cultural system he was denouncing.[71] It was entirely a consequence of the foolish preference for clever wit over moral rectitude. He complained bitterly,

68. Rousseau, *Œuvres Complètes*, III, 25. The same thought was expressed earlier in his comedy "Narcisse" in *Œuvres Complètes*, II, 965–967.
69. Rousseau, *Œuvres Complètes*, III, 19.
70. Rousseau, *Œuvres Complètes*, III, 26.
71. Rousseau, *Œuvres Complètes*, II, 973–974.

remarking that "society rewards the qualities over which man has no control: for our talents are born with us, whereas only our virtues are of our own making."[72]

To argue in this fashion was to overturn the most cherished of values held by Fontenelle and his contemporaries concerning the respective roles of the individual and society in their pursuit of truth. The *esprit classique* assumed that progress would be made if the individual curbed his inborn instincts and complied with the group's understanding of the precepts of nature. According to this conception, the measure of greatness was the ability to channel creative urges into the discovery or imitation of the fixed rules of nature. A painter's talent was judged by his ability to conform to patterns of beauty imposed by nature, discovered, codified, and taught by the Academy of Painting and Sculpture.[73] Talent was not an innate quality at all, but a set of faculties that could be trained through proper practice.[74] The classical approach was to assign the role of establishing the norms of truth to the organized institutions of culture. The awarding of prizes was merely a consequence of their particular function in the socialization of individual taste.

But Rousseau and his contemporaries—Diderot foremost among them—were forging an entirely new concept of genius, which would deny the learned society its role as an agent of progress.[75] For them, talent or genius was no longer considered an external attribute of individuals that could be manipulated, but an intrinsic quality of the person. Man no longer *had* a certain degree of genius; he *was* a genius. Since this quality was innate, no societal encouragement could help to develop it. Genius was born and not made.

These new views had the effect of displacing all the aesthetic values upon which classicism was based, and of calling into question the socializing role of academies. If, as Dieckmann pointed out, "a work of art is no longer judged by the degrees of conformity with traditional patterns and rules, but by the degree of delight it gives, and this delight is caused, not by ra-

72. Rousseau, *Œuvres Complètes*, II, 966.
73. Batteux, pp. 256–258.
74. Helvétius, III, 14, and IV, 333–338; Matoré and Greimas.
75. Rosenthal, pp. 25–56; Grappin; Willard, pp. 27–29, 52–58.

tional structure and intellectual simplicity, but by the free play of imagination and emotion," then, judgment itself becomes a personal and subjective affair.[76] The individual rather than society becomes the ultimate source of progress.[77]

Only one more step was necessary for the formation of a new concept of romantic genius. Instead of conforming to the norms determined by society for the advancement of its own understanding, the true creator was expected to transcend the accepted modes and to lead society. His relationship to society was now inverted and he became the culture hero.[78] The learned society, therefore, necessarily took on the guise of an outmoded or even conservative institution which was either irrelevant or, even worse, detrimental to progress. When the romantic genius expresses his creative urges, he is by definition pitting them against the practices of his culture, symbolized by the learned institutions. In fact, the extent of his originality can be measured by the degree of his disagreement with the community. The authentic genius is not only superior to his fellow man, but he must suffer the incomprehension and rejection of his society. In the extreme case, the true genius must be a martyr. He is by definition that misunderstood, tortured individual infused with the fever of enthusiasm that only posterity is capable of appreciating.[79]

This new romantic concept of genius was the perfect justification for disappointed seekers of the academic stamp of approval. It will be noted that the concept applied equally to the activities of geniuses and madmen, since it denied the value of objective judgment. For this reason, the academicians totally rejected it as a repudiation of their principal function in the scientific society. For the same reason, the critics embraced it as the most effective of sticks with which to beat the Academy.

The history of public criticism of the Academy of Sciences is almost as old as the Academy itself. The earliest recorded attack on the Academy was in a satirical vein. Between

76. Dieckmann, *JHI*, II, 152.
77. Cassirer, pp. 304–307.
78. Cassirer, p. 330.
79. Monglond, II, 175–179; Fellows.

1674 and 1676, four pamphlets aimed directly at the Paris Academy were issued from a private press in Hamburg.[80] They were printed in the form of proceedings of a rival "Academy of Bertrand," whose membership was limited to men bearing the name of Bertrand. Each pamphlet contained a set of mathematical challenges, interspersed with insults leveled at the Parisian academicians Carcavi, Niquet, Roberval, and Auzout, as well as against the Academy in general. The author, printer, and director of this supposed group was a French Huguenot engineer, Bertrand de La Coste, who had suffered a galling setback at the hands of the Paris Academy in 1671 when he had failed to gain its approval for his perpetual-motion machine.[81] The pathetic story of this would-be scientist is worth telling because his plight and reaction were typical of an entire class of critics.

La Coste was a self-taught mathematician and military engineer who had gained a certain success as a mercenary in the employ of several European rulers.[82] Between sieges and battles, he dabbled in geometry, and in 1663 managed to publish a treatise offering a solution to the problem of squaring the circle. Despite the refutation of this solution by a Dutch mathematician, La Coste continued to tackle other allegedly insoluble problems of science that had stumped great minds for centuries.[83] He drew up plans for a perpetual-motion device which he dubbed "Archimedes' machine," and set out for Paris to win a reward for his invention. As was to be expected, Colbert referred the matter to the Academy which, in turn, sent La Coste to its carpenter to have a scale model built for demonstration. This was done at the inventor's expense and in his feverish expectation of great rewards. Once completed, the device was tested before other groups and then presented to

80. Consult the bibliography for the amusing full titles of these pamphlets by La Coste. For the press, see Benzing, p. 174, no. 43.

81. Some biographical details are in Haag, VI, 180–181, and *DBF*, VI, 286–287.

82. These exploits are recounted in a fictitious letter from "A. V. des Angles" to Jean Robert Chouet in the section following the "Avis au Lecteur" of La Coste's *Ne Trompez Plus Personne*.

83. Despite the refutation by Johannes Muller in 1663, La Coste persisted in his claim in his *Démonstration de la Quadrature*, printed in 1667 and translated into Dutch.

the Academy in July 1671. The moment of glory had finally come! To his dismay, La Coste was first grilled by the academicians about what he considered to be irrelevant physical and mathematical matters. His answers were so unsatisfactory that the august assembly dismissed him before he could give them a demonstration of the newly constructed model of Archimedes' machine. All attempts for a further hearing came to naught and La Coste was forced to return to Hamburg a complete failure. His hopes were crushed and he was unable to grasp what had befallen him.

It is evident there was a serious misunderstanding. According to La Coste's own account, his machine was no more than a weight-lifting device with a sizable mechanical advantage that enabled a single man to lift weights of over 1,000 pounds.[84] Yet he was offering this device as a perpetual-motion machine. Like so many other amateurs, La Coste had but a feeble grasp of the scientifically accepted definition of perpetual motion and, judging from his alleged squaring of the circle, very little understanding of geometry as well. His claims were clearly unjustifiable. The academicians felt they could dismiss his case after having assured themselves of his scientific and mathematical ignorance, without even having seen the device in operation. Too many exalted geniuses had falsely claimed to have conquered the great insoluble problems of antiquity for them to dwell on this poor soul.[85]

But La Coste's wounded pride and his investment of time, labor, and money would not let him rest in peace. In 1674, he obtained a certificate from the Elector of Brandenburg, Frederick William, referring to the potential utility of his invention. Conveniently confusing this attestation with the Academy's refusal to credit him with the invention of a true perpetual-motion machine, he set out to ridicule his Parisian judges. They were portrayed as having the qualities of animals, Niquet (the "nigaud") as a cow, Roberval as an ass. Carcavi was treated with even less respect in the epitaph La Coste composed for him:

84. "Approbation de la Machine d'Archimède," in La Coste, *Reveil Matin*, preceding the text.
85. Montucla, *Histoire des Mathématiques*, IV, 619–643; Dircks.

Cy gist le bon Caricavi
Qui dicit Pater Peccavi
Le bien d'autruy Furavi
Es Mathematiques Erravi
En mes brayettes Cacavi[86]

Because the humor was so heavy, the point was made with great force. La Coste also attacked the institution of the Academy, claiming its members were living off the Crown without carrying out their responsibilities toward the public.[87] His major attack was aimed at the very nature of the Academy, which he symbolized in the pun-like motto, "Cartel est notre plaisir."[88] It was the academicians' pretensions of setting themselves up as "sovereign Arbiters of Science" that irked him most profoundly.[89] By challenging them to solve a series of mathematical puzzles, he meant to show them that "wisdom" [*les sciences*] does not depend upon money or authority, but upon the grace of God who doles it out as he sees fit."[90] What annoyed La Coste most was the complete lack of response by the Academy, even after it had been publicly attacked and ridiculed. Silence was the height of insult.

There were many other La Costes during the course of the eighteenth century, although few took to satire for revenge. The only other entertaining attack on the Academy came in 1783 with the publication of the proceedings of a fictitious "Académie d'Asnières," filled with subtle references to the Paris Academy.[91] The legion of squarers of the circle, trisecters of angles, and perpetual-motion inventors took their efforts too seriously to resort to light-hearted ridicule. A correct solution to any of these problems would indeed have constituted a major scientific breakthrough and should have brought instant fame to the discoverer. Because he expected this instant recognition of his genius, the would-be scientist was all the more wounded by the rebuff. He could see it only as an outrageous personal insult and a profound miscarriage of justice, the result of either

86. La Coste, *Ce n'Est pas la Mort aux Rats*, following p. 14.
87. La Coste, *Reveil Matin*, p. 25.
88. La Coste, *Reveil Matin*, privilège.
89. La Coste, *Ne Trompez Plus Personne*, p. 41.
90. La Coste, *Reveil Matin*, p. 74.
91. *Mémoires de l'Académie d'Asnières*. Another, private satire called *L'Académicien Manqué* was published by Toraude, reprinted in Demachy, p. lxxxix.

the stupidity or the jealousy of his judges. As La Coste had pointed out, the Academy's refusal to admit the value of his work was also a serious breach of responsibility that had to be exposed to the public. Indeed, the normal reaction of the squarers of the circle was to break into print with cries of anguish, hoping that the public would act as a higher authority to vindicate their claims. Phrases like the following, found in a pamphlet appropriately entitled *Avis aux Plus Puissants Génies de l'Univers*, were typical.

The most (or least) learned of academicians in scientific matters cannot understand how true, exact and perfect is my figuring . . . [but] all the members of the Tribunal of Science will be forced to agree on the truth that I proclaim and the falseness of the judgments they [academicians] have rendered of my work.[92]

The academicians, of course, looked upon such self-acclaimed geniuses in a different light. Since the professionals took great pride in upholding high standards, they tended to consider the squarers of the circle and their ilk as members of a lunatic fringe of science who must be prevented from doing damage to the profession. In the early days of the Academy, claims of squarers were usually examined with some care before being rejected. In almost every instance, the errors arose either from a lack of knowledge of classical geometry or physics or from an incomplete understanding of the nature of the problem the claimants were attempting to solve. A great deal of time and effort was thereby wasted, for both the inventor and the judges. Even after the mathematician Montucla had published his *Histoire des Recherches sur la Quadrature du Cercle* (1754), exposing with great clarity the pits into which intrepid squarers had fallen, claims to solution of the problem continued to demand the Academy's attention. Montucla had described the situation accurately when he said that the squarers

do not fail to tire mathematicians and especially the academies by soliciting a judgment for their alleged discoveries. They go from tribunal to tribunal, that is, from one academy to another, from the provincial ones as courts of the first instance to the Parisian one. They complain bitterly against a denial of justice when they cannot

92. Dufé-Lafrainaye, *Avis*, p. 3.

be heard and rarely miss the opportunity to deny the competence of their judges or attack them once they are condemned.[93]

As the pace of the Academy's business quickened and its time became more precious, the demands of the lunatic fringe had to be checked. On 3 May 1775, D'Alembert requested a change in the Academy's procedures. The minutes of the meeting recorded the event as follows:

On the occasion of the paper by M. De la Frainaye [Dufé-Lafrainaye] the question was raised if we would continue to receive and examine papers on this subject and a few others whose scrutiny takes so much time and is fruitless. Upon which the Academy voted that henceforth it will not receive nor examine any paper concerned with squaring the circle, trisecting the angle, duplicating the cube, and perpetual motion, and that this decision will be made public.[94]

Although it reduced the work load of academicians, the decision also served to widen the gulf that separated the squarers from the academicians. Since the irrationality of *pi* and the establishment of the laws of thermodynamics had not yet been demonstrated so that the insolubility of these problems was not fully guaranteed by a scientific argument, the Academy could be accused of having made an arbitrary ruling. The chorus of complaints redoubled, all of them pointing to the danger this kind of "tyranny" would create if it remained unchecked. The clockmaker Romilly remarked that if the Academy could decide with a simple negative vote on what was outlawed, there was no hope for any individual who disagreed, no matter how talented.[95] The authority of the *corps* clearly outweighed the authority of logic, and science might thereby be the ultimate loser.

One squarer was so indignant that he brought suit against the Academy. He was Guillaume le Roberger de Vausenville who, much to the society's embarrassment, had once been a corresponding member of the Academy.[96] A Norman amateur

93. Montucla, *Histoire des Recherches*, p. 10. During the 1763–1773 decade, over thirty-five committee reports on attempts to square the circle were recorded in the minutes of the Academy.
94. AdS, Reg 3 May 1775, fol 125v; *HARS* (1775), pp. 61–66.
95. *Journal de Paris*, 19 Jan. 1778.
96. Morin-Lavallée, pp. 81–82; Oursel, II, 153–154.

of science who had made some minor contributions to astronomy and invented a useful device to rule music paper, Vausenville was falsely persuaded that the solution to squaring the circle was a lemma for the solution of the famous longitude problem.[97] Having squared the circle to his satisfaction, Vausenville considered himself eligible for the 150,000 livres he thought was offered by the Academy for a paper related to the question of longitude at sea.[98] In his eyes, then, the Academy's refusal to acknowledge his solution was not only a breach of its mission but also an illegal act. For, as he pointed out, the money had been willed to the Academy by one Rouille de Mêlay for that specific purpose, and the society was legally bound to observe the terms of the will since it had accepted the gift. The case was thrown out of court because it was built on a false premise, but not before Vausenville had lashed out at the Academy.

In a letter he read before at least five Parisian academies and then reprinted for public distribution, Vausenville accused D'Alembert of having "placed the liberty of opinion in chains."[99] Vausenville challenged the great mathematician in unmistakable terms: "How do you know I am in error? Who told you? No one knows, and you as little as anyone else, and yet you condemn me without a hearing, without regard for me, for the rights of individuals or for justice." Four years later, enraged by the Academy's intransigence in the face of his accusations, he struck out at the institution with a set of rhetorical questions:

Is it by oppression and tyranny that you pretend to assist in the progress of human knowledge? Tell me, please, are you the Legislator? Do you possess the authority to enslave genius to your whims, by making reason bend to your opinions? Have you the exclusive power to trace out nature's course by placing stumbling blocks against public liberty? What right have you to intellectual knowledge over anyone else? Is your genius to be taken as a mea-

97. Vausenville, *Essai*, p. 175.
98. The same misapprehension was shared by others. See Montucla, *Histoire des Mathématiques*, IV, 538–553; and Marguet, pp. 45–50.
99. Vausenville, *Quadrature*, pp. 5, 17, for this and the following quotation. The letter was read successively to the Académie Française, the Academy of Architecture, of Surgery, of Inscriptions, of Sciences, in Dec. 1773 and Jan. 1774. A copy of the letter is in AdS, dossier Vausenville.

sure of what is just, of the highest efforts of the human mind? Is it no longer permissible for others to think and act?[100]

He ended his diatribe with a ringing public denunciation that must have shaken some academicians out of their complacency: "It [the Academy] is not fulfilling the object intended by the ruler who established it; it is abusing its reputation to tyrannize the sciences by exercising its despotic powers to the detriment of the law established by the sovereign."[101] Vausenville claimed he had rights that were equal to the Academy's, even though he acknowledged that his power was puny by comparison. Only the "force of reason" was on his side, and it would prevail over injustice.[102]

But there were others, besides this fringe element of squarers of the circle and inventors of perpetual-motion devices, whose claims the Academy denied and who thereupon attacked the royal institution. The noteworthy aspect of their arguments is their close resemblance to those advanced by La Coste and Vausenville. It was the nature of the institution and not the subject matter in contention that was really at issue.

One of the Academy's major detractors in the pre-Revolutionary period was the agronomist Louis Maupin, who bombarded the public and the Academy with a host of pamphlets and letters. His troubles began in 1767 when his new fermentation theory, which was designed to improve wine-making, was found unacceptable because of the lack of conclusive experimental evidence.[103] The academic commissioners were especially skeptical that certain aromatic vapors could be retained in the wine by sealing casks tightly and refused to approve Maupin's theories until reliable evidence could be produced. They noted in passing that his assertion rested upon the views of the Montpellier chemist Venel on the solubility of gases in

100. Vausenville, *Essai*, p. 164.
101. Vausenville, *Supplément*, p. 14.
102. Letter of Vausenville dated 20 Jan. 1779 in AdS, dossier Vausenville. The Academy was sufficiently concerned about the matter to have Lalande explain its official position in the *Journal de Paris*, 17 Feb. 1777 and 8 July 1779. The legal defense of the Academy was entrusted to one of its astronomers and an important member of the Paris Parlement, Dionis du Séjour (see AdS, Reg 16 June 1779, fol. 203r). There was also a discussion of the case at the Société Libre d'Emulation, recorded in AN, $T^{160}23$, no. 224.
103. AdS, Reg 25 Feb. 1767, fols. 38v–41r.

liquids, which, it happened, contradicted those held by one of the commissioners, Macquer.[104] The first academic committee was careful not to deny the correctness of Venel's theory, insisting only that experimental evidence was essential.

Instead of satisfying the Academy, Maupin presented a similar paper two years later, this time obtaining a different set of examiners. Their report was identical in all essential points, and it reaffirmed the Academy's desire to judge on the basis of evidence rather than theories.[105] But Maupin refused to take the Academy's decision at face value. He claimed that a cabal composed of Duhamel du Monceau and Macquer was working to suppress his ideas, and that the former was named because Maupin had previously crossed swords with him in the Academy.[106] On an earlier occasion, Maupin had asserted that his ideas about the spacing of wine shoots for better production were entirely new, whereas the Academy insisted Maupin had been preceded by Tull, Duhamel du Monceau, and Lullin de Châteauvieux.[107] In this case, too, the Academy was careful not to deny the validity of the theory, but simply took issue with Maupin's claims to novelty.

His reaction was to rush into print without the Academy's *privilège* and to magnify the novelty and importance of his contributions to agriculture and to the nation's welfare in general. Maupin then submitted his printed works as "new evidence," hoping to obtain recognition of his genius from the learned society. This time the Academy refused to examine these new works because of its policy of passing judgment only on scientific manuscripts.[108] That policy long antedated the Maupin case and was based upon a strict interpretation of the Academy's role as censor. Once a work was in print it had presumably already passed through the hands of a censor and therefore another judgment was superfluous and potentially

---

104. The two other commissioners reporting on Maupin's paper were Duhamel du Monceau and Malouin. See *Gazette d'Agriculture*, 14 Oct. 1783, pp. 652–653.
105. AdS, Reg 13 Dec. 1769, fols. 387r–388r. This time Maupin's claims were examined by Fougeroux and Cadet.
106. Maupin, pp. 34–35.
107. AdS, Reg 9 Aug. 1763, fols. 303v–305v.
108. AdS, Reg 4 Dec. 1779, fol. 302v; Maupin's letter in AdS, dossier 4 Dec. 1779.

destructive of the principle of authority. The Academy's ruling was also aimed at self-preservation in the face of the increasing flood of scientific publications.

Maupin took this as further evidence of persecution and began to assail the Academy publicly in terms already familiar to us.[109] On the personal level, he claimed the Academy was unwilling to recognize his discoveries because this would confute the views of two academicians. On a more general level, he attacked the institution as also being unjust to the nation as a whole, for his theories were of great potential utility for revolutionizing agriculture. In contrasting his activities with those of the Academy, Maupin added that the institution was shirking its public responsibilities by engaging in nonutilitarian and frivolous enterprises. He also charged the Academy with exercising tyrannical action by interposing itself between a useful discovery and the public. The academicians, Maupin said, are "master of all the public and private channels and the 'doctors' of science have thousands of ways of showing off merit and talent; but they also have thousands to muffle it by closing off all its outlets. . . ."

There were others, from all walks of life, who found it difficult to live by the Academy's decisions. Father Castel of the Jesuit order, the artisan Jodin, the inventors Demandre and Fyot, the geologist Soulavie, the chemist Delamétherie, the astronomer Rutlidge, and the agronomist Morize all expressed deep misgivings about the Academy's role in their own lives and that of science.[110] Others remained silent, privately nursing their wounds and waiting for an opportunity for revenge. Some, like the botanist Descemet and the mineralogists Romé de l'Isle and Monnet, were resentful for having been denied election to the Academy. Others, among them Carra, Barbot du Plessis, and Bergasse, were profoundly displeased with the treatment of silence or disdain their favorite theories received

109. The comments in this paragraph are in Maupin, pp. 11, 12, 19, 20, 34, 35. The quotation is from p. 35. See also Maupin's letter to Macquer: BN, MS. fr. 12306, fols. 67–69.
110. For Castel, see Schier; for Jodin, his own *Examen*; for Demandre, Montucla, *Histoire des Mathématiques*, II, 779, and *Biographie Universelle*, X, 370–371; for Fyot and Rutlidge, see Rutledge, pp. i–xx, 60–61, 99–100. On Soulavie, see Aufrère; on Delamétherie, see *Nouvelle Biographie Générale*, XXIX, cols. 209–212.

from the society and were potentially its greatest antagonists.[111] Two others particularly stand out as figures who were given the opportunity to influence directly the revolutionary generation: Marat and Bernardin de Saint-Pierre.

Marat's association with the Academy dated back to 1778, when he sent in a short manuscript containing his new "discoveries" on imponderable fluids. In April of the following year, the Academy endorsed a report of one of its committees which, while complimenting the author on the merit of his experiments, remained cautious about the validity of his theoretical views on the igneous fluid.[112] Shortly thereafter, the essay appeared in print as the *Découvertes sur le Feu, l'Electricité et la Lumière Constatées par une Suite d'Expériences Nouvelles qui Viennent d'Etre Vérifiées par MM. les Commissaires de l'Académie des Sciences*, headed by the society's favorable report. Marat was overjoyed, for it looked as if he had finally found a vocation that would satisfy his pathological craving for fame.[113] Feverishly, he prepared his next scientific essay on light, filled with numerous experiments, many of which were meant to invalidate Newton's optical theories on color. So proud was Marat of having discredited the scientific hero of the era that he called attention to this momentous event in the letter of presentation to the Academy.[114] He was confident that the august tribunal of science would once more bear out his opinion of himself as a trail-blazing scientist.

This time, however, the Academy was less charitable. Because of the magnitude of Marat's claim, his work demanded close scrutiny. The academic committee noted errors of logic in his paper, which reminded them of the charlatan-like attempts others had recently made in trying to discredit Newton.[115] Moreover, Marat's ideas, even if correct, were not original. The committee took over six months to deliberate and, despite their witnessing of a number of Marat's experiments, failed to reach a common decision. After more than nine

111. For Carra, see BN, MS. fr. 12336, fols. 59–60; for Barbot du Plessis, see AN, F¹²2216, pièce 1583; and for Bergasse, see Bergasse du Petit-Thouars.
112. The report is reprinted in Cabanès, 2nd ed., pp. 511–514.
113. Raspail, pp. 234–252; Burr; Gottschalk, ch. I; Champeix.
114. Letter from Marat to Maillebois dated 19 July 1779, in Cabanès, 1st ed., pp. 299–300.
115. Champeix, pp. 246–249.

months of waiting, Marat insisted upon a judgment. It was not at all what he had expected, especially considering his long wait. The report said

since the experiments were so numerous . . . that we were unable to verify them all (notwithstanding all the attention we gave the matter) with the precision necessary; and that furthermore they do not seem to us to prove what the author imagines they do; and that they are in general contrary to what is best known in optics, we think it unnecessary to enter into details to make them known, and for the reasons just advanced do not consider them fit for the Academy's approval or assent.[116]

As could well be expected, Marat was furious. When he recounted the incidents leading up to this report, he argued that the Academy was persecuting him, not only for his anti-Newtonianism but also because certain members were angry at his arrogant refusal to stand as a candidate for admission into the learned society.[117] He considered the conclusions of the report to be a mockery of justice, and felt deeply wounded that the Academy would not even discuss the experiments which a few members had witnessed. He later insisted that the Academy continued to persecute him by blocking all normal channels of publication and preventing reviews of his book from appearing in leading periodicals. His hopes for a brilliant scientific career were dashed, and the rejection of his work became fuel for his paranoia.[118] During the Revolution, Marat was to be one of the most malevolent and powerful critics of the academies.

But before the Revolution, Marat hardly dared to vent his rancor publicly. Instead, he poured his heart out to his friend and admirer, Jacques-Pierre Brissot de Warville, who carried his message to the public. It was from the pen of this able and eloquent journalist that the most intelligent and well-articulated attacks against the academies appeared before 1789. The most extensive of these came in 1782 in the book *De la Vérité*. What

116. Cabanès, 2nd ed., p. 515. According to Marat's unverified but plausible account related in Cabanès, 1st ed., pp. 308–309, there was a bitter debate in the Academy between his supporter Le Roy and Newtonian enthusiasts, probably led by Lalande and Laplace.
117. See Marat's remarkable letter to Roume de Saint-Laurent dated 20 Nov. 1783 in Marat, *Correspondance*, pp. 23–44.
118. Cabanès, 2nd ed., ch. XXII.

made this work so powerful was the ability Brissot had of turning specific instances of academic rebuffs into illustrations of broader and more general significance. *De la Vérité*, which he once described as his "profession de foi sur les académies," carried the intriguing subtitle, "Meditations on the Ways of Reaching Truth for all Aspects of Human Knowledge."[119]

As far as we know, Brissot did not take an anti-academic position as a consequence of a personal rebuff. He had in fact been schooled all his life in anti-academic and anti-establishment affairs. Brissot's earliest academic contacts were with writers and, quite appropriately, his first public attack, made in 1777, was aimed at the Académie Française.[120] He then became friendly with the journalistic Linguet, who filled his own periodicals with sarcastic jibes directed at the entire sect of *philosophes*, particularly singling out D'Alembert for abusive comment.[121] Brissot tells us that he "quickly became a rabid partisan of Linguet" and eventually collaborated in editing his idol's *Journal* and *Annales*.[122] We can assume from this that Brissot was familiar with the Vausenville affair, which Linguet, among others, fully exploited for his anti-academic campaign.[123]

By the time Brissot met Marat, he had also read and digested enthusiastically the works of Rousseau, whom he cited and used as an inspiration for his treatise on truth.[124] His opinions on Parisian academicians in general and scientists in particular were thus already well formed when he learned of the treatment Marat had received at the hands of the Academy. It confirmed all his beliefs that the officials of science lacked feeling and "sensibilité," and that mathematicians in particular were vain, peremptory, and possessed of "good stomachs but bad hearts" and a "sécheresse de leur âme."[125] The use of such phrases by Brissot was symptomatic of his pre-Romantic sympathy with the morally pure creator struggling against a heartless society. It seemed as if the conduct of men of science was

119. Letter of Brissot to Roland de la Platière dated 23 June 1787, in Brissot, *Correspondance*, p. 143.
120. Brissot, *Pot Pourri*, pp. 11–13, 48–50, 66–83.
121. Devérité, pp. 60–74.
122. Brissot, *Mémoires*, I, 82.
123. *Annales Politiques*, VI, 148–160, 386–395.
124. Brissot, *De la Vérité*, pp. 109–112; Techtmeier, pp. 250–255.
125. Brissot, *Mémoires*, I, 188.

of even greater significance than their use of logic to reason about nature.

Brissot's argument against academies in the *Traité de la Vérité* rests upon a belief that the real seeker of truth must, by the nature of his pursuit, be simple, pure, and solitary.[126] According to Brissot, truth depends solely upon the righteousness of the seeker and his degree of genius, never upon communal assent. From this it follows that organizations like the academies, designed to bring men together, were detrimental to the intensive pursuit of knowledge. Learned societies, like society in general, corrupt the innate creative qualities of individuals by cultivating social habits that inhibit the emergence of truth. They tend to encourage intrigue, disputes, and all the basest dispositions of human nature which are irreconcilable with righteousness, meditation, and the self-confidence of true genius.

Not satisfied with having cut away the very ground on which the academic system rested, Brissot commented directly on the incongruity of allowing constituted bodies to act as judges of merit. As elective aristocracies of the mind, they contradict all of the self-evident freedoms necessary for research:

The empire of science can know neither despots, nor aristocrats, nor electors. It mirrors a perfect Republic in which utility is the only title worthy of recognition. To admit a despot, aristocrats, or electors who by edicts set a seal upon the product of geniuses is to violate the nature of things and the liberty of the human mind. It is an affront to public opinion which alone has the right to crown genius.[127]

The substance of Brissot's criticism went beyond even these cutting remarks. The practice of rendering judgments, which he correctly identified as the Academy's major professional occupation, inevitably forced habits inimical to progress upon its members. The academician's constant use of authority unchecked by public opinion fostered the idea that he was infallible, or, as Molière put it, that "nul n'aura d'esprit hors nous et nos amis."[128] Under such conditions, how could

---

126. Brissot, *De la Vérité*, pp. 90, 125, 163 ff.
127. Brissot, *De la Vérité*, pp. 165–166.
128. Brissot, *De la Vérité*, p. 175.

anyone expect the Academy to be anything but a conservative force in the scientific community? Whatever its avowed purpose, the Academy, in Brissot's view, was a guardian of accepted truth rather than the promoter of new knowledge. Innovation was the work of the isolated genius such as Marat who had to struggle against the spirit of academic conservatism to make the truth prevail.

The Marat case seemed to Brissot to be the perfect example for pressing his views. Here was an enthusiastic experimentalist daring to combat the Newtonian orthodoxy of the Academy and prevented in the most ruthless manner from demonstrating his novel views to the scientific community. Brissot recorded a long conversation he held with one of the Academy's most convinced Newtonians, Laplace, who willingly admitted his opposition to Marat's views even before having heard them.[129] This seemed to Brissot—and quite rightly—the acme of academic prejudice. It was proof that academicians were insensitive to the struggles of innovators as well as arrogant and self-satisfied. Nothing seemed more contrary to the spirit of the search for truth as Brissot saw it.

For Brissot, another striking example of the evil role of the Academy was its activities following the rejection of mesmerism in 1784. The lawyer Nicolas Bergasse, friendly with Brissot, was among the most effective advocates of Mesmer's system.[130] To Brissot, Bergasse's *Considérations sur le Magnétisme Animal* seemed a complete refutation of the Academy's criticisms. Nevertheless, every means was being employed to stem the counterattack of the mesmerists, and it looked as if academic power and intrigue would once more triumph over the weight of evidence. To offset this injustice, Brissot produced in 1786 a strongly worded attack against the academies entitled *Un Mot à l'Oreille des Académiciens de Paris*, even more sarcastic than *De la Vérité*. This was a concisely written brochure in which Brissot vented his spleen against academies in general and the literary and scientific ones in particular. He reiterated the charge that "insolent haughtiness" toward men

129. Brissot, *De la Vérité*, pp. 333–340. Laplace is identified in Brissot, *Mémoires*, I, 198–199.
130. Brissot, *Mémoires*, II, 53–54; Bergasse du Petit-Thouars, pp. 27–40; Darnton, pp. 76–87.

of genius is contrary to the best interests of science and can be explained only by the academicians' self-interest. In order to preserve their academic integrity, he continued, academicians must deny that talent exists outside their ranks. To admit that a mere amateur like Bergasse could overturn their considered opinions by rebutting an academic report would be to admit fallibility.

The Academy's intolerance and intransigence were contrary to its mission and, according to Brissot, denied the fundamental freedom of discussion among men of letters. Most shocking for him was the human injustice that such policies engendered. Having just emerged from a short stay in the Bastille for a crime he had not committed, Brissot readily empathized with men of letters such as Marat and Bergasse who had been, in a figurative sense, imprisoned by the arbitrary judgments of academic authority.

In this pamphlet, Brissot cited one more victim of academic intolerance whose outlook on mankind and society was remarkably similar to his. It was his friend, the writer Bernardin de Saint-Pierre, who was soon to become one of France's most popular authors. In 1784, Saint-Pierre had published his three-volume *Etudes de la Nature*, in which he advanced a new "romantic" approach to the study of nature and introduced some remarkable theories about tides which ran counter to the views prevailing among scientists.[131] Saint-Pierre vigorously rejected the standard methods of naturalists who catalogued dried specimens in museums, counted and described species with the aid of instruments that distorted nature, and dared to reduce living phenomena to physical mechanisms acting without the benefit of Providence. Like Jean-Jacques Rousseau whom he befriended, Saint-Pierre felt that the analytic methods employed by professional scientists blinded them to the essential beauties, sensibilities, feelings, and harmonies that nature was constantly displaying. By dissecting nature, the scientist was in effect killing its very essense.[132]

131. For a sympathetic account of Saint-Pierre's views as a naturalist, see Roule. Brissot and Saint-Pierre were in contact before 1787 (see letter of Brissot to Saint-Pierre dated 12 May 1787, in Boston, Public Library, MS. Ch. I, 10. 23).
132. Saint-Pierre, *Etudes*, 1st ed., I, 13-45, II, 1-62.

Saint-Pierre rejected the academic approach in favor of his simple, unadorned, common-sense apprehension of nature which effectively by-passed the tortuous metaphysical systems of naturalists that had obscured sublime Nature as a work of the Almighty. He drove home the importance of direct contact with nature by recounting a charming tale about the horse chestnut.[133] After deep and protracted consideration, several academies concluded that it served no alimentary purposes in nature and relegated it to the status of a substance useful only for making candles or preparing face powder. But Saint-Pierre, without the erudition of academicians, learned from a simple child goatherd that the horse chestnut served as a stimulant for the production of milk in goats. Why not follow this simple path to understanding and listen attentively to unlettered peasants living close to nature? he asked. Books, surveys, discussions, and esoteric terminology had failed to give mankind any essential piece of information. Men of learning were respectable and sincere, but they would never be in a position to transmit the important lessons nature had to teach mankind. "Let the academies accumulate machines, systems, books, and eulogies," he said. "The principal credit for them is due to ignorant men who provided them with raw materials."[134] Let Everyman be a scientist.

That such a doctrine would find little sympathy with the leaders of the scientific community, intent as they were upon establishing professional criteria, was to be expected. What they feared was that ignorance itself would become a qualification for seekers of scientific truth. Saint-Pierre was in fact a perfect example of this, a fact that the men of learning did not fail to underline. In the course of his discussions about the formation of oceans on this globe, the author of the *Etudes de la Nature* suggested that currents and tides were a direct result of the periodic melting of polar icecaps, rather than the consequence of attraction, and, furthermore, that the earth was elongated at the poles rather than flattened as partisans of Newton had demonstrated in the 1730's.[135] In his discussion of as-

133. Saint-Pierre, *Etudes*, 1st ed., I, 39–40.
134. Saint-Pierre, *Etudes*, 1st ed., I, 47.
135. Saint-Pierre, *Etudes*, 1st ed., I, 219–222, III, 522–553.

tronomical phenomena, Saint-Pierre fully demonstrated his ignorance of elementary geometry and was therefore officially ignored by the Academy, which deemed his work beneath its dignity. Only Lalande attempted to set Saint-Pierre straight, but to no avail.[136] It was a classical case of complete misunderstanding on both sides.

Saint-Pierre and his friend Brissot interpreted the Academy's silence about the book as evidence of its hostility. Like other foes of Newton, they claimed the society was preventing the appearance of book reviews favorable to the *Etudes de la Nature* and surreptitiously discrediting new genius. Such a view fitted well with what they perceived to be the pattern of persecution against simple, righteous, and well-meaning lovers of nature. In the preface to a new edition of his work, Saint-Pierre repeated the well-worn anti-academic theme, asking his readers "never to forget that the Republic of Letters must be a true republic which recognizes no authority other than reason. Moreover, nature has brought each one of us into this world to communicate directly with her. Her intelligence shines on all of us just as the sun does"[137]

Thus, on the eve of the Revolution there was a well-articulated anti-academic position that had developed beyond the normal complaints of rebuffed seekers of academic praise. The position was not only shared by all sorts of would-be scientists—squarers, agronomists, magnetizers, anti-Newtonians, artisans—but also embraced men like Brissot whose humanitarian, literary, and political sensibilities were jarred by the essentially autocratic and elitist nature of the academic system. The more perceptive among them recognized that the activities of these government-sponsored institutions created problems in the politics of science not altogether different from those encountered in Old Regime society. Because the Academy was linked in so many subtle ways to that society, it was inevitable that the growing libertarian and pre-Romantic spirit that

136. Letter of Lakanal to Saint-Pierre dated 30 pluviôse III [18 Feb. 1795], in Brussels, Académie Royale de Belgique, Stassart autograph collection. The new tidal theory was easily refuted by Suremain de Missery; Villeterque; and Hourcastremé, *Aventures*, IV, 64. See also Delambre, *Astronomie au Dix-Huitième Siècle*, pp. 561–562.
137. Saint-Pierre, *Etudes*, 3rd ed., IV, lxxviii.

gnawed away at inequities of the Old Regime would also place in question the Academy's very existence.

The academicians stood fast by their principles. Few if any of them were conscious of the magnitude of the forces of change that were slowly undermining the organization of French society. Their attention was clearly focused on the maintenance of science's forward motion by encouraging work that added to the storehouse of knowledge and by weeding out the feeble attempts of ignorant novices. In the face of the growing numbers of scientific *cognoscenti*, their mission to keep up standards even at the cost of making enemies appeared all the more essential. The Academy had been assigned the role of guardian of science and took that assignment seriously. Although personally irritated with the abuse heaped upon their institution, the academicians were fully aware of filling a vital need and were willing to suffer the consequences. None formulated this better than Condorcet who said in answer to an unnamed critic:

You seem a bit predisposed against the academies; you think them infused with an *esprit de corps* which tends to make them too demanding. I would rather admonish them for being too soft. The Marat incident is proof of this. The only mistake the Academy made was to have at first accepted as new experiments which were already known and which were novel only through the jargon the author had given them.

Academies have two incontrovertible objects. The first is to act as a barrier against charlatanism of all kinds, and that is why so many complain. The second is to maintain sound methods in the sciences and to prevent any branch of science from being totally abandoned. They have a third important function, as long as scientists do not neglect public opinion—that is to make them independent.[138]

But it was impossible for science and its major institution to achieve a greater degree of insularity in the face of the Revolution, which immersed all activities in political turmoil for decades to come.

138. BI, MS. 876, fols. 95–96, partially reproduced in Robinet, pp. 24–25.

Chapter 6

Revolutionary Context

*W*HILE there has been convincing evidence presented to support our contention that the roots of the anti-academic movement antedate the Revolution, it is also true that the closing of the Academy did not occur until 8 August 1793. The seeds for its destruction were sown and some even began to sprout in the Old Regime, but they could not grow to full maturity until the social and political setting had been altered significantly. In that crucial respect, the revolutionary context was a necessary condition for bringing about the end of the academic system as it had existed for over a century.

The national upheaval had repercussions for the Academy that went far beyond the dislocation it brought to individual academicians.[1] As might be expected, the Revolution served to accentuate tendencies already manifest before, but muffled by Old Regime practices. Century-old problems of the relationship of the Academy to constituted authority were magnified by the emergence of a strong legislature accustomed to discuss in public the value and the proper role of national institutions

---

1. These personal "dislocations" ranged from death by the guillotine or in jail (Lavoisier, Bailly, Dietrich, Bochart de Saron, and Condorcet), to unexpected career changes (Sage, Cassini IV, Lamarck), to emigration (Cornette, Chabert, Broussonet). A good number of academicians were pressed into politics, with varying degrees of success: Bailly, Broussonet, Cadet, Condorcet, Darcet, Dionis du Séjour, Fourcroy, Jussieu, Lalande, La Rochefoucauld, Le Roy, Monge, Périer, Tenon, Thouin, and Vandermonde; still others took on administrative functions they would normally have shunned in their communities: Berthollet, Cousin, Lavoisier, L'Héritier, as well as some cited above who held political offices.

like the academies. Moreover, the importance of pamphlets and newspapers forced these questions upon the public with greater impact. Hence both the movement toward reform and the expression of antagonistic criticism were considerably intensified and given a greater chance for success. In addition, new possibilities for social organization provided a political setting and specific ends around which a variety of dissident views could be grouped.

At first, though, the Revolution kindled the hope among certain scientists that drastic reforms of the academic system, heretofore impossible to obtain, could now be enacted for the institution's benefit. In fact, until the middle of 1792 there was considerable reason for academicians to expect that the coming of the Revolution offered their institution the promise of an even brighter future.

Despite some pessimistic estimates about the combined effects of political turmoil and the inflationary spiral, scientific activity progressed at a normal pace, and the Academy continued to play its role at the hub of the scientific life of France. In astronomy, Laplace persisted in reporting on his efforts to incorporate all celestial phenomena into a strict Newtonian framework, Messier and Méchain continued to scan the skies for new comets, and the Academy followed with great interest and a certain amount of jealousy the results obtained by the Herschels across the Channel using their new forty-foot reflecting telescope.[2] Mathematical research also continued its course, with the presentation of memoirs by Arbogast and Le Gendre, respectively, on the integration of differential equations and on elliptical functions.[3] The Cassini map of France made further progress during the early years of the Revolution, and geodesy scored new triumphs when Greenwich and Paris were linked by triangulation. In physics, Coulomb's pioneering work on the measurement of magnetic forces was presented to the society, but was quickly overshadowed by the controversy over the newly discovered galvanic fluid, involving Galvani himself, his nephew Aldini, and their antagonists, Valli and

2. Lalande, *Bibliographie Astronomique*, pp. 681 ff.; Delambre, *Rapport*, pp. 100–163; Cassini, pp. 33–42; and Chapin, "Astronomy."
3. Nielsen; Delambre, *Rapport*, pp. 64–88.

Volta.[4] In chemistry, Fourcroy, in collaboration with Vauquelin, presented the Academy with experimental observations on the synthesis of water, on various inorganic reactions, as well as on the chemical analysis of organic matter.[5] Van Mons's experiments on mercuric oxide were also discussed in academic sessions early in 1793 and served as the occasion for still another debate over the merits of the new chemistry, pitting Lavoisier against the old-fashioned chemist Baumé.[6] During the Revolution, the Academy was no less active in the life sciences, discussing at various intervals new theories of digestion and respiration offered by Lavoisier and Seguin, of botanical growth by Hassenfratz and Seguin, and the massive research efforts of Vicq d'Azyr on the brain and Mascagni on lymphatics.[7] If there was a slackening in academic publications, it was clearly not the result of a loss of scientific vitality in the institution.[8]

The Academy also continued to play its role as consultant. Governments of all political persuasions bombarded the society with an ever-increasing number of requests for expert advice. The circumstances of the Revolution even created additional technical problems requiring immediate solution. A vast quantity of silverware and church bells found its way into government centers when property of the Church and the *émigrés* was confiscated. The Academy was thereupon called to evaluate methods of assaying metals and to develop a cheap and effective method for separating copper from bronze alloys. The process, rapidly developed by Fourcroy, proved a great boon in 1793 and 1794, when the metal was in great demand for the war effort.[9] Indeed, following the outbreak of war in April 1792,

4. AdS, Reg 11 and 13 July 1792, pp. 213–214.
5. Smeaton, *Fourcroy*, p. 243.
6. Meeting of the Academy on 5 June 1793 as reported to the Société Philomatique: BS, MS. box 132, no. 170. The AdS, Reg for this date does not include a detailed description of these activities.
7. Brongniart's report of the Academy's meeting of 26 March 1791, in BS, MS. box 132, no. 63, and of the 23 May 1792 meeting in BS, MS. box 132, no. 127; Desgenettes, II, 174–175; Cuvier, *Rapport*, pp. 207, 321.
8. The troubles were mainly of a financial nature. The Academy's regular printer, Anisson-Duperron, resigned and his task was assumed by the Dupont press. See JT, p. 88; Guerlac, *Scientific Monthly*, LXXX, 99; and Saricks, p. 207.
9. Richard, pp. 263–271; Smeaton, *Fourcroy*, pp. 120–121; Kersaint, pp. 185–186.

there was a sharp increase in academic activity concerning weaponry, powder, ambulances, food preservatives, and other war-related matters. In keeping with its old tradition of governmental assistance, the Academy responded quickly and with a great sense of dedication, paying little or no attention to the shifts in political ideology through which administrations went. As a consultant and in the midst of turmoil, the Academy was beginning to take refuge from ideological commitment by functioning as a politically neutral bureaucratic organization serving the entire nation.[10]

The changes in governmental structure did pose problems of protocol for the Academy, particularly for its secretary, Condorcet, whose political views were well known to the world at large. The sources from which technical demands emanated had multiplied, and the Academy found itself answering not only the king, his ministers, the intendants, and the parlements, but also legislative assemblies and their committees, municipal governments, and even Parisian districts and sections. In one sense, this atomization of administrative centers was only an accentuation of tendencies already experienced before the Revolution and required no new policy decisions. Its principal effect was to increase the amount of time the Academy spent on consultative matters, tending further to disrupt the balance once achieved between pure research and consultation. Yet most of the public spokesmen for the Academy actually welcomed this change because it provided them with a politically neutral utilitarian argument in favor of the Academy's continued existence.

The most notable use of the Academy's special talents was in the role it was assigned by the legislature in the reform of weights and measures.[11] Because this task was given such a strong symbolic significance as the earliest positive reform of the Revolution, it was repeatedly paraded as a prime example of science's potential value to the nation and a concrete instance

10. Maury, *Académie des Sciences*, p. 320; Bertrand, pp. 420–424. This does not contradict the fact that individual academicians had pronounced and often deep political feelings.
11. Hourcastremé, *Dissertation*, p. 19; Bigourdan, *Système Métrique*; Favre, pp. 97–153.

of the Academy's proper function in society. As we shall see later, that argument was not entirely free from danger.

The idea of standardizing the shockingly large variety of weights and measures in the kingdom long antedated the outbreak of the Revolution. As the *cahiers de doléances* and numerous pamphlets underlined, commerce and industry had been seriously handicapped by the chaotic conditions created by this diversity. To the revolutionaries this was an outstanding example of an irrational, feudal system that had to be uprooted. Although none of the proposals to alleviate this condition offered before 1789 had been effective, the new political atmosphere seemed propitious. A month after the storming of the Bastille, the academician Le Roy suggested to his colleagues that they demand that the National Assembly establish a single standard for weights and measures.[12] For tactical reasons, the Academy preferred to introduce the matter through the good offices of a respected representative, and they found their champion in Talleyrand, who was briefed on the topic by several academicians.[13]

The proposal was all the more appropriate since a similar motion had been introduced in the House of Commons by Sir John Riggs Miller on July 24, offering the hope of a truly international reform.[14] Talleyrand and the academicians at first agreed to follow the Miller suggestion for determining the unit of length by pendulum measurements at latitude 45°, leaving the technical details in the hands of the learned societies on both sides of the Channel. In his speech before the National Assembly, Talleyrand made this point explicitly: "I presume that the National Assembly will not take any decision on this matter without first consulting the Academy of Sciences to whom, by all rights, belongs the authority to settle all opinions on such matters."[15]

On 8 May 1790, the Assembly adopted the general princi-

12. AdS, Reg 14 Aug. 1789, fol. 207r.
13. Talleyrand, *Proposition*, in the preface of the printed copy. This quotation is curiously left out of the version published in *AP*, XII, 104 (9 March 1790).
14. Hansard, XXVIII, 297 (24 July 1789), 315–323 (5 Feb. 1790).
15. *AP*, XII, 104–108 (9 March 1790).

ples proposed by Talleyrand, requesting the scientific society's cooperation.[16] The following month, Condorcet brought back the Academy's official acceptance in a ringing speech extolling the virtues of science and its leading institution. He did not miss the opportunity to commend the Assembly for having exercised sound judgment in looking to the Academy for guidance. He said: "You have shown that the wise representatives of an enlightened nation could not fail to recognize either the value of science or the utility of the associations concerned with accelerating its progress and multiplying its applications."[17]

Successive legislatures continued to demonstrate their confidence in the Academy's handling of this monumental reform, giving it the authority to set standards, to decide upon the method of division of the unit, to draw up tables of conversion from older standards, and to communicate directly with all the municipalities of France to implement the reform. The Assembly even went along with the Academy's costly decision to base the unit of length upon the geodesic measurement of an arc of a meridian rather than the simpler and cheaper pendulum method earlier advanced by Miller and Talleyrand.[18] There are grounds to suspect the true motives for the Academy's decision to favor the more expensive and complicated method for carrying out the reform. The new plan gave the Academy an increased budget and considerably lengthened the period of the legislature's sponsorship. Moreover, it was expected to provide an important verification of previous meridian measurements, important for astronomical and geographical purposes and to test Borda's new repeating circle.[19] Whatever the real reasons for its particular choice of method, the support of the weights and measures project is a clear indication that the learned society enjoyed the legislature's confidence and could continue to press home its utilitarian value. These signs augured well for the future.

In its appearances before the National Assembly, the

16. *AP*, XV, 43–44 (8 May 1790).
17. *AP*, XVI, 200–201 (12 June 1790).
18. *AP*, XXIV, 379 (26 March 1791). The appropriation was for 300,000 francs.
19. Biot, *Essai*, p. 36; Favre, pp. 121–130.

Academy exuded confidence. The academicians, realizing that the legislature's printed proceedings were read with avidity by the public, took care to present a utilitarian and humanitarian face, at the same time reminding the nation of their more traditional accomplishments. Lalande, for example, was specifically instructed by the Academy to offer a full set of its memoirs as a gift to the Assembly, and to remind the nation that the scientific organization "has always been careful to direct its concern towards objects of public utility."[20] By this tactic, Lalande followed the lead given earlier by Condorcet, whose speeches also linked the pursuit of science to the destruction of intellectual and social prejudices. In his 10 June 1790 address before the legislature, Condorcet had explained that the search for truth in which the Academy was engaged had a profound liberating influence on the minds of men and that it tended to replace distinctions based upon irrelevant social criteria or political power by more noble intellectual qualities such as talent and knowledge.[21] A short while later, he seized upon the occasion of Benjamin Franklin's death to echo these sentiments in a ringing eulogy read at a *séance publique* on 13 November 1790, ending it with a flourish extolling the role that science played in the preparation of the liberal ideology of revolutionaries:

Always free in the midst of servitude, science transmits to those who cultivate it something of its noble independence, or it flees from countries subjected to arbitrary rule, or it slowly prepares the revolution which must destroy them. . . . In the natural order, political enlightenment follows in its [the cultivation of science's] wake and is buttressed by it . . . an ignorant people is always enslaved.[22]

The Academy also congratulated itself publicly upon having one of its members, Bailly, chosen as delegate from his district in Paris, then as president of the Third Estate and subsequently of the National Assembly, and later elected as the

20. AdS, Reg 24 Nov. 1792, p. 200.
21. *AP*, XVI, 20 (12 June 1790).
22. Condorcet, *Œuvres*, III, 423. This meeting was described by Halem, pp. 341–342, and in *Journal Gratuit*, pp. 36, 72.

first mayor of Paris.[23] It was clear that Bailly's qualifications for political office derived principally from his prominence as a savant belonging to three academies and his reputation as a trenchant critic of charlatanism and a champion of humanitarian reforms, both demonstrated in committee reports on mesmerism and hospitals that he had helped to draft. Before violent political factionalism in clubs and in the legislative assemblies emerged, such self-adulation seemed in order.

The Academy was careful to remember its connections with the royal administration. Delegations of academicians still met and even dined with ministers of the Crown, presented the Dauphin with gifts when he visited the Academy, and continued to offer publications to the King and his brother at court.[24] Condorcet was often forced to write two sets of letters of almost identical content to the Assembly and to the king whenever the Academy officially expressed its gratitude or made a formal statement to governmental superiors.[25] The scientists' allegiance was to the nation, and while there were two centers of authority the society was forced to appear submissive to both the Crown and the legislature.

The difficulty of remaining neutral in the face of mounting political factionalism was aggravated by the beginnings of serious internal dissensions that threatened to split the scientific organization. That there were deep rifts opened early in the Revolution was a direct result of the actions of politically minded academicians who expected that the revolutionary atmosphere in Paris would assist them in pushing through academic reforms that had failed under the Old Regime. It is of considerable importance to note here that, when partisan politics finally made its way into the academic sanctum, it was introduced by way of disagreements over the organizational structure of the institution and its relation to the two centers of governmental authority rather than by its occupational

23. E. B. Smith, ch. III; Hahn, *RHS*, VIII, 351. The Academy made several official visits to congratulate Bailly: AdS, Reg 4, 8, 22, 25 July 1789, fols. 183v, 184v, 194r and v, 4 Aug. 1790, p. 149; Bailly, *Mémoires*, II, 149–150; Sage, *Supplément*, pp. 187–188; *AP*, XVI, 200 (12 June 1790).
24. AdS, Reg 29 July and 1 Aug. 1789, fols. 195v–196r, 6 and 20 March and 21 April 1790, pp. 69, 80, 93; Sage, *Tableau*, pp. 79–80.
25. AdS, Reg 21 May and 13 Sept. 1790, pp. 126, 210–211; BI, MS. 888, fols. 52–53.

concerns. As will become increasingly clear, the Academy's troubles with the Revolution derived principally from the institution's inability to cope with the changing societal environment and not from difficulties originating in the developing scientific life in France.

The demand for academic egalitarianism which had so often failed to win decisive support before 1789 was hard to ignore in the new revolutionary setting. At the time it was reintroduced for discussion in November 1789, Paris was filled with excitement over the drafting of a new constitution and the possibility of a new regime that would remove the inequities of the old.[26] It was auspicious that the Duc de La Rochefoucauld d'Enville, who broached the subject in a speech immediately after the Academy reconvened from its autumn recess, was a highly esteemed liberal representative of the nobility and a member of the scientific society's most privileged class. As an honorary member of the Academy, he commanded respect from the Crown and power within the scientific society altogether out of proportion to his scientific accomplishments. It was just this sort of incongruity that he typified that made his proposition to do away with his own class of *honoraire* so palatable. In the National Assembly, the nobility had spontaneously renounced its privileges on the night of August 4; now the same magnanimous gesture was expected from the Academy's honorary members.[27]

La Rochefoucauld made great capital of the similarity in the incongruous structures of the body politic and the body scientific. He referred pointedly to the "prerogatives" of each academic "order" and to the "inequalities" it sanctioned among a group of intellectual peers, and to the need for "liberty" within the society which could best be secured by the drafting of a new "constitution." He pushed the analogy with the August debates in the National Assembly even further by suggesting that arbitrary ministerial authority was, along with the "prejudices of the century," initially responsible for the Academy's structural troubles. When a new and more equitable

26. For the documents concerning this reform, see Hahn, *RHS*, XVIII, 20–28.
27. This demand was phrased in bombastic terms in the anonymous pamphlet *Suppression de Toutes les Académies*, a copy of which in BN bears the manuscript inked date of 2 Sept. 1789.

constitution was finally adopted, he expressed the hope it would
be presented to the National Assembly as a "patriotic offering."
Only as an afterthought did he suggest that the monarch should
be consulted to ratify the changes.

His specific proposals were in line with the desire to see the
minister stripped of academic control and the National Assembly recognized as a new center of authority. Elections of members and officers were to be kept entirely in the Academy's
hands and the king was to act as a figurehead, rubber-stamping
its decisions. Social differences among academicians were to be
overlooked so that all would enjoy equal voting rights. The
only distinction La Rochefoucauld was willing to retain was
the pension and a special class for scientific amateurs who, by
the nature of their occupation, could not be in regular attendance. In his scheme, there were to be only two classes, the
"Ordinary" and the "Free" academicians.

The immediate reaction of the assembly was predictably
in favor of liberty, equality, and fraternity. Who could have
opposed such noble sentiments at this juncture? Even the astronomer Cassini IV, who expressed some objections to La
Rochefoucauld's scheme, chided the speaker for choosing a
terminology that suggested some academicians were to be free
while others were not.[28] He agreed that it was time to open the
franchise to all academicians, though he felt that new members
should not be given full voting powers until they had attuned
themselves to academic habits with a few years' practice. Cassini
was joined by other speakers in urging that the radical nature
of La Rochefoucauld's proposal be toned down, so that it was
eventually decided a constitutional committee should be appointed to draft new statutes. This committee was elected on
16 December 1789, and included the secretary and the treasurer, Condorcet and Tillet, and the senior academicians Laplace, Borda, and Bossut, all three of whom were presumably in
sympathy with the changes advocated by La Rochefoucauld.[29]

There could be no open argument against equality, but
men like Cassini and Le Gentil cautioned the Academy against

28. Cassini, "Mes Annales," pp. 4–6, 8–9, 16–21; Hahn, *RHS*, XVIII, 25.
29. AdS, Reg 16 Dec. 1789, p. 239.

being swept off its feet by the fever of reform. They urged small changes in the existing regulations rather than sweeping transformations and, above all, recommended that the issues be given slow and dispassionate consideration. In their arguments, one can detect the same reticence toward change expressed decades earlier by Duhamel du Monceau and Nollet. On one point, the group opposing quick and radical reforms was explicit and, at times, emotional in its stand. Whenever the role of the Crown was discussed, voices were raised to fever pitch. Cassini and Le Gentil, among others, expressed their deep attachment to the monarchy and insisted the Academy persist in its traditional relations with the king. They clearly resented the implications in La Rochefoucauld's speech that the Academy ought to turn its back upon its benefactor and usher in a revolution in science to parallel the political events of the day.

That sentiment was sufficiently strong to force La Rochefoucauld to soften his antiministerial line in a follow-up address a few days after he delivered his original proposal. In an effort to find a middle ground between those who shared his reforming instincts and the more conservative academicians, he brought himself to praise the Crown and to list evidence of past royal munificence. By taking this stand he cut the ground from under the more traditionally minded academicians. As a tactic, it proved effective, even though few of his colleagues were fooled by this sudden, forced demonstration of royalist sentiment. La Rochefoucauld's association with the Société de 1789, a new liberal political club, was too well known to be overlooked. All its members, including the academicians Condorcet, Lamarck, Monge, and Vandermonde, shared optimistic views concerning the power of rationally conceived plans for reform to solve political dilemmas at all levels of society.[30] They were willing to concede minor points in order to obtain reform.

The conservatives in the Academy were temporarily mollified, and a constitutional draft containing seventy-four separate articles was prepared and submitted to the whole assembly

30. Desgenettes, II, 172–173; Challamel, pp. 400–414; Guerlac, *Scientific Monthly*, LXXX, 95; Baker, *AS*, XX, 214–218.

on 10 March 1790.[31] Serious discussion was not taken up until June, and then only once a week, article by article. It would seem as if the presiding officers for 1789 and 1790, who happened to be of conservative temperament, did little to speed up deliberations.[32] These dragged on for months, and would probably not have been completed for another year had it not been for the National Assembly's explicit demand on 20 August 1790 for a new academic constitution. Immediately and unanimously the Academy agreed to hold four extraordinary meetings a week to complete its discussions.[33] These special sessions continued until September 13, when a final version emerged for presentation to the National Assembly.

The constitutional debates were time-consuming because so many of the academicians took the issues seriously. There were seldom fewer than thirty-five academicians present for the extraordinary meetings, and at times over fifty attended regularly scheduled sessions. The length of the debates and the results of votes indicate not only a keen interest in these internal affairs, but also a serious division of opinion over a few crucial issues. Judging from contemporary accounts, feelings ran high and left a permanent impression on members consistently holding views that lost by a small margin.[34] In emotional intensity these meetings duplicated the bitter factionalism that the revolutionary era was fostering throughout French society.

We need not dwell at length on those aspects of the new constitution that sparked little debate, except to indicate how far-reaching some of the changes were. Equality was accomplished by following La Rochefoucauld's suggestion of retaining only one type of class division, separating the ordinary or active academicians from those whose attendance was irregular. In the process, the distinctiveness of honorary members and *associés libres* was lost and their total number reduced. One of

31. Lavoisier, *Œuvres*, IV, 597–614. The editor has mistaken this draft for a proposal connected with the 1785 reforms. There are several manuscript drafts of the committee report in AdS, dossier Règlements, and AdS, Lav 934–936. A new version of the proposal in Lavoisier's hand is in AdS, Lav 878.
32. Sage was director of the Academy in 1789 and Le Gentil in 1790. For Sage's conduct in this affair, see his *Tableau*, pp. 79–80.
33. AdS, Reg 21 Aug. 1790, p. 166.
34. Cassini, "Mes Annales," pp. 3–6; Lalande, "Collection," pp. 7, 129; Sage, *Opuscules*, p. 75, and *Mémoires Historiques*, pp. 74–75.

the consequences was also to reduce the number of presiding officers from four to two and to raise the issue of their appointment. Both the functioning of the standing committee system and the election procedures for new members were streamlined by the creation of an executive committee and the disappearance of complex processes for nominations that had previously been followed by each disciplinary section. In fact, there was a general effort to reduce the role of these individual sections. Although new academicians were still to be elected according to their interest in a particular science, they no longer held status because of their sectional assignment. In the old regulations, for example, the three oldest members of each section were given pensions. Under the new proposal, seniority without respect to section was to be the determining factor for reaching the pension stage. Such an attitude was a repudiation of Lavoisier's contention of 1784 that the Academy should be a society composed of a set of small societies divided according to disciplinary interest.[35] The spirit of the times was to rationalize procedures and to create greater equality by eliminating special interest groups within the body scientific.

Only one other discernible trend caused major difficulties and deeply touched political sensibilities. The proposed constitution radically diminished the Crown's effective control of the institution in favor of academic autonomy. What authority remained was entirely honorific and in all major instances was to be shared with the National Assembly. No longer was the king to select new members from a list of three candidates, nor impose his will in the choice of presiding officers or the secretary and treasurer. In this way, the academicians were attempting to guard themselves against arbitrary ministerial interference. But some members wanted to cut ties with the Crown completely and attach the society only to the legislature. The idea, radical as it seemed, was not out of place, since control of the national budget had already been transferred from royal authority to the National Assembly and its finance committee. In point of fact, the legislature rather than the Crown now controlled the Academy's life line.

The most serious debate broke out over the adoption of

35. AdS, Lav 887.

article 67, which originally stated: "The secretary of the Academy will inform the minister in charge of academies of elections in order to obtain the consent of the king. . . ." The article was opposed by Fourcroy, who proposed that confirmation of elections should be left to the National Assembly alone. He was supported by a number of other antiroyalist academicians, probably including Monge, Vandermonde, and Broussonet. On the other side, and rallying considerable support, there was a group who insisted that the Academy's choice of new members must be ratified first by the Crown. This group was led by Jussieu, and probably included Le Gentil, Dr. Le Monnier, Brisson, and Coulomb. The first vote on these two proposals ended in deadlock, but a second vote at the next session favored Fourcroy's substitute motion. On another motion of similar import, Vandermonde mustered enough votes to empower the National Assembly to determine the exact formal relations the Academy was to entertain with other established institutions within France. On the other hand, the conservative faction took some consolation from the Academy's vote to retain the adjective "royal" in the official name of the institution.

With our hindsight, it is easy to see that these debates were wasteful and even injurious exercises. Not only did they politicize academic sessions and thereby foster factionalism, but the new constitution was rapidly rendered obsolete in its details by the changing political situation. Although submitted to the legislature or its committees in September 1790, it was in fact never ratified or put into practice.[36] Indeed, during the three years between its completion and the dissolution of the Academy, there was constant expectation that a new set of regulations would be decreed from above. In the meantime, the Academy chose to function under the old procedures until new ones were formally approved and promulgated. The incongruous result was to promote a feeling of uncertainty despite the Academy's progressive decisions and to place the society in the embar-

36. I have been unable to locate the final version, which should normally be filed with the reform projects of other learned societies in AN, F[17] 1350, dossier 2, or F[17] 1310. Already in 1790 it was difficult to find (see *CIPCN*, II, 247). Only the drafts of Condorcet's covering letters are extant in AdS, Reg 13 Sept. 1790, and BI, MS. 883, fols. 52–53.

rassing position of following obsolete practices it had formally denounced.[37]

Even if the Academy had wanted to keep its difficulties private, the revolutionary penchant for exposing the weaknesses of Old Regime institutions would have prevented it. Knowledge of these internal disagreements and procedural incongruities became public from the start and served to undermine the Academy's prestige. The original La Rochefoucauld reform proposals were immediately discussed in the press and seized upon as the occasion to propose further changes in the academic system. One journalist suggested that the Academy should be abolished because it was a nest of counterrevolutionaries.[38] Another claimed the Academy, because it was supported by national funds, was accountable to the public for its activities, while still another journalist urged that the institution should become self-supporting.[39] Marat, now turned journalist himself, agreed wholeheartedly with the latter, adding that under the Old Regime, despite large salaries given to academicians such as Condorcet, Tillet, Rochon, and Sage, the nation had failed to receive a single noteworthy invention in return.[40] There was a mounting chorus of critics whose views, while often differing one from the other, made the academic reformers' task increasingly complicated. Any plan initiated by the institution was bound to meet with public criticism. As we shall see in the next chapter, it was partly for this reason that the academic reformers abandoned hope for a self-transformation and rallied around a new and more general reform plan elaborated in the National Assembly.

While the Academy was struggling to adjust to factors that touched it directly, other changes were taking place that affected it indirectly. Freedom of the press and of association,

---

37. The problem was repeatedly raised in the Academy's meetings: AdS, Reg 12 Jan. and 6 Sept. 1791, pp. 267, 278, 441; and Lalande, "Collection," p. 129.

38. Morize, in a letter entitled "Abus à Réformer," sent from Evreux on 22 Dec. 1792, in AN, F^{17}1004^A, no. 363. He had expressed displeasure with the academic system on numerous occasions previously.

39. *Chronique de Paris*, 27 Nov. 1789, pp. 382–383, probably written by Millin; Aubert de Vitry, pp. 44, 162–63.

40. Marat, *L'Ami du Peuple*, 17 Aug. 1790, pp. 5–7.

both established in the early days of the Revolution, particularly helped to erode the Academy's virtual monopoly over
scientific and technological matters, contributing to the waning
of the institution's authority.

Freedom of the press seriously undermined the stature of
the *Histoire et Mémoires* as the foremost means of scientific
communication. As soon as the restrictions imposed by censorship were lifted and the *privilège* system disappeared, a flood of
new periodicals made their appearance.[41] Specialized journals
serving the interests of the newer sciences were particularly successful. Prominent chemists joined to found the important *Annales de Chimie*, Bertholon published his private *Journal d'Histoire Naturelle*, while another *Journal d'Histoire Naturelle*
edited by Lamarck, Bruguières, Olivier, and others also made
its appearance for a short time. Here were the true beginnings
of a specialized scientific press in France, which was at last independent of the Academy. In addition, there were new medical
and agricultural periodicals: the *Ephémérides pour Servir à
l'Histoire de Toutes les Parties de l'Art de Guérir* of Pelletan
and Lassus, the *Journal de Chirurgie* of Desault, and the *Journal d'Agriculture* or *Feuille du Cultivateur*, edited at one time
or another by Tessier, Reynier, Dubois de Jancigny, Parmentier, and Cretté-Palluel. A smaller—and possibly less successful
—group of specialized journals devoted themselves to utilitarian
matters connected with science, including Bertholon's *Journal
des Sciences Utiles*, the *Journal Gratuit*, the *Journal du Point
Central des Arts et Métiers*, and the *Journal des Sciences, Arts
et Métiers*—the latter two edited by artisan groups. Each of
these new journals appealed to a particular interest group that
had found the existing media of communication either too general in its scientific concerns or too closely under academic
control.

Paralleling the emergence of specialized publications was
the creation of review journals or newsletters that would keep
readers abreast of the current state and progress of science. The
idea was by no means new, having previously given rise to the
short-lived *Avant-Coureur* and, initially, to Rozier's *Observa-*

---

41. On the scientific press of the period, see Ranc, "La Presse Scientifique";
Genty, *Progrès Médical, Supplément*, XI, 33–37; Festy.

tions sur la Physique.[42] Three new publishing ventures were begun after censorship was lifted. Fourcroy founded *La Médecine Eclairée par les Sciences Physiques*, which was meant to bring together news of all the scientific discoveries of interest to those engaged in the healing arts.[43] Another important review journal was the *Bulletin des Sciences*, at first circulated in manuscript for the mutual instruction of members of the new Société Philomatique, and eventually printed and sold on the open market. Still another type of general periodical was the naturalist Millin's new *Magasin Encyclopédique*, which reported titles of new books and announced coming events of scientific interest.

While it is true that this proliferation of scientific publications could not have occurred in the Old Regime because of tight academic control, it would be a mistake to picture them as overt rivals of the *Mémoires*, the *Connaissance des Temps*, or the *Journal des Savants*. They functioned in fact as much needed complements aimed at satisfying needs of the growing scientific community already felt before the Revolution began. A large number of academicians individually supported their existence, often serving on their editorial boards and generously contributing papers. Without their being direct competitors of academic publications, their very existence nonetheless destroyed the uniqueness of the Academy's journals and reduced the effectiveness of its control over the scientific and technological affairs of the nation. As might also be expected, articles appeared which, before 1789, would not have received the Academy's approval, thus noticeably lowering standards of publication.

What the Academy was losing as a consequence of this new revolutionary freedom was its traditional power to make judgments which had a legal or moral force over the entire community. With the end of the era of the *privilège*, the Academy was forced to recognize a decline in its over-all authority. Academicians able to perceive this change therefore suggested that subcommittees reporting on scientific or technical papers sent

42. Smeaton, *AS*, XIII, 219–230; McKie, *AS*.
43. In Fourcroy's letter to Antonio Savaresi dated 16 Dec. 1790, he explains the importance of freedom of the press for the new enterprise (BN, MS. fichier Charavay, vol. 75).

to the society should refrain from the traditional use of the word "approbation" in their concluding paragraphs.[44] On one occasion, the Academy acknowledged the new realities by emphasizing that its principal role was as a "consultative" body rather than an "authoritarian" one.[45] To say this publicly was to admit that a fundamental change had occurred since the outbreak of the Revolution.

The other important area in which the Academy's uniqueness and prestige were seriously diminished was in the proliferation of independent scientific societies, commonly referred to as the *sociétés libres*. Like freedom of the press, freedom of association was cherished by revolutionaries as a major accomplishment of the new era. Such was the attachment of some radical revolutionaries to this new freedom that even the legally required formality of registering group meetings was feared to be contrary to this ideal.[46]

Just as no journals were consciously set up to rival the Academy's publications, there were no voluntary associations formed specifically to combat the Academy. Yet their very existence drew attention away from the government-affiliated organization and offered a visible alternative to Old Regime patterns. As criticism of the Academy mounted, the scientific community—including some academicians themselves—began to look toward such associations as a sensible option to the traditional Old Regime patterns. It seemed to many as if these private groups were better suited to the pursuit and propagation of science than were the old-fashioned learned societies.

Of the numerous voluntary associations that dotted Revolutionary France, three stand out: the Lycée, the Société Philomatique, and the Société d'Histoire Naturelle.[47] They had two noteworthy features in common. All three had links with prerevolutionary groups that lived in the shadow of the Academy. In addition, each was committed to the propagation even more than to the advancement of science. In contrast to the Academy, which had never considered teaching or "haute vulgarisation" as one of its roles, most of the voluntary associations

44. AdS, Reg 30 March 1791, p. 314.
45. *Rapport Fait à l'Académie*, p. 1 n. 1.
46. Robespierre, in his speech in *AP*, XXVIII, 749 (5 July 1791).
47. A list of other voluntary associations is given in Appendix I.

were engaged in educational ventures, from which they derived both membership and financial backing. Beyond this practical need, their interest in the diffusion of science corresponded to the revolutionaries' profound desire to democratize science by making its fruits understandable and applicable to all members of society. Since most of the successful *sociétés libres* derived their sustenance from and served social groups that had been shut out of direct participation in the scientific life of France by the Old Regime practices, they came to represent the antithesis of the elitist and professional traditions forged for over a century by the Academy.

The Lycée had been founded in 1785 as an offshoot of the Musée de Monsieur, and maintained an excellent reputation as one of the more enlightened and popular centers of public scientific education in Paris.[48] Like the Collège Royal which it resembled in many ways, the Lycée was principally concerned with offering high-level adult education in all subjects of general interest to the upper classes of Parisian society. By the time the Revolution came, it had already developed a respectable tradition of teaching which was consciously aimed at lifting what was presented to the gullible public above the less than reputable lectures of the day given by men such as Comus, Famin, Filassier, Miollan, and Perrin.[49] Against these demonstrators of dubious quality, the Lycée pitted scientists of academic quality, including Condorcet, Lacroix, Deparcieux, Fourcroy, Ray, and Süe. After the outbreak of the Revolution, the institution, faced with serious political and financial difficulties, underwent a drastic reorganization that enabled it to carry on its own respectable traditions in the midst of great political turmoil. One major aspect of the institutional transformation was the elimination of aristocrats from the list of financial backers and their replacement with republicans willing to invest money in this public enterprise.[50] In keeping with the times,

48. Smeaton, *AS*, XI, 257–259; Taton, *RHS*, XII, 130–153. Contemporary descriptions of the Lycée are found in Darnton, pp. 178–179; Isambert, pp. 163–164; and Maugras, pp. 74–82, 195–200.
49. Taton, *RHS*, XII, 140–141.
50. Fourcroy's role in this affair has been discussed with differing interpretations by Smeaton, *AS*, XI, 259–265; Hahn, *AIHS*, XII, 285–288; Kersaint, pp. 46–50.

the Lycée also took the name Lycée Républicain in 1793. The
directors' ability to respond promptly to the changing political
winds was clearly one of the institution's greatest assets. Such
flexibility was indeed one of the major reasons it survived, and
constituted the principal advantage of private, voluntary associations over government-affiliated societies. The contrast, as
we will see later, was not lost upon a number of influential academicians.

The Société Philomatique had its origins in the gatherings
of students who met informally after a set of lectures on chemistry given at the Invalides by young Alexandre Brongniart
starting in the summer of 1786.[51] Their first recorded meeting
as a "Société Gymnastique" took place on 10 December 1788
and, in addition to Brongniart, included several medical students, the agronomist Silvestre, and the naturalist Riche de
Prony, who was an intimate of the Fréminville salon at the Invalides.[52] It was a sense of congeniality and a desire to share
scientific information that originally brought these men together and led to their adoption of the name "Philomatique."
From the very beginning, the group fell into the habit of reporting and commenting on scientific discoveries currently in
the news or discussed in meetings of learned societies. Their
plan to finish the Academy's task of preparing a complete
Description des Arts et Métiers was, of course, too ambitious,
and often led them into discussions about applied science, particularly that part derived from chemical knowledge. In fact,
interest in chemistry seems to have provided the main focus of
the group, which expanded steadily in the early years of the
Revolution.[53] It grew so quickly that by 1791 a monthly manuscript bulletin summarizing the latest scientific news was pre-

51. In a letter to Silvestre dated 14 Oct. 1793, Brongniart recalls there were
two possible dates for the society's foundation (BMHN, MS. 1989, no. 876).
The manuscript text of Brongniart's lectures is in BMHN, MS. 2322, 1.
52. Duveen, *AS*, X, 340, considers Silvestre the most important founder, while
Crosland, *Society of Arcueil*, p. 169, credits Riche de Prony also. The latter's
connection with the Invalides salon was through his brother, the academician
Prony who was married to a Fréminville (see Paris, Ecole des Ponts et
Chaussées, MS. 2703, and Dupin, pp. 203–215). Equal attention should be
paid to Brongniart as a founder. His connection to the Invalides circle was
through his father, Théodore, the resident architect (see Launay, pp. 23–24,
50–51).
53. BMHN, MS. 647; BS, MS. box 128.

pared by the secretary for its members. It was this newsletter which in 1792 turned into the *Bulletin des Sciences*, printed and sold by the Dupont firm. Of all the societies of the era, the Philomatique seems to have been the least affected by contemporary politics, steering clear of thorny discussions about its members' political views and of composing elaborate "constitutions" for itself. Its ambitions were limited, and since it was a private association sustained by membership dues it could operate in relative isolation from the turmoil of the times.

In this respect, the new Société d'Histoire Naturelle stands in sharp contrast. Its prerevolutionary background was considerably less benign, originating as it did with members of the ill-fated Société Linnéenne. The naturalists who regrouped on 27 August 1790 began their association in a vastly different mood, eminently aware of the political implications of their association.[54] They met on the heels of public debates in the National Assembly about the future of the Jardin du Roi, which was threatened with a cut in public funds unless it reformed its organization in accordance with less autocratic patterns.[55] The founding members of the Société d'Histoire Naturelle—particularly Louis Bosc and Millin—were older than those of the Philomatique and more politically oriented. From the very beginning, they seem to have attracted naturalists of a republican stripe who felt it essential to introduce a political dimension into their carefully articulated constitution. Delamétherie, the editor of the *Journal de Physique*, Creuzé-Latouche, a member of the National Assembly, and the academician Fourcroy, who joined the naturalists before the year was out, were all enthusiastic associates of the reform movement and operated in this voluntary association almost as if it were a political club.[56] As early as 22 October 1790, Fourcroy was advocating the adoption of five articles in the newly framed regulations concerning "the patriotic sentiments that members who compose it or are

54. The contrast is noted by Brongniart in his letter to Silvestre dated 7 frimaire an II [27 Nov. 1793], in BMHN, MS. 1989, no. 883.
55. *AP*, XVIII, 176 (20 Aug. 1790).
56. The subject of a constitution for the new society was broached at the first meeting by Millin (BMHN, MS. 464, p. 1). Creuzé-Latouche, though never listed as a member of the society, took part in a considerable number of meetings. His political views are discussed by Marion.

to be admitted must possess."[57] The minutes of the meetings even record the lengthy discussions held in March and April 1792 to determine if one of its members, Ramond de Carbonnières, whose "conduct in public affairs was blameworthy," should be ousted.[58] A number of letters of resignation from the society—including one from Ramond himself—were clearly prompted by the increasingly republican tone of the meetings.

The naturalists were also more concerned with public gestures and influence than the philomaths. Exploiting their contacts with ministers, legislators, and the press, they mounted a successful campaign to gain national sponsorship for a new expedition to circumnavigate the globe in search of the La Pérouse voyagers, who had not been heard from since March 1788.[59] After the National Assembly passed the decree authorizing this venture, both the Academy and the Société d'Histoire Naturelle were consulted to organize the scientific details of the expedition, implying that this new voluntary association possessed as much right to make professional recommendations as did the venerable Academy.[60] Whenever public ceremonies called for the presence of naturalists, delegates from the society made themselves quite conspicuous. Thus the press made much of their role in the erection of busts of Linnaeus and Buffon at the Jardin du Roi and in the celebration honoring Jean-Jacques Rousseau held at Montmorency on 25 September 1791 —both events conveniently possessing a politico-scientific character.[61]

In such an atmosphere, it was inevitable that the *société libre* should be extolled as the typical product of the Revolution, in contrast to the academies, which tended to be portrayed

---

57. BMHN, MS. 464, p. 9.
58. BMHN, MS. 464, pp. 104, 107. Besson and Pelletier resigned in sympathy (BS, MS. box 133, no. 40). Ramond's political "crime" was his stubborn support of the constitutional monarchy and opposition to the legislature's treatment of priests. See Girdlestone, pp. 193–198.
59. BMHN, MS. 464, pp. 25–26, 43. For details, see Hamy, *Débuts de Lamarck*, pp. 140–145, and Martin-Allanic, pp. 1519–1521.
60. *AP*, XXII, 457–458 (22 Jan. 1791); XXIII, 78–81 (9 Feb. 1791); XXVII, 93 (9 June 1791). Lalande was opposed to this expedition (see *Moniteur*, 3 Feb. 1791, p. 137).
61. *Chronique de Paris*, 31 July and 25 Sept. 1791, pp. 856–857, 1081; Fourcroy, *Médecine Eclairée*, I, 48; BMHN, MS. 464, pp. 82–83, 103–104, 109; BS, MS. box 128, for the proceedings of the Rousseau festival.

as obsolete creatures of the Old Regime. One symbolized freedom, liberty, and *laissez-faire*; the other constraint, oppression, and *dirigisme*. The voluntary association emerged as the embodiment of democratic science, just as the Academy had stood for what revolutionaries termed "aristocratic" science. Pamphleteers and journalists found that contrast easy to exploit. Millin, as editor of the *Actes de la Société d'Histoire Naturelle*, reminded his readers of the machinations of the Academy allegedly initiated to destroy or suppress the Société des Arts, the Société Libre d'Emulation, and the Société Linnéenne.[62] It was easy for him to caricature the Academy, representing it as the source of scientific despotism, and to cast doubts on its reliability by reminding the readers of the institution's close connection with the much maligned ministerial governments. The same was true of Delamétherie, who was particularly articulate in his tirades against all official learned societies.

Delamétherie's first opportunity to lash out at the Academy came in the eulogy of the deceased crystallographer Romé de l'Isle, which he inserted in the April 1790 issue of his periodical, the *Journal de Physique*. He remarked bitterly on the difficulties Romé de l'Isle had faced trying to gain access to the Academy and on the profound injustice of his failure to win a seat.[63] According to Delamétherie, the crystallographer was immensely talented, but lacked the proper connections or the personality to muster the necessary votes for admission. As always, the claim was that intrigue triumphed over merit.

Nevertheless Delamétherie saw some hope for the future of science if it followed new directions:

We like to believe that under the rule of liberty such abuses will be repressed. The gentlemen of the Academy must not forget that they are salaried at great expense to the nation . . . to the extent of 117,780 livres. But that is the least part of their remuneration. They have been able to take over all the numerous positions open to scientists, such as those in the Jardin du Roi, the Observatory, the Collège Royal, the Mint, the School of Mining, etc. To refuse

62. *Actes de la Société d'Histoire Naturelle*, p. xiii.
63. *Journal de Physique*, XXXVI, 319. Lalande tried to refute Delamétherie's article in *Journal des Savants*, June 1790, pp. 443–444. The reasons for the Academy's refusal to elect Romé de l'Isle are given by Burke, pp. 62–63, 80. This refusal seems in retrospect to have been poor scientific judgment on the Academy's part.

to elect a deserving scientist . . . is therefore a crying injustice . . . a real theft.

In voluntary associations, such as the Royal Society of London, no injustice is committed in rejecting an otherwise qualified candidate, since the association—being a purely voluntary one—has the right to reject anyone whose moral qualities do not correspond to those shared by the majority of members, as is the case in political clubs; but such is not the case with associations salaried by the nation.[64]

He returned to this contrast on every occasion, reminding his readers that science had flourished in other epochs and in other lands without the existence of paid academicians, that Descartes, Fermat, Gassendi, Pascal, and others had been capable of great feats without academic appointments. When there was an Academy of Sciences, he continued, it was despotically ruled by a Fontenelle, a Réaumur, a D'Alembert, who continually prevented talented scientists such as Romé de l'Isle from receiving proper recognition. The academies were "powerful corporations, full of intrigue and cabal . . . crushing all true merit unwilling to bow down to Balaam's ass."[65]

The case in favor of voluntary associations and against the academies was easily transposed into a preference for liberty over despotism, which in effect left no option for the true patriot. Gradually the vocabulary used to describe the inequities of the Old Regime was appropriated to depict the academies as "imperial," "gothic," despotic corporations and *jurandes*, as part of the "order of academic nobility," or, to follow Marat's phraseology, as modern charlatans.[66]

As we saw in the last chapter, none of these ideas was completely original with the revolutionaries. But in the revolutionary setting, they struck many more resonating chords and swayed more opinions toward an anti-academic position. Brissot's *De la Vérité* meant more in 1789 when paraphrased in a

64. *Journal de Physique*, XXXVI, 319 n. 2.
65. *Journal de Physique*, XXXVIII, 46–48. Similar comments are in *Journal de Physique*, XXXVII, 318–319; XXXVIII, 74; XXXIX, 82, 235. Condorcet answered these and other charges by Delamétherie in a letter reprinted in *L'Amateur d'Autographes*, 1 Sept. 1865, pp. 257–258. See also BI, MS. 861, fols. 398–399.
66. Cloots in *Chronique de Paris*, 8 Aug. 1792, pp. 882–883; Villette in *Chronique de Paris*, 28 May 1791, p. 345; Restout, p. 6; Marat, *Charlatans Modernes*.

pamphlet than when originally published in 1782.[67] Marat now found a sympathetic audience for the denunciations of the Academy which he had prepared before the Revolution by publishing them in his fiery political newspaper.[68] But the most devastating of public utterances belittling the academic system came from the pen of Rousseau's most popular disciple and academic critic, Bernardin de Saint-Pierre. In a style more pleasing than Marat's, his message was just as clear:

> If the sciences and letters have an influence on the prosperity of a nation—which we cannot doubt—it might be proper to have the nation elect the members of its academies, as they do members of other assemblies. New knowledge must be held in common, as are the other riches of the state. When the academies elect their own members, they become aristocracies harmful to the republic of science and letters. Since one can only be admitted by courting their leaders, one must tie himself to their systems of thinking. Errors are thus maintained by the authority of corporations, while isolated truth finds no partisans.[69]

Saint-Pierre's most influential comment against the academies came in the form of a short novel called *La Chaumière Indienne* which, through its association with *Paul et Virginie*, was destined to become a best-seller. It was a charming fable, based in large part upon Sonnerat's *Voyage aux Indes Orientales et à la Chine*, and describing the trip of a learned academician sent by the Royal Society to gather information on the greatest problems of mankind.[70] His world tour ended on the Bay of Bengal at the Jagannath pagoda of the chief Brahmin priest. Our erudite doctor had reduced the 3,500 questions he was investigating to three essential ones which constituted the key to knowledge itself: How can one find truth? Where is it to be found? Should it be shared with other men?

To a man of erudition associated with the august Royal Society, the Brahmin's answers were devastatingly insolent. The high priest asserted that truth was open only to the Brah-

---

67. Aubert de Vitry, pp. 126–127, 142–143. Brissot's book appeared in a second edition in 1792.
68. Marat, *L'Ami du Peuple*, 18 Sept. 1789, p. 193; 7 Aug. 1790, pp. 3–4; 17 Aug. 1790, entire issue; 16 March 1791, pp. 1–3; 4 June 1792, p. 2; and *Journal de la République Française*, 15 Dec. 1792, p. 8; 14 Jan. 1793, pp. 4–5.
69. Saint-Pierre, *Vœux d'un Solitaire*, p. 208.
70. Bird.

mins themselves, because it was contained in ancient books that they alone could read. Moreover, it was the sacred possession of the sect and should be withheld from the common man. Those wishing to gaze upon the eternal truths would have to be initiated into the order. Our doctor was outraged and attempted to refute the chief priest, but was totally frustrated by the Brahmin's complete self-assurance. Since his very title insured that "every word he speaks is a ray of intelligence," the high priest could refrain from engaging in argument. The interview was abruptly ended, and the European left, a totally crushed man.

Saint-Pierre wrote the fable as a parody of the academies, who, he suggested, also arrogated truth to themselves alone. Like the Brahmins, they claimed the ability to read the book of nature, which was denied to the common man. It was the title "academician" that guaranteed the purity of their judgment and justified their unwillingness to debate with commoners. Academicians were the high priests of knowledge, and their arrogance was as shocking as the conduct of the Jagannath priest.

To make certain the parody was understood by his readers, Saint-Pierre included in the introduction to the fable the story of his own misfortunes suffered at the hands of the Academy of Sciences. He was careful to disassociate himself from anti-intellectualism by insisting that "it is not science itself that I am blaming; but I wanted to show that because of their ambitions, jealousies, and prejudices, these learned societies all too often act as obstacles to progress."[71] He also wanted to contrast the vain erudition of academicians and the arrogance of priests with the wisdom of the simple man who, like himself, communed directly with nature. This is why he ended the fable on a favorable note.

In it, the disappointed English scholar is driven to take shelter by a tropical storm and seeks refuge in the humble home of a pariah. It is in this "chaumière indienne" that he finds the secrets the Brahmins were unwilling to share, and symbolically from the lips of an outcast. The moral could hardly have been

71. Saint-Pierre, *Chaumière Indienne*, pp. 7–8 of the in-18° edition. In other editions it appears on p. xx.

missed.[72] Truth is not the property of a titled, intellectual elite, possessing special qualifications for grasping it. To be understood, nature must be approached in a direct, unaffected manner, with a heart filled with simplicity and virtue. Every man has access to nature, and, in fact, the lower classes are more likely to comprehend its deepest meaning because they habitually commune with it. In the search for truth, erudition is unnecessary and even detrimental.

Saint-Pierre's message was especially appreciated by a legion of less educated artisans who had long felt oppressed by officialdom but had rarely had the opportunity to vent their sentiments in public. The broadsides, pamphlets, and newspapers of the Revolution opened new channels of expression and advertisement, and the emergence of voluntary associations gave the artisan greater hopes of finding a secure political platform from which to demand an end to injustice.[73] Political circumstances were turning the dream of artisans into a reality. In 1783, they had timidly asserted that "among the almost infinite number of [learned societies of scientists] ... there should possibly be a few intermediary companies, where a new-born invention might be exposed in all its nudity, ... where an as yet unformed talent could present, without blushing, the attempts of his innovative Muse."[74] The inhibiting effect of academic domination and arrogance was now being replaced with a self-confidence that produced the most detailed and telling criticism against the learned societies. It was the newest element in the anti-academic campaign of the Revolution.

A bewildering chorus of voices called for the creation of voluntary associations for the artisans, free from the clutches of academic domination. Duchesne tried to revive the Société d'Emulation, hoping to merge it with the Lycée; Chardon and Perny demanded the creation of an Académie des Arts dedicated to the furtherance of technology; the Baron de Servières headed a group called the Société des Inventions et Découvertes; Desaudray created a Lycée des Arts as an "assemblée libre et primaire des artistes" and took a leading role in another

72. *Moniteur*, 10 Feb. 1791, pp. 167–168.
73. Gillispie, *Critical Problems*, pp. 270–275.
74. *Discours sur les Découvertes*, p. 23.

association called the Société du Point Central des Arts et Métiers.[75] The sheer quantity of pamphlets, addresses, petitions, and announcements that emerged from these groups demonstrates the great ferment unleashed by the Revolution. Yet few of the artisans possessed sufficient experience or the capital necessary to sustain their grandiose plans, which often called for the creation of newspapers or scientific journals. As might well have been expected, they were generally unable to transform their elaborate projects into realities. Much greater success was achieved by lobbying for short-range goals that spelled the dismantling of the Academy's accumulated power. Artisan pressure groups were in fact directly responsible for two entirely new creations of the Revolution, the Bureau de Consultation des Arts et Métiers and the Bureau des Brevets et Inventions, both of which were designed to protect the interests of artisans by taking over some of the functions previously exercised by the Academy.

The origins of the Bureau de Consultation go back to 10 July 1790, when the National Assembly decided as a general principle to abolish the royal pension system and to replace it with rewards given to individuals as they made substantive contributions to the nation.[76] Three weeks later, La Revellière-Lépeaux's proposal that such national rewards should be given as well to "hommes de lettres, savants et artistes" was enthusiastically endorsed following a number of speeches proclaiming the national value of literary, scientific, and technological pursuits. The bill establishing this as a principle was adopted on 3 August 1790. But it took over a year before the Assembly's Committee of Agriculture and Commerce brought in a concrete proposal and appropriated the funds necessary to implement the principle.

In the meantime, another pressing matter retained the attention of artisans and the committee. There were no satisfactory laws guaranteeing inventors the right to own and exploit their intellectual "property" in complete freedom. The old

75. Bibliographical comments on these societies will be found in Appendix I.
76. *AP*, XVII, 36–37 (10 July 1790). For the rest of the paragraph see *AP*, XVII, 444–446 (31 July 1790) and 574 (3 Aug. 1790); and Gerbaux and Schmidt, II, 569, 633, 636–637.

privilège system protected some artisans whose inventions were judged new and useful, but did nothing for the rights of other artisans. There were claims that the *privilège* arrangement lent itself to major injustices, since the choice of which artisans were to be protected lay exclusively in the hands of the Academy or its members. An effort to by-pass the alleged arbitrariness of the academicians and to recognize the rights of the inventors culminated in the adoption of France's basic patent laws of 7 January 1791.

This piece of legislation was the product of adept lobbying. In August 1790, an association of inventors, headed by the Baron de Servières, originally proposed the legislation by petitioning the National Assembly. Their demand was an unusually well-documented brief for the adoption of patent guarantees similar to those in operation in England and the United States.[77] It was accompanied by a dozen carefully researched appendices providing documented precedents and offering models for the framing of new legislative decrees. The patent question was referred to the Committee of Agriculture and Commerce, which in turn assigned it for study to the Chevalier de Boufflers.[78] With the help of the petition from the inventors' association, he was able to frame a comprehensive bill eventually presented to the National Assembly on 30 December 1790.

Boufflers was an author of considerable repute, sufficiently distinguished to have been elected to the Académie Française. In the speech accompanying the bill, he waxed eloquent on the importance of establishing as a constitutional principle that inventions were individual property just as much as real estate or other tangible assets.[79] Given this assumption, the artisan held an inalienable right which had to be protected by law. In addition, Boufflers insisted, the artisan had an obligation to share his invention with society. Boufflers therefore pictured the patent as a contract between the inventor and society in which both parties benefited from the arrangement.

To secure passage of the bill, it was essential that the lan-

77. *AP*, XXIV, 642 (7 April 1791).
78. Gerbaux and Schmidt, I, 524.
79. *AP*, XXI, 721–729 (30 Dec. 1790).

guage used to justify it be judiciously chosen to differentiate patents from exclusive corporate or commercial privileges, which were about to be outlawed as contrary to free trade. For this purpose, the committee had done well to select a man of literary talents. Boufflers also capitalized on his command of the French language to paint a touching portrait of the artisan pleading before an Old Regime bureaucrat or judge whose decision it was to award him a *privilège*. The plight of the simple inventive genius, unable to express himself clearly, intimidated by the petty bureaucrat anxious to demonstrate his social superiority, was dramatically drawn. Boufflers claimed that this humiliating procedure was responsible for the loss of countless inventions to foreign countries. He added that even the most impartial of judges, the savant, normally failed to appreciate the efforts spent by the artisan or the potentialities of his invention.

The answer to these inequities, Boufflers said, was to eliminate completely the judging procedures by creating an impersonal registry of inventions open equally to all claimants. For this purpose, a Bureau des Brevets et Inventions would operate as part of the executive branch of government as a depository to record and eventually to publish inventions, but would refrain from judging the utility of any of them. That would be determined on the open market of commerce and industry by laissez-faire competition.

The principle was quickly adopted by the National Assembly but its implementation proved more complicated.[80] It dragged on for months, in large part because of the legal objections raised by Dionis du Séjour, a member of the Academy of Sciences and a representative of the nobility, who succeeded in having the bill's adoption postponed several times.[81] But pressure from the newly formed Société des Inventions et Découvertes and remonstrances by Villette in the press pushed the legislation through on 14 March 1791, leaving the administrative details in the hands of the Minister of Interior.[82] By July,

80. *AP*, XXI, 730 (30 Dec. 1790).
81. *AP*, XXIV 462 (30 March 1791); 482 (31 March 1791); 633 (7 April 1791).
82. *AP*, XXIII, 54 (8 Feb. 1791), and XXIV, 641–644 (7 April 1791); *Chronique de Paris*, 18 and 28 April 1791, pp. 430, 469–470.

the Bureau des Brevets was in operation, with the Baron de Servières serving as director.[83]

The significance of this episode must not be underestimated. It was an important abrogation of the Academy's statutory powers, removing one of its primary Old Regime functions and representing another phase in the dismantling of academic power. Beyond this, the emergence of patent laws and the appointment of a lobbyist as director of the new patent office were a measure of the strength that voluntary associations could wield, if they applied it properly. The success of the Société des Inventions et Découvertes was undoubtedly an inspiration for other groups, such as the Société du Point Central des Arts et Métiers, which was apparently founded a week after final action was taken on the patent bill.

Another feature of the patent debate was also important for arousing the antagonism of the artisan against the academician. The major critic of the successful patent law, Dionis du Séjour, was himself an academician. It seemed plausible to suppose that he was blocking the legislation in the interests of the Academy, hoping the next legislature might be less inclined to favor the bill. Taken as an isolated event, the intervention might have been overlooked, but there were other disturbing signs pointing to machinations on the part of the Academy. They were quite naturally interpreted by the artisans as signs of the resurgence of academic power at the expense of the artisans.

One such indication was in the numerous obstacles that delayed the implementation of a system of rewards for inventions. As a principle, it had been agreed upon in August 1790, and referred for action to the Committee of Agriculture and Commerce, where it was assigned for preparation to the same member who had framed the patent bill, Boufflers. The Committee had to be prodded into action by artisans. In August 1791, Servières met Boufflers to discuss the drafting of a bill and, early in September, the Société du Point Central des Arts et Métiers requested, through the Parisian electors, that attention be given to needy artisans.[84] The language was polite, but the message unmistakably clear.

83. *AP*, XXVI, 76, 79–80 (14 May 1791); AN, F[17]1136, no. 90.
84. AN, F[17]1136, no. 95; *A Messieurs les Electeurs*.

The response was immediate. Three days later, Boufflers offered legislation to put the reward system into operation and obtained an appropriation of 300,000 francs for that purpose.[85] He proposed several levels of awards, corresponding to the degree of importance attached to each invention. In order to determine who should receive these awards and what sum they deserved, Boufflers further suggested creating a panel of judges, to be called the Bureau de Consultation des Arts et Métiers, which, like the patent bureau, would fall under the control of the Minister of Interior. By contrast to the newly created Bureau des Brevets, which served an administrative recording function only, this Bureau de Consultation would fulfill an important adjudicatory role. Because of its nature, the composition of the panel forced Boufflers to discuss the relative importance of the Academy and its artisan critics as judges of rewards.

The drafting of the legislation was a delicate affair. On the one hand, Boufflers as an academician had a marked respect for his fellow scientists. On the other hand, he was also responsive to the artisans' plight and sensitive to their demands for a share of power in the Bureau de Consultation. It was inevitable that this new body would reflect both the elitist traditions of the Old Regime and the more popular demands of the artisans' associations.

In his speech to the National Assembly on 9 September 1791, Boufflers spoke out politely against exclusive academic control. He asserted that it was no longer possible to maintain previous traditions because:

1° The details of this kind of work will henceforth multiply and be more demanding. If entrusted solely to members of this eminent society, it could distract them from even more important occupations.

    2° The arts, which flourish only in a state of liberty, would be alarmed at [the idea of] a perpetual judicatory commission which is inherent to any corporate group. From this point of view, the most enlightened body could appear as the most fearsome.[86]

Boufflers, feeling his second point required elaboration to save the individual academician from criticism, explained:

85. *AP, XXX*, 402 (9 Sept. 1791).
86. *AP, XXX*, 397–398 (9 Sept. 1791).

I know that . . . since the virtues of each member of the Academy of Sciences are almost the equal of their knowledge, the artisans ought not to worry about having such judges. But since it is always better to depend on laws rather than on men, and since the mere possibility of abuse should suffice to awaken the attention of this legislative body, we feel it necessary to transmit to the Assembly the drawbacks that artisans seem to perceive in the censure under which they have so far operated. They fear the indifference and habit engendered by a long repetition of the same acts and the same functions; they fear a kind of *esprit de corps* which, unnoticed, creeps into all discussions of the same men always meeting together; they fear a kind of arrogance which, sooner or later, is associated with the constant exercise of authority [carried out] without any interruption, contradiction or sharing; finally, they fear, so they say, even the errors of superiority itself, which might look down upon them from too great a height. . . .

Furthermore, he suggested, the Bureau should be composed of academicians and artisans, in keeping with the sacred principle of equal representation. But there was ambiguity in his flowery language. What was meant by equal representation? In the original bill presented on 9 September 1790, Boufflers had said:

The members of this Bureau will be chosen in the associations concerned . . . with matters related to the exact sciences, arts and industry; in the civil or military bodies devoted to the study of mathematics; and in the voluntary associations of artisans and citizens. . . .

Each of these groups will elect by ballot three members annually. . . .[87]

But when it was finally adopted, less than three weeks later, the composition of the Bureau overwhelmingly favored the Academy. Boufflers casually explained the change with these words:

Your committee has thought that the Academy of Sciences was quite capable of fulfilling the task we envisaged in creating the Bureau de Consultation; but in order to act in conformity with the principles often adopted by the Assembly in similar circumstances, we have felt it advisable to add to members of the Academy a number of learned experts taken from outside its ranks and selected by the Minister of Interior.[88]

This arrangement, adopted without any discussion, set the com-

87. *AP*, XXX, 401–402 (9 Sept. 1791).
88. *AP*, XXI, 368 (27 Sept. 1791).

position of the Bureau de Consultation for its first year at fifteen academicians and fifteen others to be selected by the minister from the membership of the several other associations. In theory, the principle of equal representation was maintained, but in practice the Academy was assured of the major voice in the jury's operations. Any concession to the artisans' groups was in reality nominal.

It is difficult to know for certain why Boufflers changed his mind. It is possible that he considered this bill to be a provisional measure to tide over the inventors until the matter of rewards was permanently settled in the broader context of educational reforms being prepared by Talleyrand. A similar approach had been adopted on September 17 to create a jury entrusted with deciding on fine-arts rewards.[89] It is more likely that the Academy brought is influence to bear on Boufflers by convincing him of the artisans' incompetence and lack of experience in the carrying out of bureaucratic functions. Whatever the real reasons for the change, it was interpreted by the artisans as a betrayal of their rights.[90]

The net effect was to mobilize the more radical artisan groups in protest, and to give them a fresh opportunity to rail against the Academy. A general outcry denouncing academic cabals and the injustices of both the patent and reward systems and demanding a repeal of the laws was heard on all sides. The voice of the artisans was mixed in with the voices of dissident painters and sculptors, who were also suffering from academic domination.[91] Consequently, the newly elected Legislative Assembly was showered with petitions and with delegations from the Commune des Arts and the Société du Point Central des Arts et Métiers on October 19, from the Société des Arts and the Société des Neuf-Soeurs on December 4, all demanding new laws.[92] The *Journal des Sciences, Arts et Métiers* echoed their sentiment and printed anti-academic articles which placed the blame for the inadequate legislation squarely on the learned society's shoulders. Everyone adopted the same slogan of total

89. *AP*, XXXI, 58 (17 Sept. 1791).
90. Ballot, pp. 23–26.
91. Dowd, pp. 28–34.
92. *AP*, XXXIV, 281–284, 288–289 (19 Oct. 1791); XXXV, 391 (27 Nov. 1791); and XXXV, 570 (4 Dec. 1791). Gerbaux and Schmidt, II, 613–633, 636.

freedom for the artisan by the liberation of genius from the shackles of academic and ministerial control.[93]

An attempt was even made to create a united front of artisan groups. Charles Desaudray, who was to found the Lycée des Arts later that year, drafted a "Constitution Nouvelle des Sciences, Arts, et Métiers," and obtained signatures from delegates of several societies, including that of Trouville, a bitter inventor who had run afoul of the Academy early in 1791.[94] The details of Desaudray's proposal are not as important as the fact that this provocative document was printed and widely circulated. It was becoming clear that the artisan groups were not satisfied with a small share of power. They demanded exclusive power or, to adopt their phraseology, complete freedom. Their experiences with the attempts to modify Old Regime practices by the framing of compromise measures proved that halfway measures were unsatisfactory. Moreover, they had reason to suspect that the academicians, while appearing to accept cooperation with their artisan counterparts, were in fact preparing a plan to regain control of the scientific community. Condorcet was actually in the throes of framing a comprehensive program for national education that in many ways would have given a new Academy even greater control of the scientific and technological life of France than the old one had in the Old Regime. When Condorcet unfurled his thoroughly elitist project in April 1792, the artisans were prepared to oppose it with an effective barrage of counter-arguments.

Without the experience of legislative victories in the creation of the two bureaux and the taste of disappointment that came from compromise measures, the artisans would never have been prepared to mount a collective attack against, and eventually to defeat, Condorcet's project. By 1792, they had de-

93. The Minister of Interior, Lessart, was also blamed for the arbitrary manner he used to select delegates for the Bureau de Consultation. See Paris, Académie de Médecine, MS. box 139, letter of Langlade to Vicq d'Azyr dated 19 Nov. 1791, and London, Wellcome Medical Historical Library, MS, Molard folder, "Du Choix de Quinze Membres pour le Bureau de Consultation."

94. The Constitution exists in manuscript form with annotations by Desaudray in AN, F¹⁷1002, no. 156. Its printed version is reproduced in *AP*, LXXIX, 262–263. Another, earlier effort to organize resulted in a petition cited in Gerbaux and Schmidt, II, 636 n. 2, probably presented to the Legislative Assembly on 4 Dec. 1791 (see *AP*, XXXV, 570).

veloped viable organizational structures and experienced a measure of success as political lobbyists. The feeling of self-confidence was essential for persisting in their fight against academicians who, through their emphasis on the utility of their scientific endeavor, had initially managed to secure the National Assembly's protection. Yet as the Revolution proceeded along its radical course, the Academy's prestige and effective power were gradually shrinking and their support in the legislature weakened. By contrast, their antagonists were gaining strength and finding reasons for declaring a war on all academies that would not end until they were completely destroyed.

Chapter 7

Legislative Debates

*A*CADEMIES were mentioned in the legislature and in the press long before Condorcet's report on public education officially placed them on the agenda. But until April 1792, very little direct attention had been paid to the role learned societies might assume in the new regime. It was only incidentally, with reference to a specific project or a specific bill, that they were cited at all. After the formal presentation of Condorcet's proposal to reorganize the nation's educational system, the tone and frequency of the discussion about the academic system changed completely. To a certain extent this change was a consequence of the emergence into the public domain of private discussions that had been taking place for several years around the Academy and in artisan circles. It was also a reflection of the changing political mood of the times, now more willing to sever ties with the Old Regime than it had been in the early days of the Revolution. But above all, it resulted from the nature of Condorcet's plan, which helped to polarize attitudes about academies into irreconcilable camps. As we will see in detail his proposal was precisely what might be expected from an Enlightenment scientist committed to elitist values and deeply attached to the Academy of Sciences.[1]

The emergence of the Condorcet report in the Legislative Assembly has a complicated prehistory that began early in the Revolution and developed irregularly for over two years. It is necessary to follow it carefully, for arguments later to emerge in the final debates were rehearsed in these legislative skirmishes.

1. Baker, *Studies on Voltaire*, LV, 150–156.

In January 1790 the subject of academies was first broached in a report submitted to the National Assembly by its own Finance Committee. The committee's major task was to propose a sensible budget that would put France back on the path of fiscal responsibility. To accomplish this, it had severely trimmed traditional expenditures, especially those included as part of the "Maison du Roi," under which academic budgets fell. In contrast to reductions proposed in other parts of the budget, the committee recommended that appropriations for academies be maintained, and even increased by a small amount. Thus in the budget presented for discussion on 29 January 1790, the Academy of Sciences was listed for its usual sum of 93,158 livres and 10 sols, with an additional 1,200 livres to be awarded as a new prize for the most useful discovery made or the best book published for that year.[2] The committee spokesman, Lebrun, justified the item by heaping lavish praise on the Academy's accomplishments and by referring to it as a useful institution "enjoying the respect of all Europe."[3]

It was not until August 14 that the section of the budget concerned with academic expenses came up for discussion. The first serious opposition to the committee's recommendations came as Lebrun proposed a 14,000 livres allotment for the publication of manuscripts by members of the Academy of Inscriptions and Belles-Lettres. A representative from Brittany, Lanjuinais, whose uncle had been a fervent partisan of the anti-academic journalist Linguet, came right to the point:

I would like to speak up against fees and salaries for the academies. ... They are *corps* and all *esprit de corps* is to be feared. ... Look at England; its academies receive neither tokens nor stipends; nevertheless they flourish, and their academicians are just as learned and, I dare say, even more useful than ours.[4]

Lanjuinais was seconded by his Parisian colleague Martineau, who also objected by saying, "One does not have to be paid to become a learned man. True writers and great men are born

2. *AP*, XI, 379–390 (29 Jan. 1790). For details on the Finance Committee, see Bloch, pp. 42–43.
3. *AP*, XVIII, 92 (16 Aug. 1790).
4. *AP*, XVIII, 69 (14 Aug. 1790). A variant of this speech and the next one are given in the *Courrier Français*, 15 Aug. 1790, p. 356, and *L'Ami du Roi*, 15 Aug. 1790, p. 308.

and come forth despite all adversity." The implications of their dissenting voices were ignored in the legislative hubbub, and the item was approved for one fiscal year.

The next day, as Lebrun continued to present individual budget items, another legislator, Dedelay d'Agier, interjected a thought into the discussion that served as the basis for the solution the Assembly was eventually to choose. He remarked that even though the Finance Committee was confined to a discussion of budgetary matters, there were sufficiently important issues raised to warrant the appointment of another committee to make a long-range study of "all public establishments in the kingdom concerned with science and the arts."[5] Such a committee would be entrusted with reorganizing public institutions in order to remove all the persistent sources of despotism that had allegedly prevented genius from properly flourishing. For the time being, Dedelay d'Agier's suggestion was also overlooked, but, like the comments of Lanjuinais and Martineau, it was revived at a later date in the debates.

These two sets of criticisms and Lebrun's laudatory remarks constitute the three major themes repeatedly expressed in the Assembly's chambers, echoing of views held before the Revolution. Lebrun represented the pro-academic faction, pleased with its accomplishments and satisfied merely with the transfer of academies from royal to national control. Dedelay d'Agier accepted the principles behind the academic system but wished for a transformation of its organization to maximize its effectiveness in supporting the progress of science. Lanjuinais and Martineau took a decidedly antagonistic point of view, questioning the value of government-supported learned societies altogether. Each one symbolized a position that had a direct counterpart in the Old Regime and even among academicians themselves: Lebrun was the conservative optimist, Dedelay d'Agier the reformer, and Lanjuinais the critic.

Throughout the debates that followed, Lebrun maintained his unqualified support of academies. If France was held in high esteem and if Paris was the focal point of the Enlightenment, he

5. *AP*, XVIII, 87 (15 Aug. 1790). The principle was supported by another delegate, Camus. A variant of Dedelay d'Agier's speech is reported by the *Journal des Etats Généraux*, XIV, 392.

insisted that in no small part this was the consequence of the work accomplished by academies. Beyond fostering France's intellectual hegemony over Europe, academies in general—and that of science in particular—also performed a valuable utilitarian function: "How much do the navy, the artillery, the engineering corps, the arts, and manufactures owe to the Academy of Sciences, whose reputation still overshadows that of all other academies of Europe?"[6] Isolated genius without support, he maintained, would never have been able to realize such formidable accomplishments. Consequently, it was the nation's patent duty to sustain the academies and to provide them with funds for the continuation of their work.

The argument was countered squarely by Lanjuinais himself, who developed the anti-academic position with logical force rather than the rhetorical flourish usually characteristic of the debates. Along with most of the delegates in the Assembly, he shared an important part of Lebrun's assumptions—the value of science, letters, and the arts for the nation. But he insisted that government-supported academies were neither necessary nor compatible with the principles of the Revolution. Echoing the views expressed by Delamétherie in his *Journal de Physique* and Marat in the August 17 issue of *L'Ami du Peuple*, Lanjuinais reminded the legislators that paid academicians constituted a "literary aristocracy" wielding a despotic authority that could not be tolerated in an age of liberty.[7] He cited the Academy of Sciences' suppression of the Société d'Emulation, the Royal Society of Medicine's monopolistic practices, and the Academy of Painting and Sculpture's alleged persecution of nonacademic artists as examples of the evils their tyranny fostered. To disarm his opponents who might claim that the pursuit of learning would collapse without government support, he held up the examples of England and Germany, the existence of the new Société des Inventions et Découvertes and the Société d'Histoire Naturelle, and the success of commercial ventures such as the *Encyclopédie* as evidence of the vitality of

6. *AP*, XVIII, 173 (20 Aug. 1790). Another speech by Lebrun supporting the same principles is in *AP*, XVIII, 91–92 (16 Aug. 1790).
7. For Delamétherie's views, see *Journal de Physique*, XXXVI, 319 n. 2. In his speech, Lanjuinais referred to the Société des Arts even though he meant the Société d'Emulation. See *AP*, XVIII, 174 (20 Aug. 1790). The same mistake was made by Pidansat de Mairobert, VI, 187.

private enterprises. What he argued above all was that "in the arts, letters, and sciences, there must be neither corporations nor monopolies."[8]

The speech had a powerful effect upon the legislators and could have led to a rejection of Lebrun's budget had it not been answered. The cutting of funds from the academies might have spelled their doom or, at the very least, their transformation into different (and less productive) institutions. Fortunately, there were delegates ready to counter Lanjuinais' argument, and one with a convincing compromise proposal. The Abbé Grégoire acknowledged that inconsistencies between the revolutionary regime and academic practices had to be rectified, but he believed that these could be ironed out if each learned society were asked to re-examine its regulations so it could align itself with the new political regime.[9] Undoubtedly, he was familiar with La Rochefoucauld's initiative in reforming the Academy of Sciences' statutes.[10] He may also have been aware of the discussions at the Academy of Painting and Sculpture and the Academy of Architecture about a total organizational revision, provoked by the pressing demands of nonacademic artists.[11] He was likewise abreast of the efforts of his naturalist friends to transform the old Jardin du Roi, with its oppressive system of a royal intendant who governed its administration, into a new institution which would be governed by scientists and in which they would share equally in the running of their own affairs. It was probably no coincidence that, at the same time as Grégoire was offering his compromise measure, the naturalist delegate Creuzé-Latouche circulated a twenty-page brochure urging structural reforms of both the Jardin du Roi and the academies.[12] Evidently, the scientists and their friends, alerted by

8. *AP*, XVIII, 175 (20 Aug. 1790). Comments inspired by Lanjuinais's speech are in the *Révolutions de Paris*, V, 395, and *Gazette de Paris*, 23 Aug. 1790, p. 2.

9. *AP*, XVIII, 175 (20 Aug. 1790).

10. As a naturalist and member of the Royal Society of Agriculture, Grégoire was friendly with La Rochefoucauld himself, Tillet, Vicq d'Azyr, Broussonet, Daubenton, Fourcroy, and other academicians directly concerned with the reform of statutes at the Academy or the Jardin du Roi.

11. Hautecœur, pp. 598–603; Lemonnier, *Revue de France*.

12. *Opinion de M. J. A. Creuzé-Latouche, Membre de l'Assemblée Nationale, au Sujet du Jardin des Plantes et des Académies*. It is reproduced in *AP*, XVIII, 182–184 (20 Aug. 1790). For a detailed discussion of the reforms at the Jardin du Roi (Jardin des Plantes), see Hamy, *Derniers Jours*.

Lanjuinais' attacks, had prepared themselves in advance to meet renewed objections.[13]

Having admitted that institutions of higher learning were not beyond reproach in their organization, Grégoire and his fellow deputy Camus nonetheless strongly advocated their continuation and supported the passage of the Lebrun budget as a temporary measure. An important rider to the bill was attached, demanding that each society submit a project of new regulations within the month. Apparently it was a sufficient compromise to satisfy critics such an Lanjuinais, and the bill passed.

Action was now in the hands of the learned societies. Their response to the Assembly's demands was of critical importance for the future of their existence. Had they each been able to carry out internal reforms voluntarily and quietly, it is likely that their fate would have ceased to be an issue in the assemblies. In point of fact, and as we have noted for the Academy of Sciences, a number of them encountered strong internal resistance to the changes demanded by critics. Within a number of societies, the conservative factions mustered sufficient strength to slow down, temper, and even block the reformers' efforts to draw up satisfactory new statutes. Rumors of these internal squabbles spread quickly and provided fresh evidence for the critics who claimed that the Old Regime institutions were harboring "aristocrats" unwilling to move with the times. As we saw in the last chapter, the Academy of Sciences was not free from such criticism.

Several of the academies responded satisfactorily by forwarding revised sets of regulations without any fuss.[14] But their ready compliance was generally forgotten in the wake of the controversies that erupted in the two fine-arts academies

13. On the same day, a delegation of naturalists from the Jardin du Roi presented an address to the National Assembly, which was read by its presiding officer, Dupont. See *AP*, XVIII, 177–178 (20 Aug. 1790).

14. The Royal Society of Agriculture sent its proposal with a covering letter by Broussonet to Talleyrand on 26 Sept. 1790 (AN, F^{17}1310, nos. 142 and 145). The day before, the Academy of Inscriptions submitted its project (*AP*, XIX, 239). The Académie Française complied before 6 Sept. 1790 (*Registres*, III, 635–636, and AN, F^{17}1350 dossier 2 no. 1). The contents of its new constitution were discussed by La Harpe, who helped draft it, in *Mercure*, 16 and 23 Oct. 1790, pp. 107–119, 139–155.

and in two of the learned societies concerned with the healing arts. The architects and the surgeons each submitted two different constitutional projects for their academies, which visibly demonstrated irreconcilable divisions of opinion.[15] Painters and sculptors led by Renou and David were so exasperated by their own academies' failure to alter their procedures that they banded together to form a rival association, the Commune des Arts, from which they launched pamphlet after pamphlet decrying academic tyranny.[16] Public outcries against the Royal Society of Medicine also erupted, and numerous projects for reorganizing medical research and education emerged from local presses.[17] As secretary of the Royal Society of Medicine, the academician Vicq d'Azyr found himself defending the existence of his institution against vicious pamphlets that accused it either of wielding too much power or of being ineffectual.[18] Such public manifestations of criticism and internal dissension convinced the legislators there were in fact serious organizational deficiencies within royal societies and academies with which academicians were unable to cope.

This kind of reasoning undercut the effectiveness of the Grégoire compromise and led the Assembly to adopt a new policy toward academies. Picking up the suggestion made months earlier by Dedelay d'Agier and repeated by Murinais on August 20, Talleyrand urged the legislature on October 13 to place a moratorium on all decisions relating to public institutions of learning until an over-all plan for their organization could be framed by his Constitutional Committee.[19] The mo-

15. Sections of the architects' official proposal are in Lemonnier, *Procès-verbaux*, IX, 362–367; the dissidents' constitution is in a printed *Adresse* dated 14 Feb. 1791: see *AP*, XIX, 121–135 (21 Sept. 1790). The squabbles among the surgeons are discussed by Dubois, and Lenormant. For the statutes, see AN, F^{17}1310, no. 141, and F^{17}1350 dossier 2 no. 24; and *AP*, XIX, 239 (25 Sept. 1790), and XX, 741 (25 Nov. 1790).
16. Tourneux, *Bibliographie*, III, 879–883; AN, F^{17}1310, nos. 152–155, and F^{17}1350, nos. 15–26; *AP*, XXIV, 282 (22 March 1791).
17. The most interesting proposal was made by Retz. Gallot's writings are discussed by Merle, p. 38, and Ingrand.
18. For details about the printed attacks, see Tourneux, *Bibliographie*, III, 664. The Society's reformed constitution is in AN, F^{17}1310, no. 127, and F^{17}1350 dossier 2 no. 5. Vicq d'Azyr wrote a more comprehensive plan, reprinted in *Histoire de la Société Royale de Médecine, 1787–1788*, pp. 1–201. Drafts for both plans are in Académie de Médecine, MS. box 114. See also Genty, *Progrès Médical. Supplément*, XIV, 49–56.
19. *AP*, XIX, 588–589 (13 Oct. 1790).

tion was adopted without discussion, and the subject of academies was temporarily laid aside.

The adoption of Talleyrand's motion marks a turning point in the history of academies during the Revolution. By placing the fate of the academies in the hands of legislative committees, the bill forced debates behind closed doors and off the public podium where it had been constantly under the watchful eyes of journalists such as Marat, who never missed an opportunity to carp at Old Regime institutions. Temporarily at least, public display of internal dissensions was submerged, and the societies were able to continue their activities without completely losing face. Academic reformers who despaired at the possibility of voluntary transformations within each society must have welcomed the transference of this responsibility to legislators. They cherished the hope that effective lobbying would help them achieve their ends more easily than would fruitless debate within learned societies themselves. Furthermore, the Assembly's decision to consider public institutions of learning as a whole put it on record as favoring wholesale reform, thus repudiating the stand adopted by conservative academicians who claimed that change was superfluous. Talleyrand's academic friends—Condorcet, La Rochefoucauld, Vicq d'Azyr, and others—must have been delighted at this turn of events. Now the only question to be answered was what direction and degree of change the Constitutional Committee would suggest.

Talleyrand worked on the project for almost a full year. He collected large quantities of information about the state of institutions of learning, perused the hundreds of suggestions for reform printed in pamphlets and books, and studied the proposals for reorganization emanating from institutions such as the academies themselves. To digest all this material and to sift it for use in his *Rapport sur l'Instruction Publique*, he sought the help of dozens of experts, among them several prominent academicians. In his *Memoirs*, he recalls that

I also made myself responsible for the report on public education for the Constitutional Committee. To accomplish this great task, I consulted the most educated men and the best known scientists of the age, Lagrange, Lavoisier, Laplace, Monge, Condorcet, Vicq d'Azyr, La Harpe. All helped me. The reputation this report has acquired demands that I name them.[20]

The Talleyrand report was remarkable for its comprehensiveness and in the detail of its specific proposals, if not in the novelty of its philosophical premises.[21] It aimed at a complete transformation of existing institutions of learning by their absorption into a state-organized and state-supported system of education. The conception put forth by Talleyrand was that of a thoroughly national and comprehensive system that would bring order into the chaos of Old Regime practices, and that would institutionalize the links that tied the diffusion of knowledge to its progress. He based his views upon traditional Enlightenment assumptions about the need to conquer ignorance through the advancement of knowledge and to provide opportunities for universal education to safeguard citizens' liberties.[22]

In its structure, the new system of education was supposed to reproduce the country's political organization, with levels of instruction paralleling cantons, districts, departments, and the capital. It was in Paris that the apex of this hierarchical scheme was to be located. There, Talleyrand suggested, a National Institute would be created, "which, enriching itself with knowledge from all parts of France would continuously create a combination of the best means for the teaching of human knowledge and its indefinite growth."[23] Its function would be both to educate the best students selected from the lower echelons of the system and to oversee the continued progress of letters, sciences, and the arts. Toward that end, Talleyrand planned to staff it with the intellectual elite of the nation and to provide it with all the laboratories, libraries, museums, and other means necessary for the advancement of learning. To insure that it would be a truly national institute he elaborated an intricate system of communication with provincial institutes of learning that would make the Institute in Paris a "correspondence center," where information would be constantly gathered and, after it was digested, distributed to all local in-

20. Talleyrand, *Mémoires*, I, 163–164. See also Lacour-Gayet, *Talleyrand*, I, 137–140.
21. Vignery, p. 23.
22. La Fontainerie; Clifford-Vaughan, *British Journal of Sociology*, XIV, 135.
23. *AP*, XXX, 451 (10 Sept. 1791).

stitutes. He said it could be "considered either as a tribunal where good judgment prevails, or as a hearth where truths could be assembled."[24]

In the detailed organization of the National Institute, which would be divided into two main classes and twenty sections, the importance of the Academy of Sciences is paramount. Structurally, it was to be subsumed under a more "encyclopedic" organization. But its membership was to remain intact and its organization, practices, and "philosophy" were expected to become the model for the more inclusive Institute. In the Talleyrand plan, the National Institute was a "super-Academy" meant to take over the functions or the direction of all the academies, royal societies, and major public institutions of higher learning, including the Jardin du Roi, the Collège Royal, and the Mining School. In essence, Talleyrand was suggesting the creation of a mammoth central organization for higher learning, "a sort of [living] encyclopedia, always studying and teaching; and Paris would see within its walls the most complete and magnificent monument ever raised to knowledge [*sciences*]."[25]

How important the Academy of Sciences loomed in the Talleyrand plan is measured by the numerous references made to it in the project. Not only were the Academy's disciplinary arrangements copied as the basis for the larger organization, but its communal practices were also singled out for praise. The value of having a meeting of scholars of different subjects, Talleyrand claimed, had been demonstrated for over a century in the Academy where it had "the great advantage of communicating the spirit of calculation and method to all classes of the Academy, forcing everyone to be exact in his research, clear in his exposition, and pertinent in his reasoning. . . ."[26]

In other ways as well, Talleyrand drew upon past academic practices to fashion the new system. First of all, he adopted most of the principles held by the eighteenth-century intellectuals concerning the Republic of Letters. In his proposed statutes for the Institute, he explicitly abolished the

24. *AP*, XXX, 462 (10 Sept. 1791).
25. *AP*, XXX, 465 (10 Sept. 1791).
26. *AP*, XXX, 464 (10 Sept. 1791).

honoraire class, fixed stipends according to seniority, and stipulated complete equality among members. Talleyrand also accepted the principle of cooptative elections and acknowledged the criteria of talent and intelligence as the only ones valid for admission to the Institute. In every section of his report, he demonstrated complete adherence to the principles of equality within the Republic of Letters and to its intellectual superiority over those without. Like previous spokesmen for an elitist Republic of Letters, he warned of the dangers ignorance begets:

Are we not aware that, even under the freest of constitutions, the ignorant man is at the mercy of the charlatan and much too dependent upon the learned man; and that a well-rounded general education alone can prevent, not the superiority of intellects, which is a fact and beneficial for all, but the excessive domination that this superiority would permit if any class of society were condemned to ignorance.[27]

Like Condorcet, he emphasized the importance of intellectual elites for combating "charlatanism which, in free states is always more abundant, and needs to be strongly repressed."[28]

Talleyrand's debt to the reformist ideology is more extensive than he ever acknowledged, so great is his dependence upon it. There is, for example, a marked resemblance between Talleyrand's project and both the botanist Adanson's "Project for a Universal Philosophic Academy" and some undated jottings in Lavoisier's hand.[29] One might also suspect that Talleyrand took his inspiration from the "Detailed System of Human Knowledge" that prefaced Diderot and D'Alembert's *Encyclopédie*.[30] But the most likely source from which Talleyrand drew was the series of articles on education published from January 1790 to June 1791 in the *Bibliothèque de l'Homme Public* by Condorcet. In the fifth of these, entitled "On Education Concerning the Sciences," Condorcet outlined

27. *AP*, XXX, 448 (10 Sept. 1791).
28. *AP*, XXX, 463 (10 Sept. 1791).
29. For Adanson, consult Pittsburgh, Hunt Botanical Library, MS. AD 304. For Lavoisier, see AdS, Lav 86, 409, 866. Lavoisier's interest in Talleyrand's project is discussed in Guillaume, I, 354–379. See also Baker and Smeaton.
30. Talleyrand makes a veiled reference to this in *AP*, XXX, 45 (29 Aug. 1791). An article in the *Chronique de Paris*, 25 Sept. 1791, p. 1080, specifically links Talleyrand's report to the *Encyclopédie*.

his thoughts on the importance of scientific societies and their role in the progress of learning. We may assume that the author offered a copy of this memoir, or at least expounded his views on the subject, to Talleyrand.[31]

These comments about Talleyrand's failure to be original are not meant to deny the importance of his work. Rather, they can serve to explain why there was so little substantive difference between his published views on the super-Academy and those formulated later by Condorcet in his April 1792 project. Both men shared similar ideologies about the necessity for state-supported encyclopedic institutions of higher learning, and both based their plans on the assumption that the elitism exemplified in the Academy of Sciences was necessary for the public good.

If there was a difference between the Talleyrand and Condorcet plans, it was in the reception each received. In contrast to the delayed but heated debates that followed the publication of Condorcet's project in 1792, Talleyrand's proposal for a National Institute passed almost unnoticed.[32] It was read before the Assembly on the 10th, 11th, and 19th of September 1791, barely three weeks before the expected dissolution of the legislature. It is clear that the recent capture of the fleeing monarch at Varennes and the coming elections for the new Legislative Assembly absorbed public interest and pushed the long and detailed Talleyrand report into the background. To put the plan into operation also required a considerable expenditure, which few delegates were willing to underwrite. It was therefore decided "to leave something for our successors to do."[33]

The new Assembly showed its determination to legislate in the area of education by creating a special new Committee of Public Instruction. At the outset, the newly elected delegate Condorcet was chosen to preside over this committee. We

---

31. Garat, who knew both Condorcet and Talleyrand, referred to the latter as a "disciple" of the former in *Décade*, IX, 216.
32. *Chronique de Paris*, 25 Sept. 1791, pp. 1079–1080; *Révolutions de Paris*, IX, 465–467; *Point du Jour*, 28 Sept. 1791, pp. 482–484; *Journal des Etats Généraux*, XXXIV, 322–323. For the reaction of the press to other parts of the plan, see Aulard, *Dictionnaire*, I, 2854–2856.
33. *AP*, XXXI, 325 (25 Sept. 1791).

know from the minutes of its meetings that he dominated the proceedings and became its favorite spokesman before the Assembly. The choice was a logical one, for of the two dozen committee members Condorcet had been the most actively concerned with educational reforms and the one whose experience as secretary of the Academy of Sciences gave him a command of relevant details few others possessed. But whereas his title and his name inspired confidence among his fellow delegates, they eventually worked against his plan, not only when it came under attack from artisan quarters, but especially when the political mood of the nation shifted away from Girondin politics and Condorcet was accused of subversive activities.

Shortly after its initial meetings, the Committee of Public Instruction appointed a subcommittee composed exclusively of scientists and academicians—Condorcet, Pastoret, Lacepède, Arbogast, and Gilbert Romme—to draft a general plan of education for presentation to the Legislative Assembly.[34] It was duly prepared and the section relating to the Academy of Sciences was approved by the entire committee on 20 February 1791 without lengthy debate. As far as most substantive issues were concerned, the draft was similar to Talleyrand's proposals, offering a hierarchical system of education paralleling political institutions. It called for a five-step system, of which the first four were devoted to teaching and the fifth occupied by a National Society of Sciences and Arts whose purpose was "to direct education and concern itself with the progress of the sciences and arts, and in general with the perfecting of human reason."[35] Like the Talleyrand report, it was to a large extent based upon the practices and ideologies evolved in the Academy of Sciences.

Although the educational plan technically belongs to the Committee of Public Instruction as a whole, it is generally referred to as the Condorcet plan. Affixing a precise authorship to a committee document is difficult. Even if the evidence were available, the exercise would probably prove to be uninformative, since the views Condorcet held about the organization of science were generally shared by his colleagues. At an earlier

34. *CIPAL*, p. 19.
35. *CIPAL*, p. 227.

date, one of them, the abbé Audrein, had even published a pamphlet urging the creation of a "French National Academy" to direct the entire system of education.[36] What is more, there was on the committee a preponderance of members of the Republic of Letters, or of men who, through their training, would naturally have shared its elitist values. Pastoret, elected vice-president of the committee, was a member of the Academy of Inscriptions and Belles-Lettres. Arbogast and Romme had been successful teachers of mathematics and physics, Lacepède was a naturalist at the Jardin du Roi, and Lazare Carnot and Prieur de la Côte-d'Or were both military engineers with a strong bent for scientific matters.[37] Quatremère de Quincy, who supported the artists critical of the Academy of Painting and Sculpture and whose brother had been stricken from the Academy of Sciences' rolls, is the only significant member of the committee who might have had reasons to object to its predisposition toward elitism.[38]

Condorcet's plan differed from Talleyrand's previous plan in only two significant ways, both of which are consistent with Condorcet's own philosophical attitudes and the particular political circumstances of the period in which it was drafted. One was to turn the National Society into an even more elitist organization, almost totally independent of political authority. The other was to emphasize the utilitarian functions of this super-Academy to justify the considerable expenditure its creation would require and to respond to the public criticism that had been previously leveled at the "frivolity" of certain Old Regime learned societies.[39]

The National Society was to be divided into four classes, two of which, significantly, were devoted to applied science, following Condorcet's special views on this subject.[40] First and

---

36. Audrein, p. 15.
37. Bouchard, *Prieur*, pp. 76–81, 459–460.
38. R. Schneider, pp. 151–163. Quatremère d'Isjonval was dropped from the rolls on 11 July 1786 for prolonged absence from the meetings. He was in fact hiding from creditors in Spain, following a declared bankruptcy. See Bachaumont, XXIX, 217–218, and XXXI, 211.
39. Chamfort, pp. 2, 36, and the review in *Affiches*, 5 June 1791, pp. 2082–2084.
40. More details on Condorcet's educational principles are in Compayré; Cahen, pp. 324–379, 494–497; and Delsaux, pp. 185–207.

foremost, in replica of the Academy of Sciences, was the First Class for mathematical and physical sciences, composed of forty-eight Parisian members, each assigned to a specific discipline. The Second Class was reserved for Condorcet's pet subject, moral and political sciences, which were expected to place human affairs on the same path toward truth that had been discovered by scientists in their yearning to understand nature.[41] The study of metaphysics, morals, law, legislation, political economy, and history was to be carried out by thirty Parisian members adopting the same dispassionate sense of objectivity and reliance upon undisputed evidence exemplified by the First Class. The Third Class was the largest, entitled "Application of the Sciences to the Arts" and composed of seventy-two Parisian members divided into ten different subdisciplines. The literature and beaux-arts Class, incorporating the Académie Française and the fine-arts institutions, was relegated to the fourth and least significant place in the National Society.[42] Through this conscious ordering of classes and the relative importance assigned to different subjects, Condorcet deliberately altered the Old Regime balance between academies, favoring science and its application. He was, above all, stressing the utilitarian potential of the National Society.

Condorcet's concept of applied science encompassed more than technology and the medical arts represented in the Third Class. As is now commonly known, he considered science a profound force for the moral and mental progress of mankind, as well as a source for material progress. Thus for Condorcet, the Second Class (moral and political sciences) was, like the First and the Third, meant to serve the needs of French society as well. All knowledge, including that derived from the "social sciences," was potentially of national utility.

The importance of the National Society for the commonweal was emphasized by the special role Condorcet assigned it in the realm of state-supported public education. Talleyrand had suggested the creation of a single institute devoted to teach-

41. Granger, ch. IV; Manuel, ch. II; Baker, *AS*, XX, 212–218, and *Studies on Voltaire*, LV, 129–133.
42. *CIPAL*, p. 208, where Condorcet also belittles the "arts agréables."

ing as well as to research. Condorcet was equally concerned with formalizing the links that tied the advancement of science to its diffusion, but, having a personal aversion to teaching, he found a way of separating these functions without severing their relationship. His solution was a bureaucratic one. It was the task of each class of his projected National Society to select the teachers of the *lycées*—at the highest level of education—and that of an executive committee of the Society to oversee the actual teaching in these institutions. Condorcet justified this system of appointments and supervision by insisting upon the fundamental importance of academic freedom: "The first condition of all education being the teaching of truth alone, the public institutions concerned with it must be as independent as possible from political authority."[43] He wanted to wrest education from political or religious control and place it in the hands of judges capable of basing their opinions solely upon intellectual merit.

It was for this reason that Condorcet insisted that the National Society be given a crucial role beyond advancing knowledge and providing the government with a research institute. It was to be entrusted with the guidance of higher education, and was indirectly to command all other levels of education. In his plan, the professors of the fourth level (*lycées*) were to name the teachers at the third level (*instituts*), who, in turn, would nominate the instructors of secondary and primary education. No longer would the professorial staff be subject to royal, ministerial, religious, or other inappropriate pressures. Truth, as understood and advanced by the National Society, would reign at all levels of the educational system and penetrate into the minds of all young citizens.

For all its revolutionary appearance, Condorcet's plan for a National Society was based upon highly conventional premises. The new organization was patterned on the practices of the Academy of Sciences, rationalized and extended to cover all of learning. Control of this nationalized and hierarchical system was to be vested in an elite of intellectuals who worked for the benefit of the entire society. Indeed, all of the members of the system of public instruction were to become administrators

43. *CIPAL*, p. 189.

of the state, performing well-defined apolitical tasks of national importance.[44] The plan embodied all the traditional Enlightenment principles subscribed to by the majority of French men of letters. Intellectuals were to be judged by their peers and, in return for their services, were to be paid as civil servants. Condorcet had finally articulated the ideal that members of the Republic of Letters had been seeking since the academic system was created under Colbert's direction. But the political realities of the Revolution prevented its realization.

Although it was a well-conceived and well-written document, Condorcet's project failed to stimulate debate in the chambers of the Legislative Assembly. His oratorical gifts were limited and proved to be no match for the current events that absorbed all the legislature's attention. By coincidence, Condorcet had begun the reading of the committee's report on 20 April 1792, the very day that war against Austria was declared. In fact, his presentation was interrupted by the arrival of the king and his ministers, an event that captured the fancy of the press much more than Condorcet's speech.[45] One of the few journalists who found space to comment on the new educational proposals complained about the length of the report and the fact that it was "even more abstract than Talleyrand's plan," so that most legislators failed to follow its intricacies during its first reading.[46] They were given a second chance on 25 May 1792, when it was once again presented in its entirety following its printing and distribution to the delegates. Still the full-dress debate failed to materialize. A final attempt to bring it back to the floor for discussion in August failed miserably. This time it was the August 10 insurrection that stole the legislature's attention.[47] Once more, by default, the problem of educational reform was left for the next legislative body, the National Convention.

The Convention did not turn its full attention to educational legislation until 12 December 1792. Three days later, the

44. Condorcet actually uses the term "fonctionnaire public" to describe them: *CIPAL*, p. 222.
45. *AP*, XLII, 195 (20 April 1792).
46. Prudhomme in *Révolutions de Paris*, XII, 597.
47. *CIPAL*, pp. 309, 373–375.

section to establish the National Society was dropped, thereby sealing the fate of the academies. It is tempting to explain this action by referring to the mounting Jacobin fever coupled with the declining importance of the Girondin party, to which Condorcet belonged.[48] The demise of the academies in 1793 could then be credited—or debited—to the success of the Mountain, and the reappearance in 1795 of a new Institut to replace the defunct academies could be explained as an integral part of the Thermidorian reaction. If political ideologies and parties could be shown to coincide with attitudes towards Condorcet's National Society, the swing of the political pendulum from right to left and, after the Terror, back to the right would provide a neat and plausible interpretation of the fortunes of the Academy.

Some evidence for such an alignment is available. For his part in championing the declaration of war against Austria, Condorcet was bitterly denounced by Jacobins.[49] His political associate Brissot, whom we have previously met in a different role, rallied to Condorcet's defense by painting a picture of him as the heir of the purest Enlightenment tradition:

Can you, like him, point to thirty years of campaigning with Voltaire and D'Alembert against the throne, superstition, ministerial and parliamentary fanaticism? . . . You tear Condorcet to shreds even though his revolutionary life is one series of sacrifices for the people: *philosophe*, he made himself a politician; academician, he made himself a journalist; noble, he made himself a Jacobin; . . . he devoted his work and efforts and ruined his health for the good of the people. . . .[50]

The rejoinder came from Robespierre himself, whose passionate support of Jean-Jacques Rousseau was bound to turn him against any heir of Voltaire. It was the *philosophe* in Condorcet that he denounced:

If our leaders of liberty are academicians, friends of D'Alembert, I have nothing to reply save that reputations in this new regime cannot be based upon reputations in the old; that if D'Alembert and his friends ridiculed the priesthood, they also befriended kings

48. Gillispie, *Critical Problems*, p. 257, and *Behavioral Science*, IV, 67–68.
49. Robinet, pp. 131–141, 169–181, 203–221; and the folio-sized broadside by Vadier.
50. Aulard, *Jacobins*, III, 529.

and the powerful. . . . All these great *philosophes* persecuted with malice the virtue, genius, and freedom of Jean-Jacques, who alone among the great men of that era deserves a public apotheosis.[51]

In this short exchange the battle lines between Enlightenment and romantic traditions symbolized by Voltaire and Rousseau and their correspondence with opposing political parties seemed clearly drawn. It would have taken little effort to draw these conflicting ideologies into the debate about the value of Condorcet's educational system, particularly with reference to the elitist National Society. Yet no such thing occurred. While the debate developed on the basis of specific ideological attitudes, it did not take on the sharp partisan distinctions that divided the Convention on other issues such as the war. A number of Condorcet's political allies on that issue, notably Brissot and Roland, failed to rally behind Condorcet's educational plans. The reasons are not difficult to understand, if we remember that before the Revolution Brissot was one of the Academy's bitterest and most articulate critics. Roland de la Platière, the Girondin Minister of Interior, had for years harbored a deep grudge against the academic system, suffering in silence from the injustices of which he personally felt a victim. After the idea of a National Society had been defeated, he vented his own anti-academic feelings in public.[52] On the other hand, Condorcet found an ally for his plan in the person of Gilbert Romme, who was far more radical than Condorcet on traditional political issues. It seems clear—even if a bit unfortunate for those preferring a simple explanation—that the forces that defeated Condorcet's National Society were not representative of a specific political ideology. There was growing dissatisfaction with Condorcet among the legislators, but no party platform to undermine his educational scheme.

The concerted and ideologically based attack against the Condorcet proposal in fact came from another quarter: the disenchanted artisans. They had reason to suspect a resurgence of academic power, especially because of the grievances they

51. As quoted by Robinet, pp. 174–175.
52. Roland, *Compte Rendu*, pp. 223–226; Galante-Garrone, p. 355 n. 1. Vignery, pp. 136–141, makes a similar remark about the variety of Jacobin thought on educational issues. For an important analysis of the general conflict between the *philosophes* and Robespierre, see Moravia, pp. 204–216.

had accumulated against the administration of both patent and reward systems. Their problems were psychological as well as financial and political, but in every case clearly linked to their lowly status and occupational role. Moreover, the artisans had tasted partial victory through the action of their own voluntary associations and were also encouraged by the example of the sporadic political successes of sans-culottes pressure groups in Paris. With every movement of the pendulum to the left, their expectations of political power increased.

Grievances against the newly formed Bureau des Brevets, for instance, were formulated in terms of the excessive fees that the law required for registering an invention.[53] In reality, the root of the artisan's problem lay in the deep mistrust he harbored against administrators, whom he suspected of stealing the fruit of his technological labors. The artisan was no longer forced to depend upon academicians for a patent, but the patent law seemed only to have substituted a new authority for an old one, without giving him the proper assurance of private ownership. His mistrust of authority was so pervasive that in the end the exact nature of that authority was of little consequence. In fact, the artisan continued to consider the academician as the symbol for authority, even though he had formally been removed from the seat of power.

There was trouble with the newly formed Bureau de Consultation as well. Serious objections to its composition had developed, with the critics hinting that academic maneuvering was aimed at the recovery of the Academy's control of technology. In Boufflers' original plan for manning the Bureau, there were to be an equal number of delegates from *each* scientific or technical circle, be it an official academy or a voluntary association. When the bill which temporarily created the Bureau was passed, it was changed to specify that half of the jury was to be selected by the Academy of Sciences and the other half by the Minister of Interior from among the numerous artisan groups. The unexpected change in representation and the assigning of the choice of "representatives" to the minister

---

53. *CIPAL*, pp. 279, 281–284; *AP*, XLI, 113 (3 April 1792), and XLIII, 56 (6 May 1792); and letter of Molard to G. Romme dated 26 Nov. 1792, in AN, F¹⁷1047 dossier 1.

without even consulting local societies for nominations seemed to be the grossest breach of the principles of democracy. To add to this disappointment, the artisans who were eventually chosen to man the Bureau found themselves outnumbered by academicians, and at a strong disadvantage because of their inexperience in judging the merit of inventions.

They were also deeply disturbed by the Minister of Interior's high-handed methods, reminiscent of the worst features of Old Regime arbitrariness and of academic arrogance.[54] It was the minister Roland, Condorcet's political ally and a critic of the Academy, who cast aspersions on the appropriateness of having artisans serve on the Bureau at all, and who urged the legislature not to renew the administrative unit for its second year without first altering its constitution. The artisans denounced Roland for his audacity, picturing his move as a blatant attempt to restore the Academy's exclusive control of an organization established to assure the equity of decisions concerning artisans. It seemed to them as if the clock were being turned back to Old Regime practices.

Artisan groups were incensed by what they perceived as machinations to keep them from running their own affairs. But they saved their greatest ire for Condorcet and his plan, which seemed to them even more retrograde in character. Shortly after his project was published, the Société du Point Central des Arts et Métiers printed a tract distributed to "all artisans," calling for a concerted effort to defeat the provisions of the new educational plan.[55] The author was particularly upset by the consideration given to a National Society which he correctly assessed to be the ideal embodiment of the academic system. In the flamboyant language characteristic of pamphleteers, he condemned the Condorcet project as a most pernicious effort to repress artisans' rights and to circumvent the role of voluntary associations in determining their own future. The tract was a call to arms asking the scientific proletariat to rise up against oppression and to overturn, once and for all, the arrogant pretensions of academicians by defeating the Condorcet proposal.

54. *CIPCN*, I, 105–119; Ballot, pp. 27, 73–79.
55. *A Tous les Artistes et Autres Citoyens*. In Tourneux, *Bibliographie*, III, 661, this pamphlet is described as a folio poster signed by Morin-Delairas.

The tract stated that,

If the National Assembly . . . decreed such a project, the monstrosities which are daily the object of artisans' complaints would be perpetuated and . . . they would remain subject to the despotism which they have so long bewailed and which is so harmful to the progress of the arts.

It continued by identifying the cause of the artisans with the principles of the Revolution, picturing academic domination as part of the outmoded corporate structure that had stifled "genius and the arts." Condorcet's motives for suggesting a super-Academy were exposed:

Whether this writer secretly desired to retain citizens under the academic rod and to bend them to it from infancy, or whether he —breathing the perfidious miasma of academic air that surrounds him—could not betray his academic conscience, his proposals must be rejected by the champions of equality and liberty and by those who cultivate the arts, whose genius, so long captive, is about to soar to its deserved heights.

The pamphlet ended with a request that

all artisans, scientists, inventors, and others, foreign as well as national, and even members of academies who have or have had complaints . . . against academies . . . should hasten to send them to M. Morain, secretary of the Société du Point Central des Arts et Métiers . . . to enlighten public opinion on the following question: Have the academies been more detrimental than useful for public welfare and for the progress of the sciences and arts?

The success of this specific call for anti-academic ammunition is unknown, but the pamphlet may properly be taken as the artisans' declaration of war against any form of academic domination. It was their response to Condorcet's attempts to rationalize and modernize the Old Regime academic system. Publication of his elitist scheme had the effect of polarizing sentiments about learned societies, sharpening the outlooks that in the Old Regime had already separated constructive reformers from antagonistic critics. Once these differences found the concrete issue over which to clash, organized social groups like the artisans became the most outspoken and effective enemies of the academies, not only lobbying against the passage of the educational bill before the Assembly, but also demanding the immediate suppression of all official societies.[56] Moreover, they became the ideal social unit to mobilize the motley assortment of opponents of the academic system. Artisans were joined by inventors seeking recognition, young surgeons struggling to rise in their profession, and artists battling for their right to exhibit paintings in public. The emotional link that tied them all together was a profound desire to emerge from their inferior position and to be judged in a democratic fashion by their peers. No wonder they described themselves as "artistes vrais sans-culottes."[57]

Artisans were not without constructive plans of their own. In place of the academic system that they insisted the Convention must uproot, they would have substituted a truly "republican" alternative. For the granting of rewards to inventors, they suggested the creation of a General Directory of Science and the Arts elected by the entire community of scientists and artisans from all the eighty-three *départements* of France. To promote the application and diffusion of science, they would have organized a central executive panel, composed of representatives from all existing voluntary associations, that would act as a forum to listen to original papers and to arrange for their publication. The same panel, representing a broad spectrum of interests in science, would also have been charged with establishing a teacher center.

Both plans were the inspirations of a military engineer, Charles Desaudray, whose courageous deeds on 14 July 1789 had earned him a secure reputation as a celebrated patriot.[58] His background as a cartographer, engineer, and industrial spy led him into the world of artisans, and he joined the Société du Point Central des Arts et Métiers in 1791, later acting as its representative on the Bureau de Consultation. The experience of sitting with academicians to judge the merit of artisans' efforts convinced him of the profound injustice being perpetrated by

---

56. A large number of these pamphlets are cited in *AP*, LI, 9–10, and LXXI, 8. A typical example is the letter and petition signed by fifteen artisans sent to the National Convention in November or December 1792, now in AN, F^{17}1003, no. 231.

57. *AP*, LXXIX, 256 (15 Nov. 1793).

58. Mirault, p. 20; AN, F^{12}994; BN, MS. n.a.f. 2760, fols 191–196; Sigismund Lacroix, *Actes*, 1è sér., III, 596–598; Scheler, II, 173–178.

those who composed the board. As a true patriot, devoted to the principles of the Revolution, he insisted that the panel of judges must be elected to office in much the same fashion as legislators. On this basis, he drafted a "Constitution Nouvelle des Sciences, Arts et Métiers," obtained the agreement of several other artisan groups, and presented the radical plan to the Legislative Assembly.[59] Although it received no serious attention at the time it was drafted in March 1792, it was later referred to as a possible alternative to a government-directed and -controlled panel, one which would meet the artisans' psychological needs for a more democratic jury.

Shortly thereafter, Desaudray began to plan and organize another institution which met with far greater success: the Lycée des Arts. Three major principles guided the creation of this new institution, which began to thrive early in 1793.[60] One was the establishment of a center of communication among "free artisans," devoid of academic formalities, which would serve the interests of the entire community of amateurs and raise the dignity of the artisan by having him judged by his peers. A second principle was the abolition of the barriers that separated technology from science, thereby giving them equal status in the public eye. The third was the establishment of public lectures on all subjects of general interest in the arts and sciences.

The importance of Desaudray's projects for the December 1792 debates was that they offered a potential option for legislators favorably inclined toward science and learning but unable to subscribe to the principles behind a super-Academy. When considered alongside the successful private associations of philomaths and naturalists, the General Directory and the Lycée des Arts seemed to constitute a truly republican alternative that would satisfy the demands of the increasingly belligerent artisans and perhaps even help to restore France's floundering economy. To dismiss Condorcet's National Society would not

59. This [*Nouvelle*] *Constitution des Sciences, Arts et Métiers* was written in March 1792. It is reproduced in *AP*, LXXIX, 257–272 (15 Nov. 1793). A similar idea was suggested by Molard in his letter to Romme of 26 Nov. 1792: AN, F^{17}1047 dossier 1.
60. Smeaton, *AS*, XI, 309–313, 316–319; Scheler, I, 12–16; Birembaut, *RHS*, XI, 267–269.

therefore be to condemn the pursuit of knowledge. It would merely be a concession to the current political mood. Even those who had helped Condorcet draft his project—Romme, for example—were willing to consider dispensing with a National Society, despite its having been originally presented as the keystone of the entire educational reform effort.

Debates in the National Convention on education began on December 12 with a double-barreled attack on Condorcet's National Society coming from two politically opposing sides. The first to open fire was Durand-Maillane, a member of the Committee of Public Instruction, whose dissenting opinion had not hitherto been given a hearing in the committee.[61] His position, while extreme in its formulation, was not devoid of logic. The Condorcet plan, he reminded his fellow legislators, had been drafted in different circumstances and at a time when the training of republican virtues was not a central issue. Now, Durand-Maillane argued, the sole concern of the nation must be the indoctrination of young minds with a sense of patriotism. The sole task of public education should be to teach civic virtues and to train youth for effective citizenship. For this reason, Durand-Maillane insisted that the higher levels of education should not be financed by the government and that the complex system of higher education proposed by the committee must be abandoned.

Behind this position lay a profound mistrust of science, which Durand-Maillane expressed publicly without fear. His position showed visible traces of the Stoic arguments that had been recently transmitted through Montesquieu and Rousseau.[62] Knowledge was necessary only to the extent that it supported a sense of moral rectitude. Anything beyond this was superfluous, and might even operate to corrupt the native purity of young hearts. He insisted that, outside of civics and topics concerned with the administration or defense of the state, all other sciences should be banished from public education and left to the liberty of "genius, tastes, and means."

These views fitted the democratic temper of the times so well that Durand-Maillane did not hesitate to remind the Con-

61. *CIPCN*, I, 123–131.
62. Williams, *Critical Problems*.

vention that the *philosophes*, of whom Condorcet was the
archetype, were attempting to replace the royal and religious
autocracy of the Old Regime with a new and even more pernicious meritocracy. He returned to this point several times in
his speech, suggesting that elitist attitudes ran counter to the
temper of the times. "Here and now," he said, "political equality has become more rigorous and republican liberties cannot
be stained, even in favor of merit."[63] The great danger against
which he warned his colleagues was that talent or knowledge
rather than civic virtues would slowly become the measuring
stick for success and even for election or appointment to political office. The road to meritocracy would be paved by the
establishment of the National Society, which Durand-Maillane
pictured as a dangerous guild, even more frightening than the
religious guilds previously in control of most of the cultural
institutions of the nation. He warned the Convention that

> After having shaken off the yoke of tyrants, after having disposed
> of sacerdotal domination and destroyed all vestiges of political
> bodies, it seems odd that, under the pretext of science and enlight-
> enment, we urge . . . the creation of a special class of citizens. . . .
> There is a prejudice in favor of scholars, just as there was one in
> favor of kings and priests. . . .

   The next speaker, Masuyer, a Girondin whose political
views were opposed to Durand-Maillane's, nonetheless echoed
the latter's sentiments by launching into a wholesale attack
against the National Society.[64] Masuyer's was, in fact, the most
elaborately argued position against the super-Academy, and
the absence of detailed comment on it in the press remains quite
inexplicable. Although Masuyer coined phrases that Durand-
Maillane could also have used, the basis of his argument was
fundamentally different. Both delegates profoundly resented
Condorcet's sanctioning of an aristocracy of learning, or, as
Masuyer put it, of "the formation of a supreme tribunal and a

63. *CIPCN*, I, 126, for this and the next quotation.
64. Masuyer. The text of his speech was printed almost a year after it had
been read and includes material pertaining to events after 12 Dec. 1792. For
details on its history, see *CIPCN*, I, 131–132. In a letter to Delambre dated
28 thermidor III [15 Aug. 1795], Lalande suggests Massieu was the legislator
most influential for the Academy's demise (London, Wellcome Historical
Medical Library, dossier Lalande). It is likely Lalande confused Massieu
with Masuyer.

veritable aristocratico-academic parliament." But, whereas the rightist delegate took his ideological cues from Rousseau, the Girondin grounded his argument on the fear of a centralized administration in Paris which, like a closed guild, would be self-perpetuating and whose judgments would be final. "A society both national and unique for science and the arts, a supreme administrator? Have we regressed to the fourteenth century when thought was allowed to follow only certain authorities, when opinions contrary to those held would bring about persecution?" he exclaimed.[65]

Masuyer waxed eloquent about the dangers of corporatism when he asked, rhetorically:

Has the committee forgotten what the spirit of corporatism can be, especially when stimulated by the vanities of science, the pettiness of self-esteem and the thirst for fame? . . . Must the National Convention be reminded of the hates, persecutions, and the academic machinations to demonstrate that a man avid for glory cannot tolerate contradiction, that a teacher dreams of founding schools in order to live again through his favorite disciple; that to succeed . . . the disciple must be always a docile and submissive disciple? . . . That is the character of men, of masters, of corporations, and most particularly of academies and academicians.[66]

One of Masuyer's most novel and penetrating criticisms was on the effect that directing education rather than participating in it would have upon the character of the National Society. By turning the Society into an administrative agency concerned with the politics of appointments rather than with the propagation of knowledge, Condorcet's scheme would inevitably sap the creative energies of teachers by forcing them to adopt a uniform and monotonous program of studies. Such would be the consequence of strict centralization of education and its unified control by a small band of Parisians. Indeed, concentration in Paris seemed to be one of the greatest vices of a conception such as the National Society. Masuyer painted a dismal picture of the effects of the capital on learning:

You know that the atmosphere of Paris denatures everything that it surrounds, that the nature of urban centers is to render man a

65. *CIPCN*, I, 144.
66. Masuyer, pp. 64, 67, 69, for this and the next two quotations.

physical and moral degenerate; that one is a nonentity in Paris and gets nowhere without bowing and scraping. You will create teachers lacking courage, energy, and dignity, educating their own replicas, without character and fashioned for enslavement. . . . Do not condemn them to be passing their ideas through the sieve of a depressing institution; do not be dazzled by this supposed unity of principle from which everyone would derive this system of education. Unity . . . is the seed of death in a Republic of Letters, that can only flourish and prosper as long as the individuals who compose it dare to free themselves from the obstacles of authority . . . and are left to the impulsion of their native genius.

He summed up his argument against the super-Academy by pointing out that "I have demonstrated that the National Society is an immoral and impolitic monstrosity that would hand over science—and consequently public opinion and liberty—to a privileged class, an ostensibly philosophical priesthood. . . ."

The combined impact of these speeches, reinforced by the preceding pamphlets coming from artisan circles, was enough to prevent any serious public rebuttal. The criticism laid against the National Society was so devastating that, by December 15, the members of the Committee of Public Instruction agreed to bow to public opinion and "for the present, refrain from mentioning the National Society" in its new proposals for education.[67] None had dared to come to the defense of the super-Academy, and even Condorcet was forced to admit in January 1793 that, because of changed political conditions, it was perhaps no longer necessary for the National Society to be given such formidable powers.

In retrospect, we can record the December 15 decisions as the clearest symptom that academies were suffering from a terminal disease. The decision of the committee to yield to political pressures by separating the plan for a National Society from the rest of the educational reforms proved to be fatal. As an integral part of a program to revitalize France's cultural institutions, the learned societies' continued existence might well have been defensible. But, as isolated institutions identified with the tyranny, injustice, and arbitrariness allegedly practiced in the Old Regime, their chance of survival in a hostile atmosphere

67. *CIPCN*, I, 164.

was meager at best. In fact, it is somewhat surprising that the academies were not abolished until eight months later, on 8 August 1793.

The suggestion that the academies fell victim to the anti-intellectual mood of the day has often been advanced.[68] As we will see later, this position was formulated for political reasons by close witnesses of the legislative scene in 1792. Their charge was that learned societies were destroyed as part of a program of intellectual vandalism in which both the Jacobins and the lower classes of French society reveled. Knowledge itself is supposed to have come under attack as incompatible with true popular democracy. Cassini, for example, thought it quite natural that the sans-culottes should act as levelers and show their deep hostility for learning by burning books publicly.[69]

While such sentiments were not entirely absent in 1792, there is little evidence to suppose that this was a major factor in the fall of the academic system. As far as learned societies were concerned, the only serious instance in which arguments against the value of science and learning were advocated to justify their destruction was the December 12 speech by the conservative legislator Durand-Maillane. He is the only significant public figure who attacked the value of science as an enterprise. All the earlier critics of the Academy, including Bernardin de Saint-Pierre and Marat, were careful to limit their accusations to the personal or communal habits of men who practiced science. Individual scientists might be accused of being dishonest or avaricious and institutions despotic or obsolete, but knowledge itself was hardly ever maligned.

On the contrary, even those who favored the abolition of academies rejected Durand-Maillane's uncommon position. The delegate Chénier, who spoke in the Assembly after Durand-Maillane, ridiculed his speech by caricaturing it as an attempt to make vandals of Frenchmen.[70] Another deputy, Jacob Dupont, delivered a ringing speech, of which much

68. Delisle de Sales, pp. 85–138, and the examples given by Chapin, *French Historical Studies*, V, 371–372.
69. *AP*, XLV, 377–378 (19 June 1792), for Condorcet's motion to burn titles of nobility. Cassini alludes to this episode in "Mes Annales," p. 2.
70. *CIPCN*, I, 122.

note was made, answering all the charges leveled at science and learning by his rightist colleague.[71] Dupont's oration was described as a thorough refutation of "the enemies of science and detractors of philosophy." In particular, the passage in Dupont's rejoinder in which he reminded the Convention of the essential role *philosophes* such as Condorcet played in the victory of the Revolution over the forces of obscurantism was much applauded. It can safely be asserted that the revolutionaries were not hostile to science itself. Indeed, a repudiation of knowledge would have entailed a denial of the very traditions of the Enlightenment that helped to bring the Revolution into being in the first place.[72]

How, then are we to weigh the causes of the downfall of the Academy of Sciences? It was "academy," and not the sciences, that was the main target of the revolutionaries. In the last analysis, the objections that finally counted were directed at the institution and at the image the term "academy" evoked in the minds of revolutionaries. The idea of the academy stood as too great a contradiction of contemporary political realities. It had been a royal institution, unwilling to break decisively with the monarchy early in the Revolution. Academies met in the Louvre, in the former apartments of the king, behind closed doors, and amid the trappings of royalty.[73] The Academy of Sciences had within its membership a class of *honoraires* including ex-nobles, clergy, and a few *émigrés*, as well as a secretary who had habitually eulogized hated ministers. As an institution, the Academy did not follow egalitarian principles within its own ranks, and it assumed its members were by definition superior to the common citizen. Years of experience had

---

71. *CIPCN*, I, 149–156.

72. In fact, there were a considerable number of speeches that explicitly applauded the *philosophes* for triggering the Revolution. See *Chronique de Paris*, 2 Sept. 1791, p. 987; *Chronique du Mois*, May 1793, p. 43; and *AP*, LIII, 585 (25 Nov. 1792).

73. In keeping with the times, busts of Old Regime figures were removed from the academic meeting place on 12 March 1793, tapestries and paintings on 27 July 1793, according to the account given by the astronomer Messier, in "Notice de mes Comètes," Paris, Observatoire, MS C–2–19, p. 11. AdS, Reg 29 Aug. 1792 also records furtive attempts to remove visible signs of royal association. See also the exchange of letters between Lavoisier and the Minister of Interior on 22 June and 3 July 1793, to be published in Lavoisier, *Correspondance*.

shown that academicians were privileged and haughty and, in notable instances, rich to boot. Lavoisier, for instance, was well known as a tax collector and as an administrator of the powder industry, and had even been offered a position in the treasury. To many revolutionaries, he was principally known as the man who had set up a new wall around Paris to facilitate the collection of taxes.[74] The reputation of other academicians was no better. Laplace and Lalande were known to harbor little sympathy for the menial instrument maker unable to pass a rigorous mathematical test, or for critics of Newton. Bailly was remembered as the first mayor of Paris and was held responsible for the Champ de Mars episode in which the Paris militia fired on the populace.

Above all, the Academy was still functioning as a corporate organization at a time when special-interest groups had been outlawed. It operated as an elitist subculture in a society increasingly suspicious of any distinctions, even those based upon talent or merit. The most telling argument against its continued existence was that it was an aristocracy in a democratic world. No amount of rhetoric about its national utility could obscure that fact. Elitism was the very essence of the institution, the ingredient that had made it so successful in its mission in the scientific community. In the Revolution, the Academy was inextricably caught between its own traditions and the demands of a new political society.

74. On revolutionary attitudes toward Lavoisier, see Vergnaud; and Duveen, *Journal of Chemical Education,* XXXI, 60–62, and XXXIV, 502–503.

Chapter 8

Closing the Academy

THE SETBACK that Condorcet's National Society suffered in December 1792 should have dispelled any real hopes that the Academy of Sciences could survive. In the revolutionary atmosphere, the combined effect of artisans' resentment, anticorporate slogans, and the suspicion of intellectuals' elitist "plots" weighed more heavily than counter-arguments about contributions of the *philosophes* to the preparation of the Revolution or the utility of science. By all rights, the Academy should have given up its struggle for existence early in 1793 and disbanded. Other academies were lapsing into inactivity because of sparse attendance and the law of 25 November 1792 which prohibited new elections to fill empty seats.[1] The Academy of Sciences was not spared similar difficulties.

Attendance, which had remained fairly high at the beginning of the Revolution, dropped by about 15 per cent after the 10 August 1792 insurrection.[2] Although very few active scientists emigrated, a sizable number were distracted from academic affairs by their involvement with political events.[3] Because of their legislative and administrative commitments, Bailly and Condorcet ceased to function as academicians, and Broussonet, Dionis du Séjour, and Monge were forced to stay away from meetings for long stretches of time. A number of other academicians were far too busy with their administrative occupations to attend meetings regularly, while a few others

1. *CIPCN*, I, 87–90.
2. The average attendance dropped from above forty to slightly below thirty-five.
3. In addition to Chabert, Cornette, and Dietrich, usually cited as *émigrés*, Broussonet and L'Héritier also left the country.

were sent off on scientific missions by the Academy.[4] Other academicians missed sessions because they sought refuge from the Parisian turmoil by spending several months a year in the country. Berthollet was often in Aulnay, Bougainville and Cassini in Normandy, Coulomb near Blois, Laplace in Melun, Rochon in Brest, and Tessier at Angerville.[5]

To add to the hardships of dwindling attendance, there emerged a series of annoying problems related to academic finances. They stemmed in part from the disorder discovered in the accounts kept by the Academy's treasurer, Tillet. He had died unexpectedly in December 1791, and payments to academicians were held up while an equitable settlement between the Academy and his heirs was arranged. Lavoisier, the newly elected treasurer, had barely concluded these negotiations when another issue that blocked salary payments arose. A new law prohibited the payment of double salaries to any government employee. A number of academicians who held positions as teachers were thus in danger of having their personal revenues cut in half. Thanks to Lavoisier's skillful advocacy of their case, an agreement was reached favoring the academicians, who were treated as exceptions. Immediately thereafter, another bureaucratic squabble between national and municipal governments blocked salary payments. Lavoisier advanced his own money to replenish the Academy's coffers until the fiscal mess could be straightened out. By May 1793, he was sufficiently alarmed to warn the deputy Lakanal about the effects of the Academy's financial difficulties:

Citizen, time is pressing. A large number of academicians are suffering, several have already left Paris because their finances no longer permitted them to live here. If help is not on its way, the sciences will slowly fall into a state of decay from which recovery will be difficult.[6]

4. Cousin, Jussieu, Thouin, and Vicq d'Azyr had spotty attendance records during the Revolution. Le Paute Dagelet, Delambre, and Méchain were often away on missions.
5. In addition to printed biographies listed in Appendix II, see the Berthollet-Van Marum correspondence in Haarlem, Hollandsche Maatschappij der Wetenschappen, and the Berthollet-Senebier correspondence in Geneva, Bibliothèque Publique et Universitaire, and BI, MS. 2003; for Rochon, see JT, p. 106; and for Tessier, Tessier-Vicq d'Azyr correspondence, Paris, Académie de Médecine, MS. box 120.
6. Letter of Lavoisier to Lakanal dated 13 May 1793, Archives de Chabrol, M, to appear in a forthcoming volume of Lavoisier, *Correspondance*.

The penury of academicians and their dwindling attendance were also aggravated by the passage of a law on 25 November 1792, intended for the Academies of Painting and
Sculpture and of Architecture, but which applied to all learned
societies, temporarily halting the election of new members.
Several vacancies caused by death or emigration remained
unfilled.[7]

Despite these annoying material difficulties and the prevailing political mood of the Convention, the Academy managed to carry out its functions and hold regular meetings. It was
not merely a sense of duty toward society that kept it alive. A
few academicians, whose lives had been closely entwined with
that of the institution, found enough resilience to work for its
continuation in spite of the formidable handicaps. Stress seemed
to strengthen their resolve. Some believed the Academy was
still the most vital institution for the advancement of science
and worked to keep it alive as the symbol of internationalism in
a world edging toward war. Even Cassini, Lalande, and Sage,
who were bitter or cynical about the follies of their politically
minded colleagues, maintained a firm allegiance to the institution and its principles.

None was more articulate or persistent in justifying academic traditions publicly than Condorcet himself. Perhaps
because the National Society had been his personal dream, he
refused to let the accusations leveled by his fellow delegates
against the Academy go unchallenged. Instead of seeking a
compromise which might have won him political support,
Condorcet pursued his ideology with the philosophical fanaticism that had characterized his turbulent life. He clarified and
amplified elitist and encyclopedic convictions, underlining the
benefits he was convinced would emerge from his educational
projects.

Condorcet spelled out his views in two publications in the
early days of 1793. One was the article "On the Necessity of
Public Education," the other the lengthy footnotes to the second edition of his original *Rapport*.[8] In both, he spoke directly

7. *AP*, LIII, 578–579 (25 Nov. 1792).
8. Both were written at about the same time. The article appeared in
Chronique du Mois for Jan. 1793. For the notes to the second edition, see
CIPAL, pp. 188–246, and the critical edition by Compayré.

to his critics, admitting that they were correct in their understanding of the self-contained and self-perpetuating nature of the National Society, but entirely misguided in assessing its significance. He said, in the most candid way:

There is a fear of learned corporations. But if one examines carefully the charges made against those that once existed, one will notice that they are founded upon either religious or political intolerance which no longer exists, or the old vices of these institutions which sensible men always felt, and that are easy to avoid.

To judge these corporations, one must consider the ones concerned with science . . . since they are the only ones that have enjoyed some independence. If one glances at the annals of these corporations, one will see how, by taking in men whose modest means would have prevented them from devoting themselves wholly to science and by giving others means to publish their work promptly, they have served the progress of knowledge.[9]

Condorcet readily admitted that his National Society would bring together men of superior talent and intelligence and give them control over the masses of their intellectual inferiors. But to rail against an elite because it did not correspond to the current political image of democracy was to fly in the face of reality:

Equality of mind and education are dreams. We must therefore make the best of this inherent inequality. . . . But if public education is general and widespread and if it embraces the totality of knowledge, then this inequality can be put to use for mankind by the work of men of genius.[10]

He insisted because he was certain that the progress of humanity depended both upon the perfection of the human mind, a task to be accomplished by men of superior intellect, and upon the propagation of sound knowledge among the masses. By associating the institution designed for the advancement of knowledge with the educational system, Condorcet felt he had devised a foolproof system to promote progress and to make sure that the elite would pass on its findings to the rest of society. This was the central reason for the National Society presiding over the entire edifice of public instruction.

9. *CIPCN*, I, 611. Compare to a different version of the same thought in *CIPAL*, p. 222 n. 1.
10. *CIPAL*, pp. 202–203 n. 2, and *CIPCN*, I, 610.

Condorcet may have been convinced of the logic of his argument, but no amount of repetition would make it politically palatable to the legislators. During the eight months that preceded the August 8 decree, the same slogans coined earlier denouncing aristocracies of learning were heard. The deputy Bancal des Issarts claimed that "all corporations tend toward aristocracy ... [which] even penetrates the temple of the muses and the republic of letters," and would in time usher in the return of a political aristocracy.[11] Instead of a truly egalitarian system of education, the sanctioning of this "new clergy" of intellectuals would legitimize the superiority of talent.[12] To these well-worn arguments were now added warnings of the danger of concentrating control in urban centers which, according to a Rousseauist doctrine, corrupt the innate goodness of man by introducing him to luxury, vice, and intrigue.[13] It was obvious that Condorcet's rebuttals were not having any effect, and the suppression of all academies, suggested by Romme in late November 1792, seemed to be a foregone conclusion.[14]

One hope remained. Lavoisier formulated a new tactic for the preservation of the Academy which did not involve the same doctrinaire stand taken by Condorcet. It was an approach he developed privately on the basis of his personal perceptions and a deep sense of devotion to the Academy. Lavoisier's close contacts with a new member of the Committee of Public Instruction, Lakanal, gave him a fair chance for implementing the idea in the political arena. We know that in the end he failed, and that the death warrant of the academies was delivered by the Convention on 8 August 1793. Yet this decree, which came as a surprise to most academicians, had almost been averted by Lavoisier's maneuvers, which sustained the scientists' hope for survival to the very end.

11. *CIPCN*, I, 252–253.
12. Saint-André compared the "hierarchy of scholars" to the "hierarchy of priests" in *CIPCN*, I, 278. Similar comments by Serre and Duval are in *CIPCN*, I, 286–287, 639–640.
13. This theme, already present in Brissot's *De la Vérité* and in Masuyer's *Discours*, recurs in many of the speeches combating proposals of the Committee of Public Instruction. See also Delandine, *De la Conservation*, pp. 18–31.
14. *CIPCN*, I, 87–89. The Committee of Public Instruction had tried to temper Romme's belligerent language against the academies.

As the Academy's only permanent and active officer for its last year and a half, Lavoisier identified himself more completely with his institution and its problems than he had done previously. By the time he became treasurer, he had left the tax-farm, given up his post as powder administrator, and abandoned the political activities he had once pursued in Blois and at the Société de 1789. With the publication of his essays on public finance and political economy in 1791, he withdrew from public affairs and turned to preparing a new edition of the *Traité de Chimie* and to initiate other scientific projects. His ultimate refusal of a post as minister to Louis XVI in June 1792 merely reaffirmed his resolve to devote the full measure of his energies and talent to the pursuit of science.[15]

Lavoisier's perception of the Academy's difficulties differed from that of Condorcet, in part because of their different activities during the early years of the Revolution. While Condorcet had been operating in legislative committees and on a public rostrum, Lavoisier had been involved in administrative matters, serving as a member of the Bureau de Consultation, where he came into daily contact with artisans. He assessed their complaints about academies as a tactic to carve out a legitimate place in the governmental structure next to the already established scientific profession. Even before the Revolution broke out, Lavoisier had recognized that artisans were inadequately represented. This explains his welcoming of the creation of a Corps d'Ingénieurs en Instruments and the ephemeral formation of an "auxiliary society of the Academy of Sciences" composed principally of young scientists concerned with the *Description des Arts et Métiers*.[16] He was genuinely predisposed toward applied science, as is well shown by his plan to devote a substantial section of the revised *Traité de Chimie* to the chemical arts.[17]

His strategy for saving the Academy was founded upon the belief that the attacks of the artisans arose principally from their natural desire to participate in governing science and the arts. In his eyes, the creation of the Bureau de Consultation

15. Grimaux, pp. 214–216, Vergnaud, pp. 128–135.
16. Lavoisier, *Œuvres*, VI, 67; Daumas, *Instruments*, pp. 135–143.
17. Daumas, *Lavoisier Théoricien*, pp. 110–112; Jacques Cousin, p. 143.

constituted the first step in the sharing of power. Successful as it was, there was still a need to institutionalize the artisans' role in administration by enlarging the structures of scientific institutions. In this way Lavoisier hoped that artisans' aspirations to participate in the life of science and technology would be satisfied and that their sense of suspicion, bred by exclusion, would disappear. It had been his personal experience with antagonists of the Academy, such as Desaudray, Servières, and Trouville, that cooperation was the most effective palliative to criticism.[18]

Although Talleyrand and Condorcet had provided a place for the artisans in their super-Academy, they had not made it a cardinal point in arguing for the new institution. Both had been more concerned with developing a consistent philosophical position that would appeal to their legislative colleagues. The unlettered artisans, however, could hardly be expected to appreciate this high-sounding philosophy or the debates it inspired. Instead they yearned to hear how their occupations would be given special significance in the new revolutionary society, and they sought concrete signs of their integration into the scientific community.

For these reasons, Lavoisier gladly gave his support to voluntary associations and fraternized with individuals who would probably have remained outside his circle under the Old Regime. He joined the Société d'Histoire Naturelle, where he was sponsored by the artisan-scientist Hassenfratz, and continued to give his active support to the Lycée.[19] With the vigor so characteristic of his personality, Lavoisier was drawn into their administrative activities, and was selected as the Lycée's representative to the governing board of Desaudray's newly formed Lycée des Arts. There too he earned the respect of his fellow members, who elected him for a two-month term as president, an office he held in June and July 1793. In both Lycées, he advocated a close union of the sciences and the arts. The success of courses in technology, geography, and the practical arts taught at the Lycée des Arts prompted him to recommend that

18. The correspondence between Lavoisier and these artisans, to be published in Lavoisier, *Correspondance* for 1792 and 1793, will bear this out.
19. Lavoisier was elected to the Société d'Histoire Naturelle on 23 Dec. 1791, according to the minutes in BS, MS. box 123. His membership and affiliation with the two Lycées is discussed by Smeaton, *AS*, XI, 260–264, 311–316.

similar offerings be adopted by the older Lycée.[20] He saw that
public instruction in subjects of utilitarian interest could sustain
and warrant the existence of the new Lycée des Arts and re-
affirm the value of cultivating science.

Lavoisier was also impressed by Desaudray's plan to set up
a board of directors representing all the major scientific societies of Paris, meeting periodically and administering the business of the new institution. He suggested the old Lycée would
do well to adopt this formula by creating a "Société Centrale
des Sciences et des Arts," composed of about two dozen members, each held responsible for reporting to the group on the
state of science in a designated field. Periodically, this Société
would hold public meetings at which all the Lycée's subscribers
would hear lectures on the progress of science and the arts. The
new group would thus become "a center of light and knowledge valuable for professors, artisans and all those interested in
science."[21]

Two themes dominated Lavoisier's interest in these societies. One was the importance of having institutions uniting
men previously separated by their professional occupations. He
expressed this view in a meeting of the Lycée des Arts of 9 May
1793 in which he referred to

the good that must come from the fraternal meetings of scientists
and artisans—which have taken place spontaneously . . . [It] is one
of the most desirable benefits of the Revolution which, by destroying a kind of prejudice that has grown up against learned corporations, demonstrates what is and will always be their eagerness to
contribute to the progress of knowledge and to participate in all
that contributes to public welfare.[22]

Every mention of science was coupled with a good word for
the arts, and every reference to a scientist was linked to one for
artisans. The other theme Lavoisier repeated was the need for
sharing new knowledge with all of society so that it might be of
general use. Knowledge was meant to be communal property,
in both its accumulation and its distribution.

20. For details on the planned lectures, AN, $F^{17}1004^B$, no. 580. Compare with
the list of courses and number of lectures actually offered in AN, $F^{17}1143$
dossier 3 pièce 27.
21. Lavoisier, *Œuvres*, VI, 559–569; Baker and Smeaton, p. 44.
22. *Journal du Lycée des Arts*, 14 June 1793, cited in Scheler, I, 27.

The attitudes Lavoisier formed in the brief period he spent with artisans and in voluntary associations became the basis of his new effort to save the Academy, which had always been his true mission. He decided to abandon the possibility of saving the entire academic system and to concentrate on retaining only the Academy of Sciences. In 1790, Talleyrand had argued for all learned societies, without making any value distinctions among them. The following year Condorcet had placed science at the head of learning, but developed a justification that applied equally well to all disciplines. Now in 1793, Lavoisier was willing to sacrifice the useless or "frivolous" disciplines to save the useful ones.

This plan was not without merit. It was meant to satisfy the artisans who had little understanding of and even less trust in men of letters and painters, so often in the employ of aristocrats. It would satisfy critics in the legislature who, for example, had refused to appropriate funds for the Académie Française until its national "utility" had been demonstrated.[23] It would even please those who, because of their adherence to the romantic concept of genius, objected to the practice of giving writers and artists permanent salaries. Lavoisier articulated his position with great care in a letter sent to Lakanal on 17 July 1793, emphasizing the difference between letters and sciences:

Science is not like literature. The man of letters finds in society all the elements he needs to develop his talents. . . . He depends upon no one. The same is not the case in the sciences. Most of them cannot be pursued with success by isolated individuals. . . . All the sciences aid each other to advance in common the great edifice of human knowledge.[24]

Lavoisier also reminded Lakanal that academic meetings served to sanction experiments, observations, and discoveries by putting them to a communal test, a function considered unnecesary in all other disciplines. It was clear to him that the institutional form essential for science was not necessarily the same as the one needed for letters and the arts.

An important difference also existed if one were to employ the yardstick of national utility. One could well question the

23. *AP*, XVIII, 92 (16 Aug. 1790).
24. Lavoisier, *Œuvres*, IV, 618.

value of government-sponsored learned societies for literature and the fine arts, but it was evident that the Academy of Sciences performed practical functions essential for society at large as well as for the government. To underscore this point, Lavoisier painted an idealized portrait of the role of his institution with regard to the artisan.

The Academy . . . is a free tribunal, always open to whoever wishes to have his inventions judged. It is useful for entrepreneurs unwilling to invest in the development of untried machines or inventions. The tribunal is useful for the inventors who consult it because its members are intelligent guides who make it their duty to aid them with advice for modifying or correcting their inventions whenever they deem it useful.[25]

Its greatest use was as a government consultant, and Lavoisier studiously listed all the projects undertaken by the Academy at the request of the legislature or the administration, singling out for special emphasis the standardization of weights and measures.

The implications of his position were far-reaching. By making the Academy's existence principally dependent upon its utility, Lavoisier risked turning science into a discipline auxiliary to its applications. In responding directly to what he perceived as the artisans' demands, he opened up the possibility of an "assimilation of science to the arts."[26] In the long run, such dependence might well destroy that balance among the different scientific specialties that both Lavoisier and Condorcet explicitly recognized as important.

An even more important limitation on Lavoisier's effort involved his convenient neglect of the problem of *demi-savoir* in order to win over the disaffected artisans. The two Lycées catered to a wide public, composed of a large number of amateurs. Indeed, the Lycée des Arts, which had leased a section of the popular Palais-Egalité for its meetings, received its income by subletting a part of its premises to shopkeepers, who were expected to draw paying customers to where public meetings were being held. There, scientific disquisitions were sandwiched in between readings of patriotic poems, musical recitals,

25. Lavoisier, *Œuvres*, IV, 619–620.
26. Gillispie, *Critical Problems*, p. 278.

and the sleight-of-hand demonstrations that always caught the public's fancy. The scientific mission of the Lycée often suffered, as for example when a relatively straightforward discussion of measuring the meridian was curtailed for "fear of taxing the public's attention by concentrating it for too long a time on abstract objects."[27] How apt to refer to these meetings, held in the *cirque* of the Palais-Egalité, as "spectacles"!

Lavoisier could not have been totally insensitive to the dangers his stand entailed for the progress of science. There are, in fact, indications that he sought to prevent a total integration of artisans and scientists and that he found a technique to protect the public from being duped by incompetent speakers for science and the arts. On July 18 he wrote to Lakanal asking that, in the projected organizations of learning, scientists and artisans have separate organizations.[28] In the proposal for a new system of education sponsored by the Bureau de Consultation and allegedly prepared by Lavoisier, he adhered to Condorcet's separation of the pure and applied sciences into two divisions, even though joint monthly meetings were prescribed.[29] As for the maintenance of standards in the Lycées, Lavoisier was a thorough advocate of the directory system, by which a small executive council maintained control of the affairs of the society. It was the plan he adopted for the proposed Société Centrale des Sciences et des Arts.[30]

Until the very last minute, Lavoisier's lobbying seemed to have a good chance of succeeding. Basing his argument upon the Academy's utilitarian function, Lavoisier had managed to secure exceptions for it from certain new laws that were meant to apply to all learned societies.[31] On May 17, a bill was passed authorizing the Academy to fill its vacancies in personnel. Five days later, another bill allowed academicians of science to receive double salaries, despite the general rule prohibiting this practice. To engineer the passage of these special bills, Lavoisier had worked through Lakanal, whose enthusiasm for science was fortunately matched by great energy and a string of suc-

27. *Journal du Lycée des Arts*, 29 June 1793, cited in Scheler, I, 40–41.
28. Lavoisier, *Œuvres*, IV, 623–624.
29. Lavoisier, *Œuvres*, VI, 530, 537, 554–555; Baker and Smeaton, pp. 44–45.
30. Lavoisier, *Œuvres*, VI, 567–568.
31. *CIPCN*, I, 457–464.

cesses on behalf of other scientific institutions.[32] If anyone could protect the interests of the Academy, it seemed likely that Lakanal was the man.

Unfortunately, however, the Convention was unable to keep in view the fine distinctions between academies that Lakanal underlined. The legislature was reminded that despotic ~~tyrant~~ corporations still existed. Though artisan attacks against the Academy of Sciences had subsided after December 1792, dissident artists were continuing to attack the Academies of Painting and Sculpture and of Architecture. The subject was raised at the Convention on June 28, and, on July 1, a delegate demanded that the bill suppressing the Academy of Painting and Sculpture be introduced by the Committee of Public Instruction.[33] Nothing happened for two weeks, until another delegate demanded that a report be produced immediately. The tone of his request was angry: "It is time to take a stand on these monstrous institutions . . . these corporations bred by the despotism that they so often served."

The Committee of Public Instruction assigned the task of framing the bill to Lakanal, who immediately turned to Lavoisier for help. In a letter dated July 15, he asked the treasurer of the Academy of Sciences for a memoir expounding his justification for keeping the institution alive:

> The National Convention has directed its Committee of Public Instruction "instantly to make its report on the suppression of academies." I am assigned this painful task. It would be more pleasant for me to keep quiet than to speak up for barbarism. Since it is very likely that all learned societies will be proscribed, please send me your views on the necessity of conserving the Academy of Sciences. Time is short.

Lavoisier responded two days later with a brief for his institution which Lakanal could use as the basis for his argument supporting the Academy's existence despite the suppression of other learned societies.[34] His case rested firmly on the Academy's national utility, toward which he suggested all its scien-

32. Lakanal; Hamy, *Derniers Jours*, pp. 63–67.
33. For details on this paragraph, including the quotations, *CIPCN*, II, 242–246, and VII, part I, 17–18.
34. Lavoisier, *Œuvres*, IV, 616–623.

tific activities were directed. Its traditional role as consultant to the government was discussed, as well as the long-range utility of the progress of knowledge itself. On this point, Lavoisier chose to stress the importance of science for technological and industrial progress and, playing on the legislators' sense of nationalism, emphasized the importance of the Academy in surpassing England during the eighteenth century in the areas of mathematics, chemistry, and instrument-making. He ended with a rhetorical flourish, predicting the direst of consequences if the Academy were suppressed.

Judging by the bill Grégoire proposed to the Convention on August 8 in the name of the Committee of Public Instruction, Lavoisier's approach had succeeded. It contained seven articles, two of which were designed to rid the nation of learned societies, with the significant exception of the Academy of Sciences, which received a deferred sentence of death:

Article 1. All academies and literary societies established [*patentées*] or endowed by the nation are eliminated.

Article 2. The Academy of Sciences remains provisionally entrusted with the several tasks assigned to it by the Convention; consequently it will continue to receive its annual grants until otherwise stipulated.[35]

Another article laid plans for the re-establishment of the Academy:

Article 3. The Convention directs its Committee of Public Instruction to submit as soon as possible a plan to create a society for the advancement of science and the arts.

Grégoire's speech justifying the bill also incorporated most of the points Lavoisier had advanced in his July 17 brief, and further developed the favorable comparison with England.[36] Although there was exaggeration and distortion of fact which Lavoisier would not normally have condoned, it was for a worthy cause.

35. *CIPCN*, II, 240. The draft of the decree is in AN, C 263, II 596, pièce 46. The other articles of the proposed decree dealt with the right of free assembly and the disposition of prizes, laboratories, and courses under academic control.
36. *CIPCN*, II, 250–256, and analyzed in detail by Delisle de Sales, pp. 85–138. Its disorganized contents suggest it was drafted by Grégoire in great haste.

Appearances proved to be deceiving. Although the bill looked like a victory for Lavoisier's plan, it was, in fact, backed by only a small faction within the Committee of Public Instruction, and represented the last desperate and, as it turned out, futile measure to prevent the disaster foreseen by most legislators who favored the continuance of the Academy. Although the committee's minutes are unusually silent on this matter, the future of the Academy must have been hotly debated. One sign was that Lakanal, who had previously been able to uphold the Academy's side with success, was replaced by Grégoire as spokesman for the bill.[37] There is no way of telling if Lakanal was removed from that task because of his strong pro-academic sentiments, or if he resigned because he could not bring himself to share majority views. Whatever the reason, it makes Grégoire's reminiscences on the framing of this bill critically important for our story. In his *Mémoires*, written years after the event, Grégoire characterized the attitudes then current:

A general disfavor hovered around all corporations, especially those resisting the new political order. The committee expected that . . . the Convention would destroy all the academies without discrimination and thereby single out its members for persecution. . . .

All the sensible members of the committee agreed that in order to preserve men and things, it would have to appear to bend to circumstances and itself propose the suppression of academies, making an exception of those of science, surgery, medicine, and agriculture. . . . Lavoisier came to confer with me on this plan and backed it. Against my wish, I was assigned the task of presenting the bill to the Convention. But the Convention was unwilling to make any exceptions. . . .[38]

How strong the dissension was within the committee is exemplified in the fact that one of its own members, the painter David, spoke up in the Convention directly after Grégoire's presentation of his seven articles to demand that all academies be abolished regardless of the subject with which they dealt.[39] His impassioned speech, filled with resentment against the entire academic establishment, ended the discussion and carried

37. Guillaume, I, 113–126. See also the letter of Grégoire to Garat dated 6 Aug. 1793, in AN, F^{17}1097 dossier 1 pièce 1.
38. Grégoire, I, 351–352.
39. *CIPCN*, II, 256–258.

the day.[40] The final decision of the Convention was to approve article 1 of the Grégoire motion, but to put off articles 2 and 3 for discussion for three days. As we will see shortly, when the subject came up again on August 14, it was broached in a totally different context. The verdict was final. Legally, the Academy of Sciences ceased to exist on 8 August 1793.

In contrast to the legislative confusion that had surrounded the fate of academies for almost three years, the fundamental issues on which they fell are perfectly clear. They explain why Lavoisier's move to save the Academy was destined to failure despite his deft politicking. His arguments were fundamentally irrelevant to the nature of the final debate, which pitted two irreconcilable positions about the organization of French culture against one another.

On one side were the proponents of the academic system, who conceived of the advancement and propagation of learning as a national responsibility most efficiently carried out by a self-regulating and self-propagating body of talented intellectuals, free from political control and supported by government funds. On the other side were their antagonists, who rejected the bureaucratization of cultural elites because it would place a handful of "aristocrats" in control of matters that were ultimately not subject to human command. Knowledge of nature and discovery of truth were meant to be a public affair and not the exclusive province of an elite. In the political context of the anticorporate, democratic revolution of 1793, the victory of the anti-academic movement was inevitable.

Notwithstanding his political acumen, Lavoisier had seriously misjudged the ultimate source of the artisans' attacks. Although some of them shared academic values and were satisfied with participation in the bureaucratic apparatus, the majority were so distrustful of it that they preferred its complete destruction. They were more intimately motivated by the egalitarian ideology of the sans-culottes than by the cult of merit the intellectuals had always embraced. In the final analysis, the academic system was overturned because it was the

---

40. The secretary of the Convention taking minutes noted, but later crossed out, that "this proposition, strongly supported by another member and opposed by no one," was adopted: AN, C 263, II 588, pièce 9, p. 3v.

embodiment of a profoundly aristocratic mentality implicit in the idea of a Republic of Letters.

Some academicians did not consider the August 8 decree a final sentence. Diehards read the adjournment of articles 2 and 3 as a sign pointing to the effectiveness of Lavoisier's efforts and seized upon it as a last hope for salvaging the Academy from eternal destruction. For several weeks they clung to this hope and tried to act as if the end had not yet come. The last convulsions of the moribund institution are worth recounting, not merely for their anecdotal value, but as an episode revealing the extent to which internal dissensions in the society had imperceptibly weakened the Academy from within. This episode brought into the open the political discord that had been growing inside the Academy ever since La Rochefoucauld had introduced his reform measures in 1789. Moreover, the events of August 1793 also provided concrete evidence for revolutionaries outside the Academy who had always claimed there were prominent scientists resisting the new order.[41] Such ex post facto evidence was always welcomed. Finally, the episode is important because it marks the beginnings of the disintegration of a sense of community among Old Regime scientists.

Saturday was a regular meeting day for the Academy but, because it was the anniversary of the August 10 insurrection, plans had been made to gather on Friday. Members assembled at the Louvre as they had done for over ninety years, ratified the minutes of the previous meeting, and offered various scientific papers that were intended for the *Mémoires* for 1793. The presiding officer (probably Darcet) then declared the session closed because of the Convention's bill dissolving the Academy voted upon the previous day. There followed a vigorous discussion which brought to the surface both the resentments that had been building up for years against the political regime, and the charge by some academicians that some of their colleagues were, in a "traitorous" fashion, more attached to the political order than to the Academy.

According to the only available eyewitness account, a few

41. Lakanal, p. 224, "it was a popular opinion that they [academicians] were opposed to the new order, and unfortunately there was some truth in it."

academicians—probably Fourcroy and Monge—attempted to prevent their colleagues from submitting their scientific papers for the record because the Convention had officially suppressed the Academy the day before.[42] Cassini recounted the event in the following terms, giving the episode a melodramatic character by placing speeches in the mouths of his colleagues:

> In our last session, immediately after the publication of the decree of suppression, those members of the Academy who had memoirs prepared for reading, or some work begun, rushed to record them and to offer them as the last homage to science and to the expiring Academy. . . . It was to be our swan song. But, can it be believed, our zealots were opposed to this last act of fervor, I would almost say of academic piety. They were so impatient to see our dissolution, to savor their handiwork [the dissolution], that they parried everything that would defer its demise.
>
> "As soon as you have knowledge of the decree destroying the Academy," they said, "you may no longer perform any academic act, or concern yourselves with memoirs or their printing."
>
> "At least," it was answered, "we must be allowed to close our minutes and to insert in them, as a kind of last testament, our ultimate words—that is our last productions."
>
> "No, no," they cried out with that imperious tone that success engenders, "we must obey. Let us separate. There is no longer any Academy of Sciences."

Lavoisier recalled that this debate also dealt with the letter and spirit of the law, the conservatives claiming the meeting was legal since the decree had not officially been served on them, and the "zealots" insisting the Convention had made its desire to close the Academy's doors crystal clear.[43]

    This sort of argument was not new to the Academy. The same bitter controversy had exploded behind closed doors in 1790 when the new constitution was drafted, and more recently had broken out again over the problem of *émigrés* within the Academy. As each issue was debated, the factionalism that had begun to loosen the fraternal bonds linking academicians broke them completely, leaving permanent rifts between opposing groups within the society. There was developing a genuine and

42. Cassini, "Mes Annales," pp. 12–13, and for the next quotation.
43. Letter of Lavoisier to the Committee of Public Instruction dated 10 Aug. 1793, in *CIPCN*, II, 312–313. The existence of a printed copy of this letter (Duveen, *Supplement*, pp. 42–43) suggests Lavoisier thought it important as a public document to justify the Academy's continued existence.

mutual mistrust among extremists in the Academy, which ended with an exchange of accusations of disloyalty. Conservatives like Cassini and Sage charged their colleagues with having turned the sanctuary of science into a political debating society by introducing irrelevant issues into the sessions, thereby destroying the Academy's occupational integrity and bringing about its demise. At the other end of the political spectrum, revolutionary enthusiasts like Fourcroy and Monge accused the conservatives of lending evidence to the popular charges that academies harbored aristocrats and of closing their eyes to the political realities of the day. Each side eventually held the other responsible for the destruction of the Academy.

Debate has been particularly acute after 10 August 1792, when Fourcroy had suggested that the Academy strike those members known to be *émigrés* from its lists.[44] Cassini's account, although admittedly biased, recaptures the tense atmosphere of the occasion:

I will never forget one of the most indecent of these scenes . . .

Several members of the Academy had been absent for a long time and were presumed to be *émigrés.*

"We must purge the Academy of these traitors," said our zealots, "we must strike them from our lists without delay."

"This is a matter for civil authorities."

"Why is that?"

"Because as long as our old regulations are in effect, the administration of our company belongs to the minister to whose care the Academy is assigned."

"No matter, it is better to give an example of patriotic zeal than to be scrupulous in observing regulations."

"It is important for a body like the Academy to temper its zeal with prudence, and to do only what is prescribed."

"Are you protecting the *émigrés,* do you approve of them?"

"We protect or approve of no one, but we have neither the leisure nor the mission of checking to see if X is in the kingdom or not, and even less to judge the motives for his absence. But above all, we do not want to denounce our colleagues."

"In such a case, denunciation would be a virtue."

"It will never be ours!"

"We take pride in it, and we will take official notice of your op-

---

44. These events are discussed by Grimaux, pp. 225–227; Bertrand, pp. 425–427; Maindron, *Académie,* pp. 65–66; Smeaton, *Fourcroy,* pp. 46–48, and Kersaint, pp. 51–52.

position by announcing it in the newspapers listing the names of those who side with the traitorous *émigrés*."

"It will be your responsibility if we are placed on the list of outlaws and if your colleagues are thereby delivered to the assassins' knives. This patriotic act is worthy of you and of a sort to bring you glory!"[45]

While the accuracy of these quotations may be questioned, the tenor of the arguments and the issues they raised were real. Academic attitudes were polarized by such outbursts, and Cassini, for one, never forgave his radical colleagues for their disloyalty to the Academy. Fourcroy, on the other hand, interpreted the punctilious concern with rules of Cassini and his friends as a cover for their disapproval of revolutionary fervor, and branded them in his mind as scientists unworthy of holding national offices.

In making such judgments about his colleagues at the Academy, Fourcroy must have been struck by the different reception his political attitudes encountered in academic circles and in the voluntary associations to which he also belonged. In 1790, he encountered no difficulties in making membership in the Société d'Histoire Naturelle dependent upon a political test.[46] The Society of Agriculture, over which he presided in 1793, removed twenty persons from its membership list, some of whom were known *émigrés*.[47] Similarly, at the Society of Medicine, Fourcroy faced little opposition to having eleven former members stricken from its ranks. By contrast, the Academy resisted such moves, and the opposition forced the decision of purifying its membership upon the Minister of Interior, Roland, who gladly ruled out thirteen names, some of whom also were *émigrés*.[48] This experience was undoubtedly important in weakening Fourcroy's resolve to save the Academy as an institution.

Unlike Cassini, who was a political novice, Fourcroy wielded considerable influence in legislative circles, and played

45. Cassini, "Mes Annales," pp. 6–7, and reproduced in part in Devic, pp. 172–173.
46. BMHN, MS. 464, p. 9.
47. Smeaton, *Fourcroy*, pp. 45–49.
48. Cassini, "Mes Annales," p. 8. The only active academicians removed from the list as *émigrés* were the chemist Cornette, the astronomer Chabert, and the mineralogist Dietrich.

a crucial role in determining the fate of France's scientific life from July 1793 until his death in 1809. He was elected an alternate deputy to the Convention, and took his seat there on 25 July 1793. Before the week was out, he had been appointed to the Committee of Public Instruction, in time to participate in the discussions on the suppression of the academies.[49] In that capacity, it is likely that he supplied Grégoire with data for his speech supporting the utility of the Academy of Sciences, and kept in contact with Lavoisier about its administrative affairs.

Contrary to what has often been asserted, there was no visible animus between the two chemists. Nevertheless, they reacted quite differently to the August 8 decree, in a manner consistent with their past personal experience and allegiance. Both were thoroughly devoted to the cause of science and convinced of its importance for the progress of mankind; but whereas Lavoisier's revolutionary experiences and the responsibility he shouldered as treasurer of the Academy drew him closer to the institution, Fourcroy's attachment to science was not inextricably linked to the continuation of the Academy. His association with it dated back only to 1786 and, as a junior member, he was less fully identified with the Academy than Lavoisier was. The difficulties he faced in imposing his "patriotic" views on his academic colleagues stood in sharp contrast to the successes he had experienced at the ex-Jardin du Roi, in voluntary associations, and at the two Lycées.[50] Thus, while willing to aid Lakanal, Grégoire, and Lavoisier in keeping the Academy alive, he was also ready to accept alternate solutions for the preservation of science.

Lavoisier's course after the August 8 decree was to urge the transformation of the Academy into a "société libre et fraternelle pour l'avancement des sciences," whose first concern would be the continuation of the tasks assigned to the Academy by the Convention.[51] In a letter sent to the Committee of Public Instruction through the good offices of Lakanal, Lavoisier listed a series of projects undertaken by the Academy

49. *CIPCN*, II, vi–vii.
50. Fourcroy was also a member of the Société Libre d'Economie Rurale, founded in March 1792, which included Roland, Romme, Thouin, and Marsillac in its membership. See BS, MS. box 128.
51. Letter of Lavoisier to Lakanal dated 11 Aug. 1793, in *CIPCN*, II, 317.

which, in his opinion, had to be continued by some agency. In addition to the project to standardize weights and measures, he listed the publication of Father Plumier's botanical works edited by Jussieu, of Desmarest's mineralogical map of volcanoes, of Desfontaines' description of the Barbary Coast, and of Vicq d'Azyr's writings on comparative anatomy.[52] His key request, however, was for appropriations to continue to pay the salaries of the academicians. In his eyes, the financial question was critical:

It is not necessary, citizen representatives, that I add that the maintenance of stipends for those who received them is a matter of elementary justice. There is no academician who, if he had applied his intelligence and means to other ends, could not have carved out an existence in society. It was through public faith that he took up an honorable though not very lucrative profession, into which the government itself had, in a way, called him, and on which he depended. Several academicians are octogenarians and sick, others have exhausted their strength and health on trips or projects undertaken at no public expense. French loyalty must not let them down, and they have at least the absolute right to the pensions designated for public officials.

Time is pressing. If you allow the scientists who composed the ex-Academy time to withdraw to the countryside or to take up other and more lucrative occupations in society, the organization of science will be destroyed, and a half century will not be enough to create a new generation of scientists. I beg you, for the national honor, for the interests of society and for the opinions of foreign nations watching us, to make provisions to prevent the collapse of the practical arts which is a necessary consequence of the destruction of science.[53]

Lakanal and his colleagues in the Committee of Public Instruction responded to Lavoisier's plea by securing the passage of a bill in the Convention on August 14 enjoining the scientists to pursue their work and assuring them of continued remuneration.[54] The academicians, if not the Academy itself, seemed to have circumvented the general proscription. In the meantime, another group of legislators was pushing a series of

52. Letter of Lavoisier to the Committee of Public Instruction dated 10 Aug. 1793, in *CIPCN*, II, 314–315. Several other items were also cited.
53. *CIPCN*, II, 316–317. As treasurer, Lavoisier was intimately acquainted with the financial plight of his colleagues.
54. *CIPCN*, II, 319.

bills through the Convention that inadvertently nullified the August 14 decision.[55] They were resolved to implement the August 8 decision by taking physical possession of the rooms in the Louvre where the academies met and by making an inventory of the statues, paintings, books, manuscripts, and other material possessions that had been confiscated by the government. By 10 A.M. of August 17, those meeting rooms had been inventoried and were officially closed by seals affixed on the doors.[56]

These conflicting decisions left the diehard academicians in a state of utter confusion. Cassini's letter to the editor of *L'Abréviateur Universel*, signed "C***," published on August 12, discussed the August 8 decree and its effects with great pessimism.[57] Four days later, a jubilant letter signed "B***" appeared, referring to the recent "resurrection" of the Academy, and chiding the "sans-culottes du monde savant" for having hoped to "break the scepter of science."[58] The writer ended on a triumphant note: "I beg these men a thousand pardons; but while they cry over it, you and I and many others will rejoice and even laugh about it. P.S. Please send my regards to citizen C***, who, in your number 224, took the matter very seriously."

But when the ex-academicians tried to reconvene as a "société libre" on August 17, they found the doors sealed and, after standing in the corridors of the Louvre for an hour, disbanded for the last time.[59] It was finally brought home to them that the majority of the Convention really favored the complete destruction of all academies, and resented the subterfuge arranged by Lavoisier and Lakanal which would, in effect, have changed the name "Academy" into "Société Libre," without altering the group's functions or academic practices. Lakanal corroborated the resentment that his August 14 decree had created in a note dated August 28, sent to Lavoisier: "This

55. *CIPCN*, II, 318–320.
56. AN, F^{17}1032 dossier 3 pièce 5.
57. *L'Abréviateur Universel*, no. 224, pp. 895–896.
58. *L'Abréviateur Universel*, no. 228, p. 911. The author of this letter may also have been Cassini.
59. Messier, "Notice de mes Comètes," Paris, Observatoire, MS. C–2–19, p. 11. For the document describing the placing of the seals, see AN, F^{17}1032 dossier 3 pièce 5.

decree, pursued at length, won me more insults than it contains words. It would have separated me from the cause of science only if mankind's sense of injustice could ever make me abandon it."[60] Once and for all, the Academy was dead and had to be forgotten.

Although Lavoisier was crushed by the prospect of science without an Academy, his ex-colleague Fourcroy immediately set out to salvage the unfinished business of the Academy.[61] He joined forces with Romme who, although in the past he had held different views on the continuation of learned societies, was a fervent advocate of the progress of science.[62] Together, they devised a new policy eventually adopted by the Committee of Public Instruction and the Convention to fill the vacuum created by the demise of the Academy. Instead of concentrating the functions of the Academy of Sciences or the super-Academy into a single organization operating from Paris, they decided to split the institution up into three parts, each of which was separately designed to carry out a special function.

Following the pattern already tested by the Bureau de Consultation, expert advice for the government was to be provided by individuals or groups specially appointed and paid by the government. Each agency or commission was to be assigned a specific task set by the appropriate legislative committee and was expected to report its findings directly to its parent organization. In this way, consultation would become wholly integrated into the bureaucratic apparatus. A second function, which Talleyrand and Condorcet had envisioned for the new learned society, was the dissemination of scientific knowledge. According to the new plan, this was to be turned over either to renewed "republican" institutions, such as the Muséum d'Histoire Naturelle, supported by the government, or, more usually, to a free-enterprise system of education. The hope was, as Fourcroy announced in December 1793, that knowledge would be disseminated by men of talent, any of whom could

60. Clermont-Ferrand, Bibliothèque Municipale, MS. 338, fol. 64.
61. Lavoisier's pessimism is expressed in his letter to Lakanal dated 1 Sept. 1793, in *CIPCN*, II, 331–332.
62. Romme's plans are revealed in the draft to a bill never presented, in *CIPCN*, II, 329–331.

set himself up as a teacher.[63] Their success would depend upon the number of auditors who chose to remain at the lectures and to pay fees. Fourcroy felt that the quality of the teacher's instruction and the impact of his intelligence on the public were sufficient guarantees to provide any capable scientist with a living and to prevent the incompetent from surviving in this profession for very long. Against Condorcet's highly structured organization for scientific teaching, Fourcroy offered a regime of total freedom and liberty that coincided well with the political mood of the times.

For this reason also, Fourcroy and Romme urged that the ex-Academy's third function, the advancement of science itself, could be carried on effectively in voluntary asociations totally independent of constituted authority. Both were confident that the *société libre* could operate as a viable organization for the progress of their disciplines. Those they had known were all infinitely more responsive to the needs of the scientific discipline than the regulation-bound Academy had been. The time that had been wasted in the Academy on electioneering, which fostered intrigues and the intervention of powerful ministerial influences, could be devoted to the pursuit of knowledge. Moreover, private associations could, without complex maneuvering, decide to initiate public lectures, raise funds, or launch new publications.

The Fourcroy-Romme policy, which both men pursued relentlessly after the Lavoisier-Lakanal plan had failed, was something of a tour de force. Politically, it was acceptable to the majority view in the Convention that had repeatedly rejected all the variations of a government-directed, pyramid-like organization of learning for France. It is a sign of the legislators' anxiety that slogans concerning the "aristocracy of science" and warnings about the creation of a system of higher education as "subversive of republican institutions" persisted even after the academies no longer existed.[64] With very few exceptions, all the important speeches about educational pro-

---

63. Speech of 9 Dec. 1793, in *CIPCN*, III, 97–102. Fourcroy's views were mercilessly criticized by Portiez, in *CIPCN*, III, 200–205.
64. The two quotations are taken from speeches by Petit and Bouquier, in *CIPCN*, II, 542, and III, 56. Similar quotations could be extracted from speeches by Coupé, Jogues, Roland, and Daunou.

posals underlined the alleged need to unshackle science from the "gothic" bonds of academism in order to set it free. The very term "société libre" exuded republicanism.

The Fourcroy-Romme plan also seemed to offer reassurance to the scientific community that it would not be abandoned. In an unambiguous manner, it would turn the scientist, who had always been a part-time public servant, into a *fonctionnaire public*, hired and paid for his special talents. In this way, the government would be forced to make a place for science in its operations. In return, the savant would be assured of a livelihood, and would remain in Paris, contrary to Lavoisier's fears. As for "pure science," every scientist would now be freer than ever before to select congenial colleagues and research problems by joining a voluntary association to his liking. There was a further, unexpected asset in the form of the *société libre*, which could narrow its scope to any specialized branch of science without being answerable to anyone but its own membership.

In short, the new scheme, forged by politicians devoted to the cause of science, was meant as a compromise between a sense of political realism and a desire to preserve a discipline and its practitioners, even though its age-old institutional structures were discarded. The physicist Biot, writing in 1803, was close enough to events to recapture the essence of the new "republican" strategy adopted by Fourcroy and Romme.[65] He remarked that, in 1793, there were two factions controlling the destinies of French culture. One was numerically dominant, but "ignorant and ferocious" and intent upon leveling its intellectual superiors. The other, more enlightened group was able to render its aims acceptable by "adopting its [opponents'] language" but with opposite ends in mind. It was by covering science with the "mantle of its enemies" that they saved it. Fourcroy and Romme were the *politiques* of science in 1793 who hammered out positive compromise measures to satisfy all the warring parties. In the process, they had been forced to organize an entirely new pattern of scientific organization for France. They initiated a system of "democratic science" to re-

65. Biot, *Essai*, pp. 46–47, 56.

place the "aristocratic science" the Revolution had completely rejected.[66] A new phase in the life of science was about to begin, this time without an Academy. How it would fare was to prove crucial to the ultimate restructuring of France's institutions of science.

66. Hahn, *XII Congrès International.*

Chapter 9

Democratic Science

HISTORIANS of the Academy, when referring at all to the two-and-a-half-year period between the Academy's demise and the foundation of the Institut, speak of it as an unfortunate intermission between two acts in the life of science in France. In fact, these years were of crucial significance for the future organization of French science, and the character of the post-Thermidorean Institut cannot be grasped without analyzing the annals of science during this period of high tension in Paris, brought about by the war and the Terror. The period opened with a bold and imaginative arrangement for the preservation of science, following a distinctly more democratic orientation, and ended with the triumphal return of government-controlled institutions of higher learning to fill the gap left by the Academy's disappearance. As will become evident, the experiment with democratic science designed by Fourcroy and Romme proved to be a miserable failure and was gradually replaced by an unplanned new pattern of organization which fitted the nation's and the scientific community's immediate needs rather than any specific revolutionary ideology. By late 1795, the original political reasons for shaping a program of democratic science were largely forgotten in the wake of more pressing demands. It appeared as if the question of science's place in the national scene would be settled by the exigencies of the moment rather than by the intensely politicized debates in the legislature.

Most visible of the three parts into which Fourcroy and Romme expected to divide traditional academic functions was the network of agencies set up for government consultation. Their numbers and organizational complexities are modest in comparison with the apparatus supported by modern industrialized societies, but, by the standards of the 1790's, they represent an intricate web in which the new fabric of French science was at first woven. To exist, this web had to conform to at least three requirements imposed by the realities of revolutionary France. Politically, consultation had to be a part of the legislative branch of government, in direct contact with and under the control of the committee system that reigned supreme until the Directory. Any lingering ties that had bound scientists to a minister, republican or otherwise, and made them dependent upon his personal favors or prejudices were necessarily to be severed. An equally important consideration was that government consultation had to relate directly to the urgent needs of a republic carrying out a social revolution within its borders and fighting off threats from foreign armies. The third reality within which the legislators who framed the new consultation pattern had to act was the shortage of available manpower possessing the requisite intellectual skills. With the destruction of the Academy, there was a real danger that the pool of talent once within easy reach would disappear from the capital, and possibly even from the nation.

It was to this last reality that Fourcroy and Romme addressed themselves first. They knew that Lavoisier's warning about the financial plight of scientists had a real basis. Before the Revolution, twenty academicians netted average annual salaries of about 2,000 livres, and another ten received grants ranging from 500 to 800 livres, including Fourcroy himself.[1] Now that these funds and the lucrative administrative offices once reserved for academicians had disappeared, most of the ex-academicians found themselves dependent upon salaries given for teaching at the few scientific institutions still in operation, such as the Collège de France and the Muséum d'Histoire

---

1. The standard budget of the Academy included 3,000 livres each for the secretary and the treasurer, eight 3,000 livre stipends for the oldest member of each discipline, eight 1,800 livre stipends for the next oldest, and eight of 1,200 livres for each of the youngest *pensionnaires*. There were also special assignments of 500 to 800 livres for some of those who had not yet reached *pensionnaire* status. See Maindron, *Académie*, pp. 115–116, and JT, p. 130.

Naturelle.[2] Before 1793, the few new governmental positions created at the Bureau de Consultation or in the Commission d'Agriculture were either nonpaying or returned inadequate monetary rewards.[3]

Two of the three organizations Fourcroy and Romme proposed in August and September of 1793 were able to compensate in part for the financial vacuum that the Academy's disappearance had suddenly created. The third, which never materialized, was a board of scholars Fourcroy suggested to select textbooks for elementary education. According to his bill, introduced in the Convention on August 30, it would have included eleven scientists, representing all disciplines.[4] The motion was sent back to committee, from which it never again emerged, probably because of the allegation that it was "a move to re-establish an academy under a more modest name." If it had been organized as Fourcroy intended, this board of scholars would indeed have exercised the very adjudicatory functions that opponents of Condorcet's plan for controlling education by an elite had repeatedly rejected.

Fourcroy and Romme's second attempt to institutionalize the practice of consultation met with greater success. On September 1, they obtained the Convention's approval to create a Commission des Arts which subsequently employed a dozen ex-academicians, paying them a stipend of 2,000 livres a year plus expenses.[5] Unlike the proposed board of scholars, the Commission des Arts was created to implement, rather than to make, policy decisions regarding science and education. Originally it was formed to inventory academic property now belonging to the government by virtue of the August 8 decree. The task of this agency was to draw up lists of "national ob-

---

2. In 1791, salaries at the Collège de France (formerly the Collège Royal) were between 1,000 and 1,500 livres a year. At the Muséum (formerly the Jardin du Roi), salaries in 1794 were around 2,800 livres. Academicians connected with these institutions included Poissonnier, P.-C. Le Monnier, Cousin, Lalande, Portal, Darcet, Daubenton, Desfontaines, Fourcroy, Jussieu, Lamarck, and Thouin. For details, see Lefranc, p. 281, and *CIPCN*, IV, 407–408 n. 1.

3. The erratic attendance of some academicians at the Bureau may well be explained by these financial problems. See Ballot, pp. 23, 69–70, 118, and *CIPCN*, IV, 265–268.

4. *CIPCN*, II, 306–307.

5. *CIPCN*, III, 327–328; Tuetey, I, xxxv–xxxix.

jects useful for public instruction," and to report its findings directly to the Committee of Public Instruction.[6] The number of ex-academicians on its staff was justified by the commission's need for persons who would be familiar with the machines, instruments, natural-history specimens, libraries, and manuscripts that fell into the national domain following the academies' demise or from property confiscated from the Church, the *émigrés*, and imprisoned suspects.[7] While this plan provided some scientists with income, it also meant their occupational roles had to follow political demands rather than scientific competence. Known critics of republicanism such as Sage and Cassini, for example, were never appointed, despite their professional qualifications. In the extreme case, ex-colleagues were placed in adversary positions, as when Berthollet, Fourcroy, Guyton, and Vandermonde were required to inventory the possessions of Lavoisier, imprisoned in November 1793 with other former tax collectors.[8]

Another direct outgrowth of the August 8 decree was the Commission des Poids et Mesures, created at Fourcroy's request on September 11.[9] Its function was to implement the legislature's decree regarding the standardization of weights and measures by completing the meridian measure and by constructing standards of length and weight. For the sake of continuity, the task was assigned to the same eleven scientists who had previously been selected by the Academy, and for whom a salary of 10 livres a day per member was decreed. Originally, the commission was considered merely as a group called together for a limited time to complete an unfinished task of the Academy, but it was occasionally consulted on other technical matters, and even entrusted with the preparation of the *Connaissance des Temps*, a job that had previously fallen on the shoulders of an academician.[10] While it would be an exaggeration to refer to this group as a surviving thread of the defunct Academy, its members behaved at first as scientists rather than

6. *CIPCN*, II, 508–510.
7. The Commission des Arts merged with the older Commission des Monuments charged with preserving France's artistic heritage. See *CIPCN*, III, 167, 169, and Tuetey, I, vi.
8. Tuetey, I, 225, 243, 547, 583, and II, 143; *CIPCN*, III, 240, 603.
9. *CIPCN*, II, 384–389.
10. *CIPCN*, III, lxxxiv–lxxxv, 257, 477.

as creatures of the political assembly. Thus, following Lavoisier's arrest, the commission requested his release on the grounds
that he was sorely needed for professional pursuits.[11] They
learned quickly that they should be operating according to a
new set of national priorities. The Committee of Public Safety
answered the petition by removing Lavoisier and five other
members as being suspect on political grounds. In the decree,
reasons for demanding a mild type of loyalty test for employment in a government agency were offered:

The Committee of Public Safety, considering how important it is
for the improvement of public morale that . . . its employees be
men worthy of trust for their republican virtues and hatred of
royalty . . . , decrees that Borda, Lavoisier, Laplace, Coulomb,
Brisson, and Delambre will cease to be members immediately.

The Academy's demise also coincided with a national
crisis of major proportions that required an even greater im-
pressment of academic talent into government employ. By
August 1793 the combined effects of allied superiority and the
crisis in authority within the army had brought France to the
edge of defeat on its frontiers, and threatened to bring royalty
back to its toppled throne. The government responded by
initiating an unprecedented mobilization of national resources
and, in a few months' time, placed the country in a state of
total warfare.[12] The problems this new policy created were unprecedented and required immediate attention. Mass conscrip-
tion had drafted over 650,000 men into military service but left
many of them without sufficient supplies, munitions, or adequate leadership. To solve these complex problems of logistics
and strategy, the Convention called on two military engineers
from its own ranks, making them members of the powerful
Committee of Public Safety. Both Lazare Carnot and Prieur
de la Côte-d'Or, who took over military commands on August
14, were conversant with science and in touch with its practitioners. Soon thereafter, they began to surround themselves

11. *CIPCN*, III, 233–242. Bouchard, *Prieur*, pp. 294–299, discusses the motives
behind the committee's response.
12. Contemporary accounts of the mobilization are given in Fourcroy, *Rapport*, and Biot, *Essai*, pp. 47–49. See also Pouchet; Mathiez, *Revue de Paris*;
Richard; Pigeire, pp. 141–154; Bouchard, *Guyton-Morveau*, chs. XIX–XX,
and *Prieur*, chs. IX–XI; and Fayet, pp. 211–257.

with a small but trustworthy band of scientists, who became the technocratic brain trust of the government's frenzied effort to bring the country to a peak of fighting strength. The story of their eventual success in "organizing victory" is one of the better-known aspects of the Revolution.

For our purposes, its importance lies in the additional role scientists were asked to perform as government consultants, and for the reputation it subsequently gave to science. The exploits of the scientists surrounding Carnot and Prieur were dramatized as an example of science's utility to the nation; these successes formed the principal public justification for the re-emergence of science as a national activity in the period when democratic science began to flounder. More than the unobtrusive efforts of those who inventoried scientific instruments or those who measured the meridian between Barcelona and Dunkirk, the exploits of this group of scientists provided a basis for replacing the public image of the scientist as an Old Regime academician by that of the scientist as an essential citizen of the new society.

Unlike the Commission des Arts and the Commission des Poids et Mesures, this small group of advisers was never formally given a bureaucratic designation, such as the "Commission des Sciences." Individuals served directly under the Committee of Public Safety, often receiving salaries as employees without title. One advantage of this lack of administrative definition was that they could move from one technical task to another without requiring special transfers. Scientists often acted as the personal agents of Carnot or Prieur, from whom they derived considerable executive authority. Whereas such practices would have been abnormal under other circumstances, they were quite appropriate to the special needs of wartime and were commonly practiced by revolutionary committees during the critical years of the Terror.

The seed for this cluster of consultants had been planted several months before the Academy disappeared by the chemist and deputy Guyton, who sat on the Committee of Public Safety before Carnot and Prieur joined it.[13] On April 9 Guyton

---

13. Fourcy, p. 13. Prieur wrote "they constituted a sort of committee within the Committee of Public Safety and Prieur served as their intermediary": Arbelet, pp. 76–77.

appointed "four citizens versed in chemistry and mechanics to carry on research and tests for new techniques of defense."[14] Three of them—Périer, Berthollet, and Fourcroy—were academicians with whom he had come into frequent contact as an editor of the *Annales de Chimie* and as a legislator. They were later joined by two other chemists, Carny and Chaptal, who were enlisted by the committee to carry out gunpowder experiments. When Carnot and Prieur were appointed to the committee by the Convention, this "congress of scientists" was enlarged by the addition of the ex-academicians Monge, Darcet, Pelletier, Vandermonde, and Vauquelin, as well as Adet and Hassenfratz, who had been closely associated with the *Annales de Chimie*. During the year and a half that this group operated, a large number of projects vital to the war effort were initiated and pursued by these scientists with the "revolutionary" fervor characteristic of the day. Their accomplishments are staggering, even after the exaggerated claims made for them are pared down to a reasonable figure. The variety of their activities is in itself remarkable.

The most spectacular of their efforts were in weaponry and munitions, where they were at first most vitally needed. Before the Revolution, small arms had been produced at the rate of 55,000 muskets a year, enough for royal needs but hardly sufficient for 650,000 men.[15] In the month of September, the Paris arsenal was processing no more than 9,000 muskets and 500 pistols, most of which were being repaired rather than manufactured. Prieur assigned Monge and Hassenfratz the task of setting up a munitions factory in Paris employing 6,000 workers and turning out 1,000 new muskets a day.[16] While never reaching the goal set by Prieur, they established a highly productive factory and, in addition, completely reorganized cannon manufacturing for both the artillery and the navy. The Périer brothers' factory in Chaillot on the outskirts of Paris became a center for the manufacture of bronze pieces for light artillery and, together with new factories in the provinces, was able to turn out more than 11,000 artillery pieces a year.[17] Here

14. Aulard, *Recueil*, III, 166; Bouchard, *Guyton-Morveau*, p. 311.
15. Richard, p. 8.
16. Richard, pp. 99–104; Arbelet, pp. 77–78.
17. Richard, p. 276.

the chemists played a major role, indicating new processes for extracting copper from church bells, in a brochure written by Darcet and Pelletier, of which 6,000 copies were printed and widely distributed.[18] They based their findings in part upon Fourcroy's work of 1791, published in the *Annales de Chimie*, and in part upon new experiments carried out in a laboratory provided by the Minister of War. At the same time Monge, Vandermonde, and Berthollet were commissioned to investigate the production of steel for swords and bayonets, for which they composed a pamphlet on metallurgy that is still considered a masterpiece of precision, conciseness, and clarity. Their *Avis aux Ouvriers en Fer, sur la Fabrication de l'Acier*, printed in 15,000 copies and sent to all corners of France, is generally credited for the great increase in production that followed.[19] Later, Monge was also directed to write a comprehensive treatise on the manufacture of cannon, the *Description de l'Art de Fabriquer les Canons*, which summarized the state of the metallurgical arts in his times and has become a classic in technology.[20]

In addition to metallurgical needs, France faced a shortage of gunpowder. Saltpeter, its most important ingredient, was normally imported from India by ship, but because of the wars it would have been foolish to expect imports to supply enough of this raw material. For the month of August, Prieur estimated that the nation had less than one-quarter of its needs filled by local production, and he projected an even greater disparity between consumption and production now that the *levée en masse* had been decreed.[21] The committee turned to Fourcroy's laboratory partner, ex-academician Vauquelin, to activate saltpeter extraction in central France. It quickly became evident that France possessed considerable quantities of unexploited raw materials. On December 4, Prieur prepared the two-page *Instruction sur l'Exploitation Révolutionnaire des Salpêtres*,

18. The pamphlet, *Instruction sur l'Art de Séparer le Cuivre du Métal des Cloches*, may have been written in part by Fourcroy. See Smeaton, *Fourcroy*, p. 271 n. 340, and Richard, p. 287.
19. Richard, p. 209.
20. Richard, p. 287. Hassenfratz, who published *La Sidérotechnie* in 1812, drawing heavily from this book, may well have assisted Monge in drafting it originally.
21. Arbelet, p. 109.

which gave simple procedures for all loyal citizens to follow in scraping their cellars, barns, stables, and soil for all the natural nitrates they could obtain, and proposed to have it bought on a national scale by government agents for 24 sous per pound.[22] It is most likely that these directions were drafted for Prieur by Berthollet, Fourcroy, or Chaptal, all of whom were familiar with the processes. In Paris, where the practice was first demonstrated successfully on a large scale, much patriotic zeal went into the citizens' search for saltpeter, resulting in an estimated production of 5,000 pounds per day shortly after the program had been launched. One of the most popular songs of the day was the "Marseillaise du Salpêtre," set to music by Cherubini.[23]

The purification of saltpeter for use in gunpowder called for large quantities of potash and soda, which could no longer be imported from Spain and the colonies. Once again, the Committee of Public Safety turned to its chemists, who urged a national program of burning trees and shrubs to use their ashes as substitutes. For that purpose, Vauquelin and Trusson wrote an *Instruction sur la Combustion des Végétaux, la Fabrication du Salin, de la Cendre Gravelée et sur la Manière de Saturer les Eaux Salpêtrées,* which was published and distributed by the committee.[24] Another group of chemists—Darcet, Pelletier, and Lelièvre—was asked to report on the preparation of soda from brine. Their report, *Description des Divers Procédés pour Extraire la Soude du Sel Marin,* not only compared the relative merits of several methods developed since the 1780's, but also exhumed the Leblanc process from government files where it had lain unnoticed for some years.[25]

In all these endeavors, the scientists were being exploited principally for their acquaintance with metallurgical and chemical processes rather than for their prowess as experimenters or innovators. Other attempts were made to capitalize on their inventive talents as well, although the results were far less impressive in terms of the war effort. In April 1793, a group of chemists under Berthollet's direction was asked to compare the

22. Richard, pp. 439–446; Bouchard, *Prieur,* pp. 277–278. The text is reproduced in *AP,* LXXX, 617–622 (4 Dec. 1793).
23. Richard, pp. 460–462.
24. Richard, p. 547.
25. Gillispie, *Isis,* XLVIII, 158–163, 166; Partington, III, 562–565.

explosive force of potassium chlorate with that of the best refined saltpeter (potassium nitrate) to determine if the former should be manufactured in large quantities.[26] Fourcroy was asked to investigate the potentialities of new "fulminating" explosives. In addition, a number of tests to improve the range of artillery pieces were being carried out with the help of scientists.[27] Until October 1793, these trials were undertaken in different laboratories at the Arsenal on the rue de la Madeleine, at La Fère near the northern borders, or at Vincennes. It was decided to create a proving ground near Paris to coordinate these tests in a locale where they would be kept from public scrutiny. The chateau of Meudon and its park were designated as the secret "établissement national pour différentes épreuves sous la surveillance du Comité de Salut Public." Monge, Guyton, and Hassenfratz often went there to observe the results of experiments with newly manufactured projectiles and incendiary bombs.

The Meudon proving grounds are better known for the role they played in the development of France's most successful new weapon, which helped to turn the tide in several battles, especially that of Fleurus. For it was at Meudon that Guyton's suggestion for using balloons as observation posts to spot enemy movements was first activated.[28] Once more, the work of ex-academicians provided the key to success. General Meusnier, who had died at the siege of Mainz, had left among his papers an extensive manuscript on his research with ballooning, which was put to quick use. After months of trials and rigorous training, a company destined to be the first airborne troop in history, the balloon corps, was finally ready for action. Guyton was dispatched to the battle front to oversee the deployment of this new weapon and returned to Paris with a triumphant report, read at the Convention by Barère. So promising was this new military invention that experiments on the strength of animal skins for use as envelopes for the balloons, and on flint glass to provide better telescopic lenses for aerial observation,

26. Richard, p. 575; Bouchard, *Guyton-Morveau*, p. 312.
27. Richard, ch. VIII and p. 669; Bouchard, *Prieur*, pp. 248–251.
28. Aulard, *Recueil*, XIX, 634; Godechot; Bouchard, *Guyton-Morveau*, ch. XIX.

were initiated at Meudon.[29] There was also established a school for training the airborne corps, whose curriculum included a good dosage of mathematical and scientific subjects, considered essential for an understanding of operations in ballooning.

Scientists were employed for a number of other projects organized by the Committee of Public Safety, ranging from attempts to extract tar from pine trees to the development of a quick-tanning process for leather.[30] The shortage of raw materials used up by warfare also led to the systematic discussion of the properties of soaps made from a variety of substances, new paper made from scraps of used newsprint, and pencils made from native minerals instead of English graphite. There is no question that the mobilization of scientific talent had been a great success for the nation, and that the frenzied efforts of Monge, Berthollet, Vandermonde, and Pelletier in the defense of the country would not quickly be forgotten.

The efflorescence of scientific activity in the name of the national emergency seems at first to belie Lavoisier's prediction that the government would seriously suffer from the destruction of the Academy. Academic talent was enlisted by the government, and many scientists now received salaries comparable to those previously handed out by the Academy. It may even be that revolutionary circumstances created more payroll positions for scientists than had been eliminated when the Academy was abolished.[31] Viewed in quantitative terms alone, one part of the Fourcroy-Romme plan was successful, for most scientists were employed in tasks related to their talents and the nation was able to continue capitalizing on their expertise. But the new practice of hiring the scientist as an employee of governmental agencies had a profound and disquieting effect upon the individual, and in turn upon the scientific community.

29. Letter of Delambre to Fréminville dated 30 prairial III [18 June 1796], AN, F^{12}2435; Richard, pp. 629–631.

30. Biot, *Essai*, pp. 81–83; Arbelet, p. 110. Many of the scientific and technological papers prepared on these occasions were published in the *Annales de Chimie*, XIX and XX.

31. An artisan named Laplace made a scurrilous attack on Hassenfratz, Berthollet, Vandermonde, and others, claiming that each was receiving the substantial salary of 10,000 livres for his efforts (AN, F^{17}4739 dossier 2). Although the figure is certainly inflated, it suggests that scientists employed by the government did not lack money.

As a result of the institutional changes that had taken place in August and September of 1793, the academician had been fully transformed into a public functionary, forcing him unwittingly to redefine his obligations and shift his allegiances. As a bureaucrat, he had to accept a set of values that differed considerably from those he had held as an academician. Whereas political opinions had been generally irrelevant for the academician, loyalty was now an understandable prerequisite for employment. Those who failed to meet the political "tests" were ignored, shut out, and even persecuted, regardless of their intellectual merits. Those who were unable to extricate themselves from past political affiliations—Bailly, Bochart de Saron, Condorcet, Dietrich, Lavoisier—were doomed to disappear, notwithstanding their potential usefulness for society. New lines of authority were also emerging. The scientific bureaucrat was now completely and directly dependent upon the lawmaker, forced to answer his every request and to solve problems in his terms. Resistance was frowned upon and could easily be misconstrued as disloyalty.[32] As one botanist put it, "One must be a citizen before being a naturalist."[33]

This transformation had a deep psychological effect on the scientists themselves, far beyond anything Fourcroy and Romme could have imagined. For all its structural defects, the Academy had been the concrete embodiment of the scientist's dual obligation: to the society that supported him indirectly by subsidizing the Academy, and to the scientific community of which he was a vital part. Now that the institution had disappeared and the scientific enterprise was "set free," the individual scientist tended to identify with the group of men he worked with in governmental agencies rather than with a community that had lost its substance. He was led to abandon his

32. Berthollet, writing to Chaptal on 28 April 1794, indicated that "resistance would be interpreted badly": Chaptal, p. 43 n. 1. Fourcroy was reported by Brongniart's mother as claiming that requisitioned scientists "belong to the Republic" and no longer to their families: Launay, p. 66. An extreme statement appeared in Darcet and Pelletier's *Instruction*: "Those who, by indolence or disinterest, will not lend their talents completely to the defense of the republic would be, by their criminal indifference, almost as guilty as the enemies of the public good": as quoted in *Journal de Physique*, XLVI, 60.
33. Letter of Bosc to J. E. Smith dated 24 July 1790, London, Linnean Society, Smith papers, 2.116.

primary allegiance to fellow scientists and to the scientific enterprise by his total absorption in governmental affairs. The breakdown of old patterns of attachment explains better than any alleged jealousy the abandonment of Lavoisier and Condorcet by Fourcroy, Monge, and Guyton. These men stopped being colleagues when the Academy disappeared.

It would be inaccurate to make the August 8 decree solely responsible for the disintegration of the scientists' sense of community. Other forces within the community itself were also at work. The growth of science had already forced subgroups of specialists to form smaller communities within the larger one. Just before the Revolution, for example, astronomers tended to congregate either at the Observatory under Cassini's leadership or at the Collège Royal around Lalande's chair.[34] Chemists, particularly those following the new system of Lavoisier, had selected the newly founded *Annales de Chimie* as their rallying point. Naturalists were gravitating to the Jardin du Roi as their gathering place. Nevertheless, all the leading members of these "subcultures" retained another identity as scientists by participating actively in the life of the Academy as a whole.

What the August 8 decree did was to shatter the bonds that had previously countered the centrifugal tendencies of specialization. Regardless of their good intentions, Fourcroy and Romme's rearrangement of the organization of scientific activity in splitting the Academy's old functions into three parts proved to be the community's undoing. Their effort to satisfy the political exigencies of the day in order to preserve normal scientific activity led to a breakdown of these activities that they never envisaged. Government consultation might provide ex-academicians with a source of income, but it was no substitute for the communal life the scientists had experienced for over a century.

Had it not been for the sense of dismay and fear that followed the closing of the Academy, Fourcroy's plan to have scientific activities carried on in voluntary associations might have worked and could conceivably have restored the bond that held scientists together in the learned society. When the Acad-

---

34. Chapin, "Astronomy," pp. 297ff., 491, and *French Historical Studies*, V, 395–400. For the naturalists, see Daudin, I, part I, ch. I.

emy was finally disbanded in spite of Lavoisier and Lakanal's last-ditch efforts to keep the institution alive, a paralyzing feeling of defeat set in. Messier, who described the final lockout of August 17 at the Louvre, referred to the way "we stayed in the corridors for about an hour" bemoaning the final act.[35] Even the prospects of meeting as a special *société libre* or a club now seemed unrealistic.

This sense of paralysis was reinforced by an element of fear that spread through the group. Lakanal had been vilified more than once for his obstinate defense of the Academy, and he was even threatened with arrest for his position.[36] Arrests of suspicious persons began shortly after the last of the Academy's meetings, and academicians were not spared. Bailly was arrested on September 8, Sage on November 11, Lavoisier on November 28, Cousin on December 6.[37] In some of his private jottings, Lalande noted that he feared for Cousin's life.[38] In his letters to Delambre, who was busy in the provinces measuring the meridian, he commented that "the Academy does not think it prudent to meet, for the time being."[39] The most encouraging sign that he could give Delambre was that "we see each other at the Bureau de Consultation, whose meetings Lavoisier, Lagrange, Laplace, Borda, Coulomb, and Le Roy attend scrupulously, and where they ask for news about you."[40]

The ex-academicians in search of an organizational structure in which to assemble had several alternatives open to them. All fell within the category of the voluntary associations that Fourcroy and Romme had advanced as legitimate and partial

35. Paris, Observatoire, MS. C–2–19, p. 11.
36. Lalande, *Magasin Encyclopédique* (1795) tome V, 442–443.
37. On Sage, see Tuetey, I, 21 n. 1. On Cousin, see Poisson, pp. 109–111. Among other academicians later arrested were Vandermonde on 20 July (Birembaut, "Précisions," p. 532, and Clermont-Ferrand, Bibliothèque Municipale, MS. 339, no. 487 bis); Haüy on 12 or 13 Aug. (Geoffroy, pp. 11–13, and letter of Tessier to Jussieu dated 17 Aug. 1793 in Avignon, Museum Calvet, MS. Requien, no. 9638); Cassini on 13 Feb. 1794 (Devic, pp. 240ff.); Poissonnier on 23 Feb. (Sedillot, p. 151); L'Héritier in July (Gerbaux and Schmidt, III, 247); and Bougainville, sometime after 20 June (Martin-Allanic, p. 1522).
38. Paris, Bibliothèque Victor-Cousin, MS. 99, p. 38.
39. Letters dated 23 Nov. 1793, New York, Columbia University, David Eugene Smith collection, volume "Lalande"; and 29 Sept. 1793, London, Wellcome Historical Medical Library, Lalande folder.
40. Letter dated 11 Nov. 1793, New York, Columbia University, David Eugene Smith collection, volume "Lalande."

substitutes for the defunct Academy. One approach was to pursue and enlarge contacts with the Lycée des Arts, which was quickly becoming one of the most popular centers of semi-scientific activity. But its strong utilitarian bent and the excessive and grandiloquent patriotism displayed by its orators were not conducive to the spirit academicians most cherished in the pursuit of science. For some of the same reasons, the Société d'Histoire Naturelle was also eliminated as a substitute academy. It was objectionable also for its narrow interest, which would not satisfy the varied concerns of many former academicians. The only group that gave promise of taking up the Academy's mission was the more sedate Société Philomatique, which had catholic tastes and ran a modest scientific journal. Indeed, there was a distinct swelling of its ranks immediately after the Academy's extinction. Vauquelin had been a member since 1789, before he became an academician, and he was joined by Lavoisier, Berthollet, Vicq d'Azyr, and Fourcroy on September 14.[41] In the next two weeks, Le Roy, Lamarck, Monge, and Prony joined the ranks, followed on November 3 by Laplace and Darcet. In due time, Pelletier and Haüy also became philomaths, and it looked as if this private association would become the haven for the lost scientific community. Its historian Berthelot has indeed eulogized it as the last refuge of science during the Terror.[42]

Whatever their initial intentions in joining the Société Philomatique, most of the academicians failed to support their new association. Lamarck and Pelletier were the only ones who attended meetings regularly.[43] The others—Prony, Monge, Vicq d'Azyr—made solemn promises to participate, but their preoccupation with governmental business prevented them from coming to meetings. In October 1793, Silvestre, the secretary of the Société Philomatique, sadly recorded that at best seven or eight members showed up—and later that one sched-

41. Société Philomatique, *Bulletin des Sciences*, I, v–viii; BS, MS. box 123. The published membership lists in the society's *Rapports Généraux* are inaccurate.
42. Berthelot, p. ix; Duveen, *AS*, X, 341; Birembaut, *RHS*, XI, 269.
43. BS, MS. box 126; letter of Silvestre to Brongniart dated 6 Oct. 1793, BMHN, MS. 1989, no. 875.

uled meeting had to be called off for lack of attendance.[44] Even the grandiose plans to publish the memoirs of the ex-Academy, which would have been the best visual sign of a true continuation of the old scientific community, failed to materialize. While the project was under discussion, one of the philomathic founders, the mineralogist Brongniart, wrote:

> The idea of having the Bureau de Consultation print the memoirs . . . is superb. . . . The *Memoirs* of the Academy have ceased to be published. Since the Société Philomatique has a large number of its members, and since it is concerned with the same subject matter, it would be natural for us to continue this series. . . . Our society would become known and would obtain the eternal attachment of the members of the former Academy whom we have just welcomed into our fold.[45]

But these projects remained dreams, and the philomaths were unable to sustain even their own modest journal. The *Bulletin des Sciences* suffered from a shortage of original articles and, even with the skillful *remplissage* of extracts and translations from printed works, the Société was unable to meet its publication schedule.[46] The issues, which normally covered one month, began to spread over two or more months in May 1793 and lacked the quality of the earlier publications. It was only in 1795, when original contributions by Cuvier, Geoffroy Saint-Hilaire, and Vauquelin began to appear, that the *Bulletin* came into its own again.

The failure of the Société Philomatique to take on the mantle of the defunct Academy has significance beyond the mere fact of the disintegration of Old Regime patterns of scientific associations. The *société libre* concept had always been advanced by all republican-minded critics of the academic system as the most viable alternative to government-sponsored scientific associations. It was the solution that Fourcroy and

44. Letters of Silvestre to Brongniart dated 30 Oct. 1793 and 24 floréal II [13 May 1794], BMHN, MS. 1989, nos. 878, 887.
45. Letter of Brongniart to Silvestre dated 8 Nov. 1793, BMHN, MS. 1989, no. 880. The suggestion for this series, tentatively named "Annales Philomatiques," had been made by Millin, according to another letter of Brongniart to Silvestre dated 2 Nov. 1793, BMHN, MS. 1989, no. 879.
46. Letter of Brongniart to Silvestre dated 8 Nov. 1793, BMHN, MS. 1989, no. 880.

Romme banked upon as the most practical way of keeping science afloat in a sea of revolution. Hence the Société Philomatique's failure to perform that prescribed task became a critical test of the very essence of "democratic science." It was later to become an implicit reason for the re-establishment of older, autocratic patterns of scientific organization. It is therefore essential that we dwell further on the fate of this experiment with voluntary associations, despite the fragmentary evidence available.

The experiment did not fail for want of *sociétés libres* or for attempts to create workable new voluntary associations. In addition to the transparent move to pack the Société Philomatique with ex-academicians, there were several other proposals for the creation of new groups. On November 26, for example, the astronomer and former academician Jeaurat urged the formation of a voluntary association of "distinguished scientists," which, in its membership, was an obvious copy of the defunct Academy.[47] Despite the fact that Jeaurat had prudently inserted the patriotic qualifier "libre" in the title of his proposed association, it paralleled the old Academy too closely to move anyone in its favor. The emphasis upon funding by the government seemed to be its key feature and reflected the precarious state of Jeaurat's personal finances more than any genuine desire to work within the framework of democratic science.[48]

A more reasonable plan was offered by Jeaurat's former colleague, the botanist Jussieu, who prepared an elaborate ninety-eight-article set of regulations for a "Société Libre des Sciences, Belles-Lettres et Arts" which apparently never went beyond the planning stages.[49] Although it bears no date, this document may well have been the response to a questionnaire that had been drafted in a meeting of the Société d'Histoire Naturelle on August 30.[50] On that day, the naturalists decided to initiate discussions on "the type of organization and regime voluntary associations must adopt . . . in order to contribute most to the advancement of the arts and sciences." The group

47. AN, F17A1006, no. 1069.
48. *CIPCN*, II, 822–823.
49. BMHN, MS. 289, fols. 29–33.
50. BMHN, MS. 464, pp. 183–187.

also decided to seek the opinions of other voluntary associations in an attempt to formulate a common, unified program of democratic science. To define the problem further, it was agreed that a subcommittee of the Société d'Histoire Naturelle would draft a set of questions about the nature of the organization of science to serve as a guide for action. These questions, even more than Jussieu's specific answers, are of critical importance, because they were drafted by a subcommittee consisting of Romme, Fourcroy, and Vicq d'Azyr.[51] They reveal the very essence of these scientists' preoccupation with the form that democratic science should take.

Unlike the aristocratic Academy, the new *société libre* they envisaged had to remain an open society, with unlimited membership and meetings open to the public. To prevent the development of any kind of oligarchy within the group, officers had to be elected by the membership for limited and specified periods of time. Romme, Fourcroy, and Vicq d'Azyr were greatly concerned with attracting members and including all potentially useful scientists and artisans rather than following the patterns of exclusiveness so long adopted by the Academy. It was most crucial that open communication with all parts of society—the other *sociétés libres*, the teaching institutions, the political clubs, and even the government—be maintained. Indeed, it was this character of congeniality—"fraternization," as the revolutionaries would have said—that guided their conception of the new shape of scientific institutions.

Inevitably, the decision to establish the new organization on a broad and rather democratic base pushed the subcommittee into examining the relationship of the voluntary organization to public utility. In turn, the problem of financing the association was also raised. Only a few general principles were set down. It was assumed that the practical arts are inextricably linked to scientific progress, but that the majority of society grasps the value of science only through its by-products.[52] It followed logically that if the *société libre* chose to consecrate

51. "Extrait des Procès-verbaux des Séances de la Société d'Histoire Naturelle des 6 et 20 Septembre 1793. L'An Deuxième de la Rép Franc.," BMHN, MS. 298. The document is probably in Vicq d'Azyr's hand.
52. BMHN, MS. 299, section 2, probably written by Millin.

itself to science alone, without emphasizing the applications, it would have to be self-supporting. On the other hand, if funds were to be solicited, either by public lectures or from the government, the utility of the society's endeavor would constantly have to be demonstrated. Such were the courses actually being pursued by the few societies then operating. At one end of the spectrum the Société Philomatique was being supported by membership dues alone, and was in theory able to devote itself to the advancement of science without paying much attention to presentable by-products.[53] At the other end was the Lycée des Arts which, through its subordination of science to the practical arts, was beginning to seek and obtain public funding.[54]

Each voluntary association took up the questionnaire prepared by the subcommittee. The Société d'Histoire Naturelle itself discussed it on September 20 and, after agreement had been reached that neither the sciences nor the arts could be cultivated in isolation, one member moved that the word "science" should be abolished altogether and a new term expressing the intimate union of science with the arts be substituted.[55] Millin, one of the most active of the naturalists, offered the words *connaissances humaines* as the proper subject of interest to be pursued by voluntary associations.[56] He later presented the Société with his "Essay on a System of Arts, or Classification of Human Knowledge," composed while in jail, where he had been held on trumped-up charges.[57] That proposal itself provided fresh subject for debate.

Ironically, the sincere efforts of the subcommittee proved to have an effect contrary to their intentions. Instead of promoting scientific activity to carry on the spirit of the defunct Academy, the questionnaire provoked a number of contradictory proposals and initiated sterile debates. The Société d'Histoire Naturelle suffered most from these discussions. Its members had been traditionally inclined toward framing rules and regulations, and this last set of questions drew them even

53. *Notice sur l'Institution de la Société Philomatique*, article XVII.
54. Smeaton, *AS*, XI, 317.
55. BMHN, MS. 464, p. 185.
56. BMHN, MS. 299, part 5.
57. BMHN, MS. 464, p. 187, meeting of 18 Oct. 1793.

further away from the business of science itself.[58] By January 1794, the presiding officer, Cels, was sufficiently alarmed to point out to his fellow naturalists that the group had fallen into a state of "torpor."[59] As one of them later remarked, the cause was very clear: "Meetings were taken up with discussions of regulations rather than with science."[60] Like the Société Philomatique, but for different reasons, the Société d'Histoire Naturelle failed to fulfill the expectations of the architects of democratic science. Despite the best of intentions, the experiment with privately sponsored scientific societies proved to be unworkable.

The reasons behind this failure are, on the surface, perfectly obvious. The political events of the Revolution and the war crisis drove the scientific community into several camps, each unable to participate in the normal life of research. Those who were politically sound were drafted into governmental service, where their talents were fully tapped at the expense of their normal scientific pursuits. Some tried to introduce political considerations into the life of voluntary associations, turning them into debating societies. Others were arrested, frightened, or simply ostracized for their "insufficient hatred of royalty." For their own safety, many withdrew to the countryside, especially after the law of 17 April 1794, which forbade ex-nobles from residing in Paris.[61] The combined result was the creation of a serious scientific manpower shortage in Paris, which precluded the continuation of the scientific mission once carried out by the Academy. The problem of optimum organizations was thus emptied of content.

Material difficulties also compounded the difficulties of the waning scientific community, contributing to the paralysis that was gradually setting in. Inflation, which had greatly reduced the buying power of scientists on fixed salaries, also created insuperable problems for the printing industry. Paper was scarce and expensive, and printers were more eager to work for

58. Letter of Brongniart to Silvestre dated 7 frimaire II [27 Nov. 1793], BMHN, MS. 1989, no. 883.
59. BMHN, MS. 464, p. 197.
60. Lelièvre, "Rapport Fait à la Société d'Histoire Naturelle de Paris, sur les Moyens d'Activer et d'Utiliser ses Séances," dated 18 floréal VI [7 May 1798], BMHN, MS. 298, p. 2.
61. Aulard, *Recueil*, XI, 620, article 6.

the government and its agencies than to publish scientific journals. Almost all periodicals suffered from these difficulties. The *Journal des Savants* ceased to appear in 1792, to some extent because it lost its previously sound financial base.[62] The *Annales de Chimie* suspended operations in 1793, just at the time its editors were being drafted into government service. The *Journal d'Histoire Naturelle* stopped appearing in 1792 when its editors, Olivier and Brugières, left on a mission to the Near East.[63] Technically, the *Journal de Physique* never missed an issue during the Revolution, even when its editor, Delamétherie, was thrown in jail.[64] There were, however, serious delays in publication. The November 1793 issue carried this apologetic note, symptomatic of the troubles facing all scientific publications: "The delays that the issues of our journal are experiencing are a consequence of the pressing circumstances of public events, which have forced the requisitioning of all printers [by the State]. . . . At this time . . . most other similar publications have been interrupted."[65]

The most telling delay was in the Academy's own publications. As was customary, the *Mémoires* were about three years behind schedule in 1791.[66] Lavoisier's physiocratic friend Dupont de Nemours, who turned printer in 1791, was awarded the job of publishing the *Mémoires*, which he expected to carry out diligently.[67] Even though the volumes covering the years 1789 and 1790 were both promised before the end of 1792, neither materialized until much later. One of the embarrassing problems for the 1789 issue was that the first 176 pages, printed before royalty was abolished, carried the running head "Mémoires de l'Académie Royale des Sciences."[68] It was too expensive to reset the type and too dangerous to distribute the issue. Even though the rest of the issue was ready for publication by August 1793, it was withheld from the public until

62. Lalande, *Décade*, VI, 70.
63. Cuvier, *Recueil*, II, 65, and III, 369.
64. Lalande, *Bibliographie Astronomique*, pp. 641–642. On Delamétherie's jailing, see AN, F⁷4762, and Guillaume, II, 403 n. 7.
65. *Journal de Physique*, XLIII, 397.
66. AdS, Reg 13 Aug. 1792, p. 424.
67. Letter of Lavoisier to Dupont dated 23 Nov. 1792, JT, p. 88; Saricks, p. 207.
68. AdS, Lav 1643.

April 1796, a full six years behind schedule.[69] The volume for 1790, edited by Lalande, eventually appeared in 1797.

One might easily conclude that, given the personnel shortage, the financial crisis, and the unreliable means of communication, it was inevitable that the experiment with voluntary associations would come to naught. The times were simply not propitious for the pursuit of esoteric activities such as science. It might even be argued further that the abolition of academic science was of little consequence since the body scientific would have been paralyzed by the direct consequences of the Revolution whether or not the Academy existed. To adopt this line of argument would be to underrate the psychological impact the destruction of the Academy had on the entire community. It would mean discounting the shattering of that intangible but nonetheless real tradition the learned society had forged for over a century. For scientists, the Academy had been the symbol of their occupation, just as Louis XVI represented the nation for Frenchmen. His execution made little difference to the daily operation of government, and yet it left a deep scar on the national conscience. In the same way, the destruction of the Academy thoroughly undermined the scientific community's psyche.

The major impact of the destruction of the Academy was the demoralization of the savants who had struggled to carry on the normal life of science despite the turmoil of the times. Throughout the early years of the Revolution, academicians had been sustained in their effort by the promise of a reorganized learned society. When discussing their occupation, they had a tendency to dismiss the political events of the Revolution as temporary "distractions" that would eventually disappear. Until 1793, most of them adopted an optimistic view of the future of science, confident that their professional world would return to normal as soon as the nation recovered its political equilibrium.[70] They may not always have been in complete agreement about the precise form postrevolutionary scientific enterprise would take, but they shared the common assumption

69. Lalande, *Bibliographie Astronomique*, p. 626. Inflationary problems contributed to the delay (see AN, F^{17}1081 dossier 3 pièce 34).
70. Letters of Bosc to Smith, dated 30 Oct. 1791, and of L'Héritier to Smith dated 9 April 1792, London, Linnean Society, Smith papers, 2.120, 5.121.

that some sort of academy would occupy a central place in their lives. When this dream was shattered, so was their faith in their future as scientists.

The end of academic meetings, rituals, and collective decisions, spelled by the destruction of the institution, was more than the symbol of the disappearance of the sense of community among scientists. The Academy had been so intimately involved with shaping the image of the savant and providing it with a concrete manifestation that its dissolution provoked a serious psychological crisis. No wonder that the naturalists were seeking a substitute for the word "science" and that the term "academician" was abandoned in favor of "artiste."[71] Both were the expression on a semantic level of a much deeper search for a new identity.

For reasons that its authors could not foresee, two of the three parts of the Fourcroy-Romme program for democratic science proved ineffective in promoting the advancement of science that both truly supported. Instead of revivifying scientific endeavor, the combination of government consultation and voluntary associations had accentuated the crisis by encouraging a change in traditional patterns of scientfic activity. In one sense, this was the immediate consequence of the turmoil of the Revolution, and of the particular circumstances that led to the Academy's destruction. From a different perspective, it was also the product of the gradual politicization of scientific institutions, which had seriously begun when La Rochefoucauld first introduced his constitutional reform plans in the chambers of the learned society in November 1789. As might well have been expected in a setting of abrupt and sweeping political transformation, the pressures placed upon the Academy and its members from the outside mounted rapidly and completely submerged the needs of the scientific community. In terms of the dichotomy presented at the outset of this study, the Academy gradually relinquished its traditional obligations to science in order to satisfy the more pressing demands of society, thereby brutally upsetting the balance it had once achieved. Its breakup in 1793 and the transformation of the

71. Grégoire, I, 350.

academician into a public servant were the logical outcome of a process set in motion by political circumstances. In the process, the continued progress of science, which had relied on internal mechanisms once guaranteed by academic forms, was deeply threatened. It was unexpectedly rescued from destruction by the third aspect of the Fourcroy-Romme plan, around which the new life of science in France gradually took shape.

The emergence of scientific education as one of the principal components of the "second" scientific revolution in the early nineteenth century is a story that has yet to be told in detail. This second revolution was marked by the eclipse of the generalized learned society and the rise of more specialized institutions, and by the concurrent establishment of professional standards for individual scientific disciplines.[72] It was the crucial social transformation that ushered science into its more mature state and, like the first revolution in the seventeenth century, cut across national boundaries. In the German states, it was marked by the rise of the university and the establishment of research institutes, in England by the curricular reforms at Oxford and Cambridge, and in France by the establishment of the "Grandes Ecoles." Everywhere in Europe, the age of professionalized science cultivated in institutions of higher learning and perfected in specialized laboratories was replacing the age of academies that had dominated the scene since the middle of the seventeenth century.

The forces behind this transformation were international in character, affecting even outlying nations such as Russia or countries in the New World struggling to set up their own national traditions. Foremost among these forces was the sheer growth of the size of the community of scientists, which, by its bulk alone, required institutional differentiation. Within each science, the increased technicality of disciplinary problems and the experimental requirements peculiar to each subject made specialization almost inevitable. Moreover, the narrowing of the gap between science and its direct applications tended to reduce the usefulness of general science in the face of the specific demands of technology. The fully educated en-

72. Merz, I, part I, chs. I–III; Taton, "Les Conditions"; Gillispie, "Science and Technology," pp. 118–131; Mendelsohn, p. xvii.

gineer or doctor had to be provided with the most advanced form of specialized education, but could not at the same time be expected to have a deep grasp of the old generalized science, natural philosophy.

While acknowledging the international character of this scientific revolution, which is ultimately linked to the broad economic, social, and intellectual movements of the age, we must note that its first concrete manifestations appeared in France. Considering the general upheaval that the Revolution provoked, it is not surprising to find the germ of new organizational patterns taking shape in France. The opportunity to effect this fundamental tranformation was more clearly available in revolutionary Paris than in any other center of scientific activity. Yet it must also be remembered that sporadic signs of these new tendencies were already apparent before the Revolution and that, in a profound sense, France was ripe for the second scientific revolution earlier than any other nation. The reasons for this have been suggested before, but it is important to reiterate them, especially since this transformation had been nourished by the Academy, the very institution it was driving into obsolescence.

In no other country had science been as professionalized or as popular as in Old Regime France. The success of the Academy had itself assured science's growth, both in numbers of participants and in the demands for specialization and rigor made for individual disciplines. It was in Paris itself that the earliest efforts at creating specialized professional societies and specialized journals had been attempted, and it was there that many of the new modes of scientific and technical education had been first adopted. The emergence of scientific education as a new force had been clearly in gestation for several decades before the Revolution broke out.[73]

Two additional factors also aided in the promotion of scientific education. Both the political preference for the democratization of public instruction and the urgent need for technically trained cadres to manage the war effort made the movement timely and appropriate. Indeed, the kind of institutions that were developed at first strongly reflected the con-

73. Taton, *Centaurus*, III, 74–78; Léon.

scious effort to broaden the base of public participation in scientific and technical matters, while at the same time directing them to the immediate needs of society. The key to the success of scientific education was, in fact, its special ability to serve the new interests of science and society, just as the Academy had been able to do a century earlier. In the new system of higher education, these interests tended to reinforce rather than to oppose each other.

Fourcroy's plans of late 1793 had been drawn up in the wake of the rejection of Condorcet's highly centralized hierarchical system of public education.[74] For that entire year, the persistent anticorporate tendencies of the legislature forced a reiteration of the resentment against any type of government-sponsored forms of higher education. Even after Condorcet's educational proposals had been "decapitated" by the removal of the super-Academy as the crowning feature of the system, legislators continued to rail against the establishment of any publicly supported educational program that went beyond the elementary level. In July 1793, Daunou, who with Fourcroy was later to be the architect of the Institut, explicitly condemned the idea of government-initiated general education in science. Instead he proposed to leave the pursuit of science to free enterprise. He said:

Instead of these brilliant and dangerous institutions, I would ask for liberty and equality and the abolition of privileges. Give the arts and sciences freedom of circulation. . . . Let this type of instruction be like commerce, honored but not directed by the state. Then, under the auspices of liberty and the protection of the law, you will see secondary schools, institutes, courses, lycées, academies establishing themselves. . . . Then instruction will spread to all parts of the Republic according to the variety of circumstances and the needs of the citizenry.[75]

Fourcroy echoed his sentiments in December 1793, summing them up by asserting that "laissez-faire is the great secret and the only road to certain success."[76] It was clear from his speech that what he and others feared above all in government-

---

74. On this point, Fourcroy's views were not identical with those of Romme. See *CIPCN*, V, 627, and Galante-Garrone, pp. 393–400.
75. *CIPCN*, I, 601–602.
76. *CIPCN*, III, 98, 102.

sponsored higher education was the perpetuation of "a Republic of letters, a Republic of science, and a Republic of the
arts" that would undermine the social unity of the "Republic
of France." The same suspicion lay behind the repeated asser-
tions of republicans representing all shades of political convictions, who insisted that the government's responsibility in
general education was limited to moral instruction in republican
virtues and "that part of knowledge necessary for the exercise
of the rights and duties of a citizen."[77] General scientific education was represented as a luxury that ought to be left in the
private sphere.

Fourcroy's assumption that the private lecture course
would replace public higher education was not borne out by
events. Judging from announcements carried by newspapers,
the number of courses offered during the Terror was a very
small fraction of those advertised in any previous year. For the
calendar year of 1789, over fifty-three different scientific
courses were publicized in the *Journal de Paris*. The same newspaper carried only a dozen such courses during the year II
(1793–1794). Subjects such as chemistry and botany, extremely popular before the Revolution, dropped out of sight almost
completely. At the Lycée des Arts, for example, chemistry
was not taught during the Terror, and botany was represented
in only twelve lectures offered by Fourcroy on vegetable physiology and eleven by Ventenat on general botany.[78] In another
domain, Lalande was the only person offering private instruction in hydrography.[79] The meager number of courses is understandable in view of the serious dislocations the Revolution
had caused and the widespread concern with the war and problems of personal survival. Behind it all was the pervasive sense
of demoralization that presided over the scientific community.

While opposing the principle of generalized scientific education at the upper levels of public instruction, both Daunou
and Fourcroy also recognized the continuing national importance of specialized professional schools, especially those that
required technical training. In his July 1793 speech, Daunou

77. The quotation is taken from a speech of Bouquier in *CIPCN*, III, 574.
78. AN, F^{17}1143 dossier 3 pièce 27.
79. *Journal de Paris*, 14 floréal II [3 May 1794], p. 1972.

had specifically urged the establishment of a publicly supported group of specialized schools for the arts and certain professions. He singled out the healing, military, and mechanical arts as areas calling for institutions of specialized learning, in which scientific subjects would necessarily be an important part of the curriculum. Yet Daunou carefully avoided calling these schools "scientific." Science was to be considered as only an auxiliary to the arts in question and was taught solely to prepare the students for their chosen profession. This was the only way Daunou could reconcile the contradictory positions he took on the government's role toward scientific education.[80] On the one hand, he maintained that science pursued for its own sake could not justifiably be supported by the nation, while on the other he recognized that it was necessary for the welfare of the country.

The same dilemma is evident in the writings of most of the public figures who spoke on the subject of scientific education in December 1793.[81] Their basic concern was to fashion an educational structure that would prevent the continuation of monopolistic control of the scientific enterprise by an aristocracy of learning. Until the legislators had become accustomed to the absence of the academic system, their obsession was with blocking its re-emergence in any form. Thus, in 1793, the ideological rejection of corporate forms of organization clearly outweighed the utilitarian preoccupation with science. But, by early 1794, the practical demands of a nation at war were beginning to tilt the balance in the other direction. It was during that year that the Convention, both before and after Thermidor, voted for the establishment or reorganization of a remarkably large number of educational ventures of which science formed a major part. By the end of the year, over sixteen different kinds of schools had been called into existence.[82]

These schools fall into two major categories, each speak-

---

80. *CIPCN*, I, 603–604.
81. Bouquier and Thibaudeau pointed out this dilemma in their speeches, in *CIPCN*, III, 105–112 and 571–581. Romme did also, but consistently placed greater emphasis on the needs of society than of the citizenry (see *CIPCN*, II, 536).
82. *CIPCN*, IV, 164–168; Aulard, *Recueil*, Tables for vols. XVIII–XXVIII, p. 89.

ing to a different group in society and fulfilling different functions. One category, which usually bore the name of "écoles révolutionnaires," was operative for a short time and was concerned primarily with the rapid diffusion of relatively elementary knowledge. Another addressed itself to the education of professionals requiring modest but advanced scientific training. It is historically significant that while the second group generally had its institutional origins in the period before the Revolution, it was the last to emerge as a conspicuous force in scientific education. Those institutions responding most directly to the egalitarian and utilitarian needs of the day were the earliest to take shape.

As might be expected, the first of the *écoles révolutionnaires* was created by the Committee of Public Safety as part of the war effort. By a decree voted on 2 February 1794, over eight hundred citizens from all over the country were called to Paris to participate in a month-long course on the arts of extracting saltpeter, making gunpowder, and manufacturing cannon.[83] Each district was required to select two men between the ages of twenty-five and thirty to be brought to the capital and given means of sustenance at government expense so that they could follow lectures outlining the theory and practice of the manufacture of munitions and visit the new Parisian factories that the "congress" of scientists had organized. The Committee of Public Safety appointed nine of the most knowledgeable and eminent scientists then available to teach in this intensive Ecole Révolutionnaire des Armes et Poudres. They included Fourcroy for saltpeter instruction, Guyton and Berthollet for powder manufacturing, and Hassenfratz, Monge, and Périer for cannon making. During the month of March, the full set of courses was given on schedule with remarkable success and great publicity. Parades, festivals, and banquets were held in honor of the committee which originated the idea, the patriotic citizens who had rushed to Paris to help save the nation, and, above all, the teachers who had so efficiently turned their wisdom to popular use. The deputy Barère, in a speech to the Convention, was exuberant about the entire

83. *CIPCN*, IV, xxii–xxiv; Richard, ch. XIII.

project, perceiving in the new mode of public instruction the joint triumph of republicanism and science. He said:

Under the Old Regime, pedantry would have taken over this institution, and would have frustrated it by delays. It would have taken three years to obtain some rather doubtful results. Today, patriotism and science will direct education; it is the love of Liberty which will speed up instruction both for those providing it and those receiving it; in thirty days' time the courses will be ended and success will be achieved.[84]

The same "revolutionary" approach and fervor were responsible for the creation or the planning of several other institutions that received government sanction and funds. The idea of the Ecole Normale was launched by Barère in June immediately after the Ecole Révolutionnaire des Armes had completed its mission.[85] Two months later, the Committee set up the Ecole de Mars on the same principle, training hundreds of teenagers to become army officers in three-and-a-half-months' time.[86] Hassenfratz, who had already demonstrated his patriotism as an instructor at the Ecole des Armes, and the Burgundian doctor Chaussier were among its teachers. On 31 December 1794, the Convention also called into existence the Ecole de Navigation et de Canonnage Maritime, organized on the same "revolutionary" principles and presumably staffed by men of equally outstanding talent.[87] It is likely that similar revolutionary schools for cartography and agriculture were founded, following the patterns already established.

The most famous of all these revolutionary institutions was the Ecole Normale, originally conceived for the education of over 1,200 persons selected by local officials to come to Paris at government expense. They were brought to Paris to learn in short order the essential elements of a number of academic subjects which, following the monitorial pattern, they would

84. *AP*, LXXXV, 209 (18 Feb. 1794). I have used a variant of this speech, quoted in Richard, p. 474.
85. Dupuy, p. 39.
86. Chuquet, chs. I–II. Hassenfratz' and Chaussier's roles are discussed on pp. 127–146.
87. Dupuy, p. 37 n. 1, and *CIPCN*, IV, xxxiv–xxxv, for this and the following paragraphs.

in turn teach to their students once they returned to their places of origin. It was the founders' intent to create, almost overnight, the entire cadre of educators for the nation, by applying revolutionary techniques to the less concrete but equally pressing intellectual needs of the nation. Copying the practices that had proved so successful for the other *écoles révolutionnaires*, the legislators enlisted the services of the most learned and distinguished savants in the land, each of whom was expected to compress his wisdom into a carefully condensed set of two dozen lectures. When it finally opened its doors on 20 January 1794, the Ecole Normale could boast of an unprecedented roster of teacher-scientists who represented the peak of achievement in their disciplines. Lagrange and Laplace were in charge of lecturing on mathematics, Monge on descriptive geometry, Haüy on physics, Berthollet on chemistry, Daubenton on natural history, Thouin on agriculture, Buache and Mentelle on geography, and Vandermonde on political economy. All but one were ex-academicians.

The germ for all of these impressive but short-lived institutions had been planted in 1794, under the pressure of acute and immediate needs. It was highly significant for the future of the scientific community that the legislators found themselves forced to call on ex-academicians to guarantee the success of these schools. It reinforced the growing conviction already expressed by the Committee of Public Safety that scientists and their talents were vital to the survival and progress of the nation.[88] A similar sentiment presided over the creation of two other important educational institutions of a different character, established in 1794, the Ecole Centrale des Travaux Publiques and the Ecoles de Santé.

Both types of schools borrowed techniques that had been found useful by the *écoles révolutionnaires*, particularly in regard to the recruitment of students. Following the now standard republican practices, each district was asked to select candidates and to send them to Paris, where they would become wards of the state. Here the similarity ended, for the Ecole Centrale des Travaux Publiques and the Ecoles de Santé were

88. This fact was underlined by Fourcroy in his *Rapport*, and by Barère in his speeches (see Gershoy, pp. 222–226).

forced by the nature of their operations to divide their charges into classes, according to intellectual merit and level of achievement. Very early in their existence, they appointed external examiners who quite naturally paid more attention to intellectual than to republican virtues for their ratings. In this way, distinctions based upon intelligence or previous training were surreptitiously reintroduced into the educational system without the need for any ideological justification. Like scientific education itself, the principle of merit was a direct by-product of the needs of French society for a professionally trained elite. Introduced in this obvious utilitarian context, meritocracy avoided a painful clash with egalitarian values.

Another characteristic factor served to distinguish these schools from the new and ephemeral educational institutions described above. Because technical and professional training was already well developed in the Old Regime, and escaped destruction because of its independence from both the Church and the University, the 1794 efforts were in part guided by the continued presence of older institutions. Indeed, the original creation of the Ecole Centrale des Travaux Publiques was based upon a conscious effort to supplant the older Ecole du Génie Militaire and the Ecole des Ponts et Chaussées by a more rational organization centered in Paris and offering a common education to engineers of all types, civil as well as military.[89] The idea had been developing since the middle of 1793, in the minds of both the deputy Lecointe-Puyraveau and the new director of the school for civil engineers, Lamblardie. It was welcomed by a number of legislators, especially Carnot, Prieur, and Barère, who recognized the necessity of maintaining standards of excellence in both schools in order to graduate the competent engineers so much in demand. By the time the Ecole Centrale des Travaux Publiques was formally established by a decree of the Convention on 11 March 1794, its scope had been expanded to include the training of architectural, naval, and mining engineers, as well as the civil and military engineers for whom it was originally conceived. Its creation signaled the thorough overhauling of several older institutions, among others the Ecole des Ingénieurs de la Marine and the Ecole des

89. *CIPCN*, V, 627–653; Artz, pp. 151–166.

Mines, and eventually the redefinition of educational norms for the engineering profession. One of the Revolution's most significant scientific creations, the Ecole Polytechnique, emerged one year later from this structural reform. The architects of all these changes and the first teachers of engineers were for the most part ex-academicians: Monge, who was the guiding light behind its institution and taught descriptive geometry; Lagrange and Prony, who taught mathematics; and Fourcroy, Vauquelin, Berthollet, Pelletier, and Guyton, who shared the teaching of chemistry.

Similar circumstances presided over the creation of the Ecoles de Santé, voted into existence on 4 December 1794 following a report by Fourcroy.[90] The need for trained doctors was perhaps even more urgent than that for trained engineers. The wars had claimed over six hundred doctors in less than a year and a half following the closing of the Faculty of Medicine on 15 September 1793.[91] Serious shortages were often met by self-styled healers, who thrived in the atmosphere of total freedom from corporate restraint. Requests came into the governmental authorities' hands demanding protection from charlatans and quacks who had been quick to seize this opportunity. The urgent need for the re-establishment of professional standards coincided with a desire to restructure all old patterns of medical education, in which a strict separation between the training of doctors, surgeons, and pharmacists had been maintained. The new Ecoles de Santé provided a common education for all practitioners of the healing arts, eliminating the senseless duplications of the Old Regime and incorporating clinical training into their curriculum. It was one of the most significant and lasting of institutional reorganizations of the period, ushering in a brilliant era of medical education and progress in Paris.

One can imagine how important the creation of all these educational institutions was for a large number of disaffected academicians, who found in them a deep psychological and financial comfort. They gained some satisfaction from the fact that their new occupations as teachers corresponded well with their previous intellectual occupations. But instead of redis-

90. *CIPCN*, IV, xxxiii–xxxiv, and V, xl–xlii; Vess.
91. Foucault, pp. 69ff.; Ackercknecht, ch. IV.

covering their identity as scientists, these ex-academicians were inevitably driven to consider themselves as mathematicians, chemists, or anatomists. The fact that a year before the Institut was in operation educational structures had emerged as dependable institutions offering ex-academicians the kind of security they had lost played a crucial role in the psychological functions the re-established Academy was able to fulfill.

By that time the savant had already found a niche in a more specialized educational institution and tended to identify himself first with a school or a profession. In the recrystallization of the scientific community after Thermidor, the precedence of the specialist over the generalist was as much shaped by the utilitarian needs of society as by the chronology of the formation of new patterns of scientific organization.

Chapter 10

The Institut

*D*ESPITE the partial failure of the experiment with democratic science, all activities of a scientific nature had not
been extinguished in the days of the Terror. Months before a
new academic organization for learning was decreed in the
Constitution of 1795, scientific life in Paris was regaining its
equilibrium. As was seen in the previous chapter, the government had provided a number of ex-academicians with salaries
and a new sense of purpose by appointing them to schools of
higher education. It also sought to stimulate the sciences by
prodding the Muséum d'Histoire Naturelle into activity and
by creating a Bureau des Longitudes to administer the country's
observatories and to plan astronomical research for the nation.[1]
In the absence of an Academy and because its membership was
almost exclusively comprised of ex-academicians, the Bureau
des Longitudes quickly took on the character of a specialized
scientific society from which astronomical discoveries were announced and evaluated. Activity was also manifest at the Collège de France, which, like the Muséum late in 1793, had re-
frained from holding scheduled *séances publiques* for fear they
would be mistaken for outlawed academic sessions.[2] A year

1. *CIPCN*, IV, 398–409, and V, 274–281, for the Muséum; *CIPCN*, VI, 317–
327, Bigourdan, *Annuaire*, and Hahn, *Enseignement et Diffusion*, p. 658, for
the Bureau des Longitudes.
2. For the Muséum's decision not to hold public meetings, see letter of Desfontaines to the Comité d'Instruction Publique, dated 23 brumaire II [13
Nov. 1793], in AN, F^{17}3880 dossier an II. On the Collège, Lalande jotted
this comment in his entry for 11 Nov. 1793, in Paris, Bibliothèque Victor-
Cousin, MS. 99, p. 38: "Although the colleges are not in session, ours was
excepted [from the law banning them], but we dare not hold our 'rentrée
publique' for fear of jealousies and arrests."

later, at the Collège's public meeting of 21 November 1794, Lalande gave a well-publicized speech, in which he spoke optimistically of the signs of renewed scientific activity in public institutions throughout the nation.[3] Even the private, voluntary associations that had been unable to sustain serious scientific discussions in 1793 were showing signs of life. Geoffroy Saint-Hilaire and Cuvier, busily engaged in zoological research, joined both the Société d'Histoire Naturelle and the Société Philomatique in the winter of 1794–1795.[4] Their scientific papers were acclaimed by these societies, discussed, and quickly published in the *Bulletin des Sciences*, which was beginning to appear regularly again. At the Ecole Polytechnique, professors and the more scientifically inclined members of the administrative council engaged in protracted scientific discussions and carried out special experiments to fulfill their obligation to "perfect the sciences and arts which are taught there."[5] Their meetings were reminiscent of academic sessions, differing from prerevolutionary ones principally because now they were legitimized by science's relation to public education.[6]

Considering the variety and vigor of these scientific activities, it is fair to ask why the Institut, with a section replicating the defunct Academy of Sciences, was established at all. In retrospect, and despite the political turmoil, we can perceive in 1794 and 1795 the formation of a new pattern of organization able to satisfy all of the scientific community's professional needs. Two forces were combining to produce this pattern. The growth of scientific enterprise inevitably had fostered differentiation and specialization, and the concurrent increasing importance of science for society was also leading to the creation of educational institutions in which the latest scientific

3. *Feuille Critique et Littéraire*, 10 frimaire III [30 Nov. 1794], pp. 25–26; *Décade*, III, 397–402. The speech was reprinted in full as "Histoire de l'Astronomie en 1794," in *Magasin Encyclopédique*, (1795) tome I, 1–35.

4. Geoffroy was elected by the naturalists on 11 vendémiaire III [2 Oct. 1794] and by the philomaths on 23 nivôse III [12 Jan. 1795]. Cuvier joined these same societies respectively on 1 pluviôse III [20 Jan. 1795] and 3 germinal III [23 March 1795]. See BMHN, MS. 464, pp. 208, 220, and BS, MS. box 127.

5. The council, which later took on the name "Conseil de Perfectionnement," was established by a decree of 30 vendémiaire III [21 Oct. 1794]. See *Journal de l'Ecole Polytechnique*, cahier III, xiv, article XXI.

6. Fourcy, pp. 65–67; Raymond-Latour, II, 63–65. The product of some of the research and experimentation sponsored by the council was published in the *Journal de l'Ecole Polytechnique*.

innovations were made an essential part of the curriculum.[7] On the one hand, the "Grandes Ecoles" were meant to answer France's need for training professional cadres.[8] On the other hand, the requirements of advanced research were met by a variety of new centers for the specialized sciences. In astronomy, there now was the Bureau des Longitudes and its observatories; in the physical sciences, the Ecole Polytechnique and its council; in mining and metallurgy, the Ecole des Mines; in the natural sciences, the Muséum d'Histoire Naturelle and the Société d'Histoire Naturelle; and in medicine, the Ecoles de Santé and several medical and pharmacological societies. In addition, activities at two different types of organization continued to embrace all sciences. One, directed to popular or adult education, was carried out with great success at the Collège de France, the Lycée des Arts, and at a new Institut Libre des Sciences, Belles-Lettres et Arts.[9] The other type of activity was being pursued at the Société Philomatique, whose active membership was still predominantly composed of young scientists anxious to complete their education by mutual instruction and to keep abreast of current research by preparing abstracts published in the *Bulletin des Sciences*. Another similar group known as the Réunion des Sciences also began to meet under Lalande's direction in February 1795.[10] It seems clear that science had survived without an Academy, whose reestablishment, therefore, could well be regarded as superfluous for the pursuit of science.

The Institut was nonetheless founded (and without opposition) by a constitutional act ratified by the Convention on 22 August 1795. To the public, the founding of the Institut was presented as an act of justice—part of the restoration of pre-Jacobin normalcy, designed to carry on a glorious tradition of French Enlightenment leadership in cultural affairs. In fact, however, the action corresponded to a set of complex and his-

---

7. For guidance here on the sociological implication, I have followed Clark.
8. For details, see Artz, pp. 151–166, 230–253; Crosland, *Society of Arcueil*, chs. III–IV; Gilpin, ch. IV; and Vaughan.
9. Blanvillain, p. 137, claims this society was formed from the "remains of the Loge des Neuf Sœurs." Its history is given in AN, F^{17}3038.
10. *Journal de Paris*, 25 pluviôse III [13 Feb. 1795], p. 185; *L'Abréviateur Universel*, 9 ventôse III [27 Feb. 1795], pp. 635–636; *CIPCN*, VII, part I, p. 70.

torically vital forces operating in post-Thermidorean France. Three of them stand out in sharp relief, combining to gain the support needed for the Institut's establishment, not only from the legislature and governing officials, but significantly from the very scientific elite of France who no longer needed a national scientific society. These three intertwined movements were the reaction to the legend of Jacobin vandalism; the resurgence of a philosophic tradition of encyclopedism; and the conviction that cultural achievements require government sanction and a public platform on which leaders of politics and the mind are seen in close alliance. Of these three movements, only the third was to have lasting importance after the Napoleonic era, forging that characteristically French marriage of cultural and ruling elites so well symbolized recently by André Malraux's tenure as Minister of Culture. For our immediate concern with the organization of the scientific life of post-Thermidorean France, it suffices to note that the creation of the Institut came after the essential needs of the scientific community had been satisfied and that it corresponded to a set of urges common to all French intellectuals. Despite all appearances, the Institut was not basically a scientific institution comparable to the defunct Academy of Sciences, but, as its full title suggests, a "National Institute of Arts and Sciences," or more precisely the public embodiment of a national Republic of Letters.

More than any others, three members of the Convention were responsible for establishing the Institut: Boissy d'Anglas, a former lawyer of royalist tendencies personally interested in the fine arts and literature; Daunou, once a teacher of philosophy and now committed to *idéologiste* principles; and Fourcroy, the veteran of legislative battles who, despite the fickleness of his political stands, continued to command respect as an adviser on science, technology, and educational matters in general. Like so many post-Thermidorean legislators with cultural concerns, they shared a desire to work for France's future, hoping that the turmoil of the Terror would never return to disturb it. Despite everything, they were still convinced that progress could be furthered by legislation, establishing institutions that would provide an enduring foundation for the nation.

More than any other evidence, this surprisingly optimistic and constructive outlook attested to the durability of Enlightenment principles.

All three legislators also helped to create and propagate the myth of Jacobin vandalism out of sporadic acts inimical to learning perpetrated during the Revolution. Along with Grégoire, La Harpe, and dozens of other figures committed to intellectual pursuits, they were quick to fix the blame for intellectual disorganization squarely on "the tyrant Robespierre" and his regime. Without intending to defend Robespierre for his anti-intellectual tendencies or for his lack of compassion toward members of the Republic of Letters, I must remind the reader that these deficiencies were neither unique among Jacobins nor uniformly supported by Robespierre's political allies. As we saw earlier, attitudes toward education and toward the importance of institutions of higher learning did not necessarily match political affiliation. What the post-Thermidorean generation of liberal authors did was to further their own goals by making a fallen political regime responsible for the eclipse French cultural life had suffered during the Terror.

This is portrayed best by contrasting the positions taken by men of letters before and after Robespierre's fall. Prior to July 1794, men like Boissy and Grégoire argued for the importance of the Enlightenment in bringing about the Revolution, without distinguishing between the various political factions that had carried the movement forward.[11] In a forceful essay entitled "Some Ideas on the Arts and the Need to Encourage Them," dated 13 February 1794, Boissy had insisted that liberty and Enlightenment were indissolubly linked and that it was the national duty to support them both.[12] When he referred to vandalistic elements in society who "wish to stop their [the arts] progress and . . . plunge France into the depths of barbarous ignorance," he was speaking directly against members of the Committee of Public Instruction opposed to state-supported higher education. He did not allude to Robespierre or his associates as vandals.[13] On the contrary, in a speech before

11. Guillaume, II, 348–381; Kitchin, pp. 113–115.
12. *CIPCN*, III, 639–640.
13. As Vignery has shown (pp. 57–69), it would have been difficult to substantiate this with conclusive evidence. Perard, in a letter to Merlet-

the Convention supporting national festivals, Boissy pointedly
defended The Incorruptible as an "Orpheus teaching men the
first principles of civilization and morality."[14] He was referring
to Robespierre's famous May 7 speech on the cult of the Supreme Being, in which the Jacobin leader had in fact denounced
the party of the *encyclopédistes* for its support of enlightened
despotism and of materialistic philosophy, while at the same
time singing the praises of scientific progress and underlining
the positive role men of letters had played in laying the foundations for the Revolution.[15] Until Robespierre's arrest and death,
Boissy's position was typical of men of learning.

A marked change of attitudes followed shortly thereafter.
On 31 August 1794, Fourcroy lashed out against the fallen
leader, charging him with persecuting men of letters:

> Our last tyrant, who himself knew nothing, who displayed gross
> ignorance, who collected evidence against some of his colleagues
> partial to knowledge and science, and who would have sent them
> to the scaffolds . . . ripped apart, calumniated, and heaped vileness
> and bitterness upon all those possessing knowledge. . . . Robespierre
> looked upon learned men with suspicion, fury and envy . . . be-
> cause he knew they would not bow down to him.[16]

Fréron and Lanthenas supported Fourcroy's assessment, adding
that the tyrant had wanted to establish a reign of vandalism
to parallel the Terror.[17] The Convention thereupon heard
Grégoire recite a list of acts of destruction of books and monuments perpetrated since the beginning of the Revolution, which
he included in his report "On the Destruction Wrought by
Vandalism." Despite the fact that many of these disreputable
acts took place years before the Jacobins were in control,
Grégoire intimated that Robespierre was responsible. He completed his brief against the "tyrant" with another list, this time
of intellectuals who had recently been incarcerated. Grégoire

---

Laboulaye, dated 7 ventôse II [25 Feb. 1794] insisted that he "would like to
convince you that the Mountain is anxious to propagate science and the arts
as much as to cherish virtue": Angers, Bibliothèque Municipale, MS. 1658 II,
dossier J. Perard.
14. *CIPCN*, IV, 797.
15. Robespierre, X, 444, 454.
16. *Moniteur*, 16 fructidor II [2 Sept. 1794], pp. 1422–1423.
17. *Mercure*, 20 fructidor II [6 Sept. 1794], XI, 56–57.

asserted that "there was organized a system of persecution against men of talent."[18]

The same accusations were echoed more forcefully the following month by Fourcroy in his speech ostensibly designed to outline the steps taken by the legislature to train engineers. He chose the occasion to charge Robespierre with a plot to exterminate science, whose failure he was later to explain by describing the heroic "mobilization of scientists" carried out by the Committee of Public Safety.[19] The fact that Robespierre was the acknowledged leader of this Committee was conveniently overlooked. On September 25, Fourcroy claimed that:

While the conspirators were attempting to abolish enlightenment from France because they feared its influence, the Convention labored with all its might against these barbarians. . . . Your committees . . . have collected too much evidence to allow anyone to doubt the existence of a conspiracy against the progress of the human mind. They have been convinced that one of the conspirators' plans was to annihilate the arts and sciences . . . and to rise to power by marching through the debris of human knowledge, led by ignorance and superstition.[20]

The coupling of Robespierre's system of political oppression with the all-too-real difficulties faced by scientists and other intellectuals was now a standard formulation repeated everywhere. On October 8, Grégoire rhetorically asked the Convention: "In what century were talented men more atrociously persecuted than under the tyranny of Robespierre?"[21] The new *Bulletin de Littérature, des Sciences et des Arts* rhapsodized over the prospect that with "liberty restored to the sciences new chefs d'œuvre will be engendered and soon . . . the Republic of Letters in France will regain its splendor."[22]

The same theme with new variations was taken up by Lalande at the opening of the Collège de France on November

18. I have followed the critical edition in Guillaume, II, 381–421. The last quotation is on p. 402.
19. Discussed in Fourcroy, *Rapport*, read on 14 nivôse III [3 Jan. 1795].
20. *Moniteur*, 8 vendémiaire III [29 Sept. 1794], p. 37.
21. *CIPCN*, V, 103.
22. *Bulletin de Littérature*, prospectus dated 11 brumaire III [1 Nov. 1794]. For other similar comments see *L'Abréviateur Universel*, 16 frimaire III [6 Dec. 1794], p. 302, and *Magasin Encyclopédique*, (1795) tome VI, 353–354.

21, at a *séance publique* that was taken as a symbol of intellectual rebirth. But the most dramatic pronouncement on Jacobin vandalism was delivered by the journalist and critic La Harpe at the Lycée on December 31. His speech, entitled "On the War Declared by Our Latest Tyrants against Reason, Morality, Letters, and the Arts," depicted in grandiloquent terms the temple of the arts overwhelmed by barbarians, debasing the refined French language by bandying "gutter" expressions, thus advertising a ferocious hate of learning and beauty that outdid the Teutonic Vandals.[23] La Harpe's lecture was widely distributed and probably established the myth of Jacobin vandalism even more securely than Grégoire's scholarly and tedious speeches on that subject at the Convention.

The rapidity with which the myth took hold and the near-uniform support given it by men of learning is an arresting fact of importance for the history of the creation of the Institut that requires explanation. For a short time following the excesses of the Revolution, there was an uncommon display of solidarity among intellectuals. The Revolution's social and political transformations, many of which began before the Jacobins' rise to power, had shaken the Republic of Letters to its very foundations. The old system of pensions doled out by royal ministers had been swept away. As teaching in *collèges* and tutoring in wealthy families disappeared, men of learning fell back on lecturing at the Lycées or earning a living from the new journalistic ventures of the day. With the disappearance of salaried positions as royal censors and the closing of the University and the academies, other traditional means of support vanished. Some intellectuals willing to participate in politics drifted into government, either as representatives or as executors of revolutionary committees' will. Others, unemployed and disoriented, wandered aimlessly in Paris or left for the provinces, where their presence aroused suspicion. Many were

23. La Harpe, pp. 4–11; *Feuille Critique et Littéraire*, 20 nivôse III [9 Jan. 1795], pp. 223–225; *Journal de Paris*, 23–24 nivôse III [12–13 Jan. 1795], pp. 457, 461–462. By 1797, La Harpe repudiated his pro-*philosophe* position and turned violently against the *idéologues*, the Institut, and the *Décade*. For details see Moravia, pp. 431–437. In *CIPCN*, Guillaume, with his republican penchant, neglects La Harpe's role.

in fact arrested and wasted away in prisons, where they naturally turned to reflecting on their sad fate, believing themselves
to be victims of a plot. In all likelihood, however, they were
victims of circumstances that destroyed their role in society
and turned most of them into citizens of marginal utility.[24]

The common bond forged by imprisonment temporarily
sublimated the academic squabbles that had divided the intellectuals before the Revolution. A noble aristocrat like Cassini
easily shared the feeling of persecution with a Voltairean like
La Harpe or with scientific journalists closer to Rousseau's
philosophy like Delamétherie and Millin.[25] Despite their genuinely differing positions on politics and on the means by which
learning ought to be pursued, they developed a sense of solidarity and willingly adopted Robespierre as the scapegoat for their
misfortune. The poet Collin d'Harleville, in his bombastic
allegory read before the first public meeting of the Institut,
proudly referred to its members as a "grande Famille Réunie,"
cemented by a common "love of work and truth and a respect
for manners and humanity."[26] He was applauded at length
when he declaimed against the isolation of intellectuals from
each other.[27] The myth of persecution was well suited to supply
them with a rallying point.

Under these circumstances, it is not difficult to understand
why intellectuals were also prone to accept a unifying doctrine
they might have shunned in prerevolutionary times. The standard interpretation for the founding of the Institut has singled
out *idéologiste* views for that role, largely because its champion
in 1795, Daunou, provided that philosophy with a visible place
in the Institut's Second Class of Moral and Political Sciences,
at the same time as *idéologiste* principles were penetrating other

24. A complete study of the Republic of Letters and its reaction to political
events during the Revolution has yet to be carried out. There are excellent
suggestions for it in Guillaume, Stein, Scarfe, Moravia, Darnton, and Baker,
"Scientism," and "The Early History."
25. All were arrested and spent time in prison. See Guillaume, II, 402–405;
and Cassini, "Mes Annales." For Millin, see *Magasin Encyclopédique*, (1795)
tome III, 21–22, and AN, F^{7}4774^{46} dossier 2.
26. *Mémoires de l'Institut National des Sciences et Arts. Classe de Littérature
et Beaux Arts*, II, 24.
27. Tissier, I, 224–225. As Baker points out in *AS*, XX, 218–219, the same
remark had been made by Garat in Dec. 1791.

parts of post-Thermidorean cultural life.[28] Commentators have correctly emphasized that the Institut became the intellectual home of the *idéologues* from 1795 to 1803.[29] There was in fact a section of the Second Class for the "analysis of sensations and ideas" specifically designed to encourage their metaphysics. But a much more general attitude shared by the *idéologues* and other intellectuals united them first.

The Enlightenment's pervasive vision of the role assumed by men of learning in civilizing society was more attractive precisely at the time their existence as a group was threatened. Early in the Revolution, political factionalism had replaced the rational procedures intellectuals naturally favored.[30] They had shared the naive conviction that widespread rational discussion and enlightenment would lead to self-evident truths and thus to a more reasonable and harmonious society. Instead the course of the Revolution had unleashed social and political discord saturated with bitterness and violence. The kind of utopian vision symbolized by Condorcet and the Société in 1789, in which civility and concord prevailed, was submerged by the display of deep passions, political opportunism, and extremism. Man the political animal replaced man the thinker, and chaos ensued.

A return to classical intellectual values typical of the Enlightenment, and strongly perfused with a dedication to rebuilding a society torn by the Revolution, was thus in order. The intellectuals imagined that "philosophy would heal the wounds suffered by humanity," expecting that a return to the cultivation of the mind would provide a strong national force for the restoration of order and social harmony.[31] Along with these deep-seated psychological motives for adopting the spirit of unity behind encyclopedism, there were corresponding so-

28. Williams, *Isis*, XLIV, 312–318; Moravia, pp. 347–430; Kitchin, pp. 121–136.
29. Simon; Van Duzer, pp. 115–133; Stein, pp. 64–79.
30. See the letter of Bailly to Charles Bonnet, dated 13 Jan. 1792, in Geneva, Bibliothèque Publique et Universitaire, MS. Bonnet 41, fols. 243, 244.
31. The quotation is taken from Grégoire's speech of 17 vendémiaire III [8 Oct. 1794], cited in *CIPCN*, V, 104. Moravia, p. 432, considers this view, which he calls "the theme of a general reconciliation," an integral part of *idéologue* philosophy. This theme was in fact shared by a larger group of intellectuals.

cial and circumstantial reasons as well. Many members of the
Republic of Letters, particularly those with reformist tendencies, had met in groups prior to 1791. Some, of course, knew
each other at the various Paris academies and societies. Younger
members had gathered in the salons of Paris and Auteuil, at
musées and lycées, in Freemason lodges, as editors of periodicals, or even on government commissions. They were the heirs
of the tradition of free and productive intellectual interchange
fostered by Diderot and D'Alembert. Without forming a
homogeneous social group or a political party with well-defined
goals, they all tended to place faith in the power of the mind,
the written word, and rational discourse to improve mankind.[32]
The more active among them actually banded together in
the Société de 1789, led by Sieyès, Talleyrand, and Condorcet.[33] Because of their common social concerns, they shared
a commitment to rationality and order, which had its political
analogue in their attachment to the principles of liberty and
property. Politically, they were adherents of classical liberalism.

Condorcet became their most prominent spokesman,
through his work in the Société de 1789, as a writer for the
Journal d'Instruction Sociale, and as the architect of an ordered
system of public education. His death while fleeing from the
"tyrant" Robespierre and the subsequent discovery of his ode
to progress written in parlous circumstances turned him into
a legendary figure of great appeal to the Republic of Letters.[34]
Daunou, who on 2 April 1795 urged the Convention to distribute throughout France 3,000 copies of Condorcet's *Esquisse
d'un Tableau Historique des Progrès de l'Esprit Humain*, presented the author's work as a "classic . . . concerned with the
perfection of society as the most noble goal of the human
mind."[35] To underline the heroic virtues of the "ill-fated philosopher," Daunou reminded his listeners that Condorcet pas-
sionately defended the true principles of the Revolution despite

---

32. Kafker.
33. The Société de 1789 claimed that it was "neither a sect nor a party, but
a group of friends of mankind . . . of brokers for social truths": Challamel,
p. 393. See also Baker, *AS*, XX, 214–220.
34. Ginguené's characterization of Condorcet as "the illustrious martyr of
philosophy and liberty" in *Décade*, 20 fructidor XIII [7 Sept. 1804], is typical.
35. *CIPCN*, VI, 11.

Robespierre's violent attack on him and the hopelessness of his personal situation. In the new *Décade*, Ginguené rhapsodized both over the meaning of the *Esquisse* and the fate of its martyred hero.[36]

The fact that Condorcet, who did not subscribe in a doctrinaire way to *idéologiste* principles, became the progenitor of the Institut, rather than Condillac, links the institution directly to a more encompassing intellectual tradition than is generally recognized. Through Condorcet's vision was transmitted the idea of the unity of knowledge in a Republic of Letters proclaimed by Descartes and the *encyclopédistes* and assumed in the successive plans for reforming education by Mirabeau, Talleyrand, and Condorcet himself. This is why the poet Collin d'Harleville struck a responsive chord when he referred to the Institut as a "grande Famille Réunie." Daunou himself underlined its fraternal and encyclopedic nature before the Convention when he spoke of the way "the Institut will once again draw together and reconcile [*raccordera*] all the branches of learning, stamping it with [the mark of] unity."[37] By adopting the philosophic principle of unified knowledge, the founders intended to promote their social idea of concord.

In the same speech to the Convention, Daunou explicitly drew attention to Condorcet's once debated and discarded concept of a National Society, now preferring to give it the name "Institut" that Talleyrand had previously adopted.[38] Not long before, in "Some Ideas on the Arts," Boissy had in fact resuscitated Talleyrand's version of a Parisian National Institute set up as a "school of intellectuals" or a "temple to glorify those having trodden the path of genius" in which, by assembling "the elite of all learned men on earth, or the entire universe of learning, . . . it would be possible to arrive at the most perfected products of the human mind, the ultimate of knowledge."[39] Like Talleyrand's Institute, this "living encyclopedia" was

36. Kitchin, pp. 110–112, 139–142.
37. Speech of 27 vendémiaire IV [19 Oct. 1795], quoted in Aucoc, *Lois*, p. 5. The philosophic idea of unity is underlined by the *idéologiste* Lancelin, I, 309; by Simon, pp. 44–48; and more recently by Buisson.
38. Naming the institution caused problems because of the connotations of the words "académie," "lycée," "université," and even "société." See Boissy, p. 37, and *CIPCN*, VI, 483.
39. Boissy, pp. 38–40.

meant to provide the nation and humanity with a perfected and universal form of instruction that would support and reinforce the genius and superiority of the French nation and its Revolution. Boissy's hyperbolic message, set as it was in the context of a plea for national support of the arts and sciences, raised once more the prospect of a government-supported elitist Republic of Letters, defined by intellectual merit and destined to lead the system of national education. On the need for organized leadership, there now was unanimous agreement among all the intellectuals, and no outspoken dissenters in the Convention. Elitism, however, was no longer principally considered as a defense against charlatanism. Instead elitism was pictured as a proof and guarantee of French cultural achievements, much as it had been envisaged by Colbert in the days of the founding of the academic system.[40] Academic institutions were once again conceived as an important element for displaying the brilliance of a civilized nation. For this reason above all others, Frenchmen of every political stripe, from Napoleon to the present, have never stopped supporting the Institut.

All the key pronouncements about the Institut accent this note. It was Boissy who, on 23 June 1795, first introduced to the legislature the proposals for a new Constitution in which the creation of the Institut was inserted. After providing still another hyperbolic version of the myth of Jacobin vandalism, he offered the new cultural body as evidence for the nation's humanitarian concerns:

This institution must honor not France alone, but all of humanity, by the striking spectacle of its power and force. It must keep watch over all the treasures of the imagination and talent, of meditation and study which Paris possesses to the admiration of all Europe.[41]

The same principle was expressed less extravagantly in Daunou's draft version of the section in the Constitution devoted to the Institut:

To maintain the renown [*gloire*] that the French people has acquired among all nations through the progress of knowledge and

40. Heubaum, 24, refers to this attitude as France's "Kulturpolitik."
41. Speech of 5 messidor III [23 June 1795], reproduced in part in *CIPCN*, VI, 335. Guillaume points out that the phrases had been used previously by Boissy in *Quelques Idées sur les Arts*.

by discoveries useful to mankind, an Institut National is created, charged with gathering discoveries, improving the arts and sciences, and which takes the place of the disbanded learned societies.[42]

One detects in the last clause, which was removed from the final version of the Constitution, a sense of the legislature's guilt at having outlawed the academies in 1793 and neglected higher culture, which world opinion recorded as evidence of French vandalism.

It was Daunou who understood and expressed best the composite motives that led to the founding of the Institut. In his measured speech at the opening of his new creation on 6 April 1796, he recapitulated the reasons for its existence by saying:

Citizens, our most pressing need is to establish the internal peace of the Republic. If there is a usefulness worthy of the arts . . . it is to restore concord and fraternity to the state, to turn national attention toward the meditations of science, the chefs d'œuvre of genius, to replace the rivalry of parties with that of talents. . . . The time has come for philosophy and letters to show its jealousy of the glories that the triumphal armies of the Republic have forced upon a stunned Europe. You who cultivate the arts and sciences have no less glorious battles to win over prejudice borne of slavery and the ravings of anarchy. The statue of Liberty stands upon innumerable trophies; crown it with the laurels of talent![43]

A curious mixture of guilt, insecurity, and national pride inspired the creation of the Institut. It will no longer do to describe the act either as the restoration of the *status quo ante bellum* or as the mere triumph of *idéologiste* philosophy. For the new structure, the architects of the Institut obviously borrowed bricks from the outlawed academies and studied the blueprints of Talleyrand and Condorcet. But the Institut was an entirely new cultural construct, as distinctive for the post-Thermidorean era as the academies had been under the monarchy and the voluntary associations were during revolutionary days.

In working out the precise structure and specific functions of the Institut, there were differences of opinion between

42. *CIPCN*, VI, 946.
43. Taillandier, p. 108.

Boissy, Daunou, and Fourcroy. Boissy, partial as he was to arts
and letters, favored a teaching Institut in which cultural leaders
would provide models to be imitated by younger and lesser
members of the cultural establishment. He therefore introduced a plan reminiscent of Talleyrand's projected super-
Academy, centered in Paris and bringing under its aegis a large
group of schools and research institutions already in existence.[44]
More impressed with Condorcet's conception of a National
Society devoted to the advancement of knowledge rather than
its immediate dissemination, Daunou cut the ties that linked
teaching to the Institut, but specified that it would supervise
and act as administrator for a number of research and educational institutions.[45] Following Condorcet's lead, Daunou also
planned to separate science from the practical arts and to emphasize the latter by assigning 40 of 128 positions in the Institut
to a Class of Applied Sciences. Fourcroy, who had been a witness since 1793 to every battle in the Convention and its committees over science and education, held still different views,
which, in the end, prevailed.[46] Remembering the fears elicited
by Condorcet's idea of a powerful National Society ruling
over the entire educational system, he insisted that all functional
relationships between them be severed. He knew that institutions like the Muséum d'Histoire Naturelle and the Observatory would resent being placed under the Institut's command.
Moreover, Fourcroy had been the major architect of France's
new system of technical education and was not anxious for a
competing or overlapping Class of Applied Sciences. He therefore eliminated a special unit for those arts dependent upon
science, and incorporated into the First Class several new sections devoted to medicine, surgery, rural economy, and the
veterinary arts.[47] By reducing the number of classes from four
to three, Fourcroy also managed to make the First Class numerically more important than any previous architect of a
super-Academy had ever proposed. Of 144 members of the

44. Boissy, pp. 28–40, 45.
45. *CIPCN*, VI, 339–342.
46. *CIPCN*, VI, 576, 645, 796–799.
47. This arrangement also coincided with Fourcroy's views of the intimate
relation between science and the healing arts, expressed in the preface to his
Médecine Eclairée.

Institut, 60 were in the sciences; and with the addition of 6 geographers relegated for an unexplained reason to the Second Class of Moral and Political Sciences, the proportion of science-linked members came to almost one half.

These successive alterations of the new cultural construct were of basic significance in setting the tone of the Institut in its first years of operation. The overwhelming predominance of the sciences over any other single branch of culture provides an excellent explanation for the scientific community's acquiescence to the creation of an institution science did not vitally need. As intellectuals, the scientists agreed that a show of cultural solidarity was desirable for psychological and national reasons. As a community of scientists, they were all the more amenable to Fourcroy's specific plan for an Institut which provided them with an unprecedented position of primacy in French culture. Without being a scientific academy, the Institut easily won the support of the scientific community.

In founding the Institut, the architects of France's cultural politics adopted principles inherited from scientists' experience preceding the Revolution. A serious attempt was made to impress upon all aspects of learning the advantages of communality and specialization that had been followed by men of science for over a century. Both the Second and Third Classes, though formed from the remnants of the defunct nonscientific academies, were patterned in their sectional organization on a model established by the sciences.[48] No longer was the functional equality of all members practiced in the old Académie Française respected. There were now to be six specialists each in morals, grammar, ancient languages, poetry, and antiquities and monuments. The Institut was meant to be, as Daunou had explained, "the epitome of the learned world, the representative body of the Republic of Letters."[49]

While it would be exaggerated to assert that the scientist thus became the cultural hero of the post-Thermidorean Republic, his resurgence in the era preceding Napoleon was a mirror of significant social and political trends.[50] The primacy

48. Details on the sectional organization of the Institut are set out in Aucoc, *Lois*, pp. 5–11.
49. Aucoc, *Lois*, p. 5.
50. Williams, *Isis*, XLVII, 380–381; A. W. Brown.

of the scientist symbolized a reordered system of social values in the Directorate. Ideally, the new elites were to be selected, like intellectuals, on the basis of intelligence, merit, and demonstrated abilities. In this way, the new hierarchy in society was to be freed on one hand from the Old Regime's dependence on birth and wealth and on the other hand from the Jacobin preference for strict political egalitarianism, which disregarded native differences of mental faculties. Because France's social and political leadership was expected to be founded on a meritocracy, it was essential, as the *idéologue* Cabanis believed, that "the nation should honor its philosophers, men of letters, scientists, and artists."[51]

Moreover, the role of the intellectual was increasingly conceived as having important utilitarian as well as ornamental functions.[52] Once again, the scientist typified the preferred image of post-Thermidorean society. On a lofty scale, the cultivation of moral and political sciences was expected to provide the nation with a more rational and better organized society, just as the natural sciences had served material progress through enlightenment. In a more concrete way, the entire Institut took over for all parts of learning the consultative functions previously assigned to the Academy of Sciences. The Second and Third Class, no less than the First, were asked by the government to give advice on literary and artistic questions in a manner that would have been unprecedented for the non-scientific academies of old. Condorcet's idea that all members of the National Society become *fonctionnaires publics* was adopted without discussion.

The same tendency to emphasize the utilitarian role of the new Institut was reflected in its close association with the government from the very beginning. In a conscious symbolic effort to show national support for the Institut, all five members

---

51. Cabanis, III, 383. I am indebted to a seminar paper by Mr. Karl E. Walther for this quotation and for a clarification of the topic of elitism among the *idéologues*. For another, more sociological view, see Clifford-Vaughan, *British Journal of Sociology*. XI, 319–331.

52. The government's appreciation of men of learning in the service of the state is noted by the contemporary British traveler F. W. Blagdon, quoted in Crosland, *Science in France*, pp. 6, 10.

of the Directorate attended its first public meeting in full dress, accompanied by other government dignitaries.[53] Subsequent public meetings were always graced by public officials amid considerable pomp.[54] On the recommendation of the Directorate, both houses of the new legislature voted to assign a salary to membership in the Institut, in recognition of its public utility and national status.[55] Another formal link between government and learning was established in the requirement of an annual accounting by the Institut before the legislature.[56] In one memorable session held by the Council of 500 on the third anniversary of Louis XVI's death, a delegation from the Institut publicly "swore its hatred of royalty," thus sealing learning's union with the Republic.[57] There was developing the kind of close, almost symbiotic association between government and culture that had been part of Colbert's dream for the academic system. In that context, the election of General Bonaparte to the Institut in 1797, at a time when he had no scientific credentials, was not surprising and should be regarded merely as another manifestation of the conscious effort to effect a firm alliance between the governing and intellectual elites of France.[58]

Without being a glorified and expanded Academy of Sciences, the new Institut was planned in the light of principles

53. Debidour, II, 67, 92–93. The opening meeting was celebrated with a great show of pomp. It has been described in the *Annales de la République Française*, quoted in Maindron, *Académie*, pp. 170–172; Millin in the *Magasin Encyclopédique*, (1795) tome VI, 559–566; Garat in *Décade*, IX (1796) 148–155; and F. J. L. Meyer, II, 37–44. An engraving of this ceremony by Girardet is reproduced as figure 3.
54. F. J. L. Meyer, II, 45–49; Bugge, quoted in Crosland, *Science in France*, pp. 88–96. The exception to this pattern is noted by Reichardt, pp. 262–263, for the 10 Jan. 1803 meeting, which he characterized as the Institut's funeral. In general, these public meetings were well attended and given special notice in the press.
55. Aucoc, *Lois*, pp. 33–40.
56. Aucoc, *Lois*, p. 8, article 6 of the law of 3 brumaire IV [25 Oct. 1795].
57. The episode is particularly noted by Messier, in Paris, Observatoire, MS. C–2–19, p. 12r and v. For the speech on this occasion by Lacepède, in which he proclaimed the eternal alliance between the principles of science and liberty, see Maindron, *Académie*, p. 163. Similar phrases were repeated periodically by Laplace, Daunou, Bitaubé, and Jussieu when they made the Institut's annual report to the legislature.
58. Maindron, *Académie*, pp. 204–207; Lacour-Gayet, *Bonaparte*, p. 11.

and experiences of the scientific community before the Revolution at the same time as it reflected the political and social
concerns of a society just emerging from a trying period of
violence, intense factionalism, and cultural disorganization. The
dedication to reason arrived at by communal efforts and the
conscious attachment to government service were not new
concepts in 1795. But their institutional expression had never
been so boldly set out as in the creation of this elitist and encyclopedic French temple of knowledge.

The paradox of the Institut was its adoption of cultural
habits developed by scientists at the very time when the organization ceased being a vital force for the scientific community.
While it helps to explain the scientists' enthusiasm for and
participation in the new organization, it raises serious questions
about the institution's actual functions. The temptation, therefore, is to write off the First Class as a mere cultural ornament,
reminiscent of its parent, the Academy of Sciences, in structure
only. While the trend toward obsolescence may well have begun in 1796, outward appearances during the Directorate and
the Consular period do not betray a lack of either activity or
distinction. On the contrary, the Danish astronomer Bugge
expressed an accurate and typical view when he wrote in 1799
that the Institut was "the first learned body in Europe."[59] Just
as had been the case in the Academy before, the leading figures
of science sat in the First Class meetings, read, heard, and discussed scientific papers, published *Mémoires*, encouraged research by awarding prizes, and acted effectively as technical
consultants to the government. Election to the limited number
of positions in the Institut was still coveted by young scientists,
and their careers were greatly enhanced by membership in the
learned society where much of the politics of science was
carried on.[60]

These outward appearances are nonetheless deceptive. The
decline in the Institut's role as a center of scientific activity was
masked for many years by the *cumul*, a French practice that
allowed overlapping membership in the Institut and the new

59. Quoted in Crosland, *Science in France*, p. 79.
60. Crosland, *Society of Arcueil*, pp. 161–168.

centers of scientific activity. Scientists like Laplace, Haüy, Biot, Berthollet, Vauquelin, Chaptal, Dolomieu, Cuvier, and Jussieu belonged to the Institut and participated in its daily affairs, but carried on their creative work independently of the society, at the Observatory, the Ecole Polytechnique, the Muséum, or private laboratories. Initial discussions about the validity and importance of their work, which before the Revolution might have occurred in the Academy's committees or in plenary sessions, were now increasingly taking place at the Bureau des Longitudes, in the classes of the Ecole Polytechnique and in its publications, in the Muséum's laboratories, at the Société d'Histoire Naturelle or the Société Philomatique, or even in private *salons* like those held in Arcueil.[61] The same individual who argued in favor of his new ideas and experiments among specialists in these research centers would eventually carry his finished products for approval to the Institut's First Class. But the occasions when unrehearsed debates took place at the Institut were significantly less frequent.[62] The Academy's most vital function as the anvil on which new science was being forged had been displaced by the inevitable growth and specialization of scientific enterprise.

Even the judicative functions once exercised by the Academy had slipped away. The establishment of a patent system in 1791 in which inventions were recorded rather than evaluated undercut the learned society's dominant role in technology. A feeble effort to restore this role was made in 1796 when the Bureau de Consultation, which distributed rewards, was in theory incorporated into the Institut.[63] In practice, however, the Institut relinquished its primacy in technology to the Ministry of the Interior—often run by scientifically trained bureaucrats—and by 1801 to the new, government-sponsored Société

61. On the *cumul*, see *Propagateur*, 6 messidor VII [24 June 1799], p. 3. Documentation to support the last contention abounds in the manuscript proceedings of the institutions named and in Crosland, *Society of Arcueil*, pp. 169–231, especially pp. 181, 213.
62. One such occasion is described by Pictet, p. 113. Significantly, the debate between Lamarck and Laplace concerned meteorology, a subject straddling several disciplines for which there were no specialized scientific institutes.
63. *Institut*, Reg I, 62–63. Despite the official transfer of duties, the Institut was still relying on the Ministry of the Interior as late as 1801 (*Institut*, Reg II, 292–293).

d'Encouragement pour l'Industrie Nationale.[64] The printing privilege once enjoyed by the Academy had also been swept away by revolutionary reforms, and the multiplication of specialized scientific journals made the Institut's publications even less useful for the advancement of science. As time passed, the *Mémoires* became more a repository than a living journal.

Special circumstances at the Institut contributed to making the *Mémoires* less attractive for the scientific community. Traditional habits adopted without question from older days prohibited the learned society from publishing articles that had already appeared elsewhere, thereby reducing the available number of contributions.[65] To compound this difficulty, delays in publication were forced upon the First Class by the law of equality among the three Classes, which was observed for publications until 1798.[66] Linguists, historians, poets, and moralists rarely felt the urgency to break into print that scientists did. Yet the principle of encyclopedism and cultural unity that had guided the founding of the Institut failed to take such distinctions into account. For this reason, even the occasional original papers read at the First Class in 1796 had to wait over two years before they were released to the public.

A crisis in the Institut's activities in 1809 illustrates with devastating clarity the extent to which the Academy's vital functions had been sapped by the new circumstances. In May, when several meetings were canceled for lack of subject matter, a committee of five—Fourcroy, Laplace, Cuvier, Le Gendre, and Lacepède—was appointed to deliberate on "ways to activate the Class."[67] Their report of July 17 realistically isolated as causes for their plight the existence of specialized scientific societies that competed successfully for original material and the large number of periodicals appearing more frequently

64. The ministers included the agronomist François de Neufchâteau, Laplace, and Chaptal. On the Société d'Encouragement, see *Histoire de la Fondation*, and Tresse, *RHS*. See also the arguments in favor of creating a Bureau d'Encouragement des Sciences et des Lettres in the Ministry of the Interior, in Ginguené's memoir of 29 brumaire V [19 Nov. 1796], in BN, MS. n.a.f. 9192, fols. 23, 102–103.
65. *Règlements Intérieurs*, p. 167.
66. *Règlements Intérieurs*. pp. 165–166; *Institut, Reg* I, 451–453; *Mémoires Présentés*, I, i.
67. *Institut, Reg* IV, 213.

Figure 3—First public meeting of the Institut held in the Salle des Cariatides of the Louvre on 15 germinal IV. Engraving by Berthout from a drawing by Girardet.

Figure 4–Costume worn by members of the Institut during the Empire.
The black uniform was set off by olive-green embroidery on the lapel
and sleeves. Engraving by Charon from a drawing by Poisson, courtesy
of the Musée Carnavalet, Paris.

than the Institut's annual *Mémoires*. In a revealing passage of the report, the committee recognized the transformation their institution had experienced:

It would be futile [for us] in the nineteenth century to attempt to return to the practices of the end of the seventeenth century, and to try to recapture the habits of a body which was [then] almost alone in cultivating science in the nation.[68]

A vote was taken that allowed the Institut to republish important scientific papers, thereby "turning our collection into a *recueil classique et choisi* of all the finest ones."

The First Class's realistic decision epitomizes the transformation of the learned society from one that had successfully guided the ship of science for well over a century to one concerned with recording science's achievements and recognizing scientific merit after it had been proven elsewhere. Statistics show that the average age of academicians was continuing its steady climb, and individual biographies indicate that admission to the Institut crowned rather than shaped scientific careers.[69] As Saint-Simon was to observe later, "the Academy no longer Commands progress; it is satisfied to Record it."[70] These were telling indications that the function of the First Class in the scientific community was markedly different from that once assumed by the Academy.

While one of its previous roles declined in importance, the Institut's service to society was considerably increased. It was as if the institution found a surrogate activity to justify its existence. In a society that recognized the primary importance of science for national prestige, cultural self-respect, and economic prosperity, the role of the Institut as the respected spokesman of learning was taken seriously. An image of the march of science and learning was projected nationally and internationally by the Institut's annual progress reports, and especially by the lengthy accounts Bonaparte requested of each

68. *Institut, Reg* IV, 228, and AdS, dossier 17 July 1809.
69. The average age of academicians in the First Class was 53.7 years in 1795. By 1803, the figure had risen to 58.0. The average age at election of the fourteen new members elected in the 1795–1803 period was 51.3. Of these, only Cuvier and Bonaparte were under 30.
70. As quoted in Iggers, p. 167. The same point was made on 30 frimaire VI [20 Dec. 1797] in *Conservateur*, p. 882.

Class in 1802.[71] In the eyes of the nation and the government, the Institut was the only unified and legitimate voice through which the Republic of Science and Letters was organized to transmit its hopes, needs, and achievements. This role, however, was increasingly limited and shaped by the Institut's association with the government. The voice of science, being endorsed by the Institut's official status, inevitably echoed government concerns and policies. Hence, when the Institut spoke out on any issue, it was difficult to tell if it was representing learning or government. So close was the relationship that members of the First Class were unable to distinguish between service to science and to state.

The practice of pressing scientists and their major institution into national service dated from the Academy's foundation under Colbert. In that respect, the Directorate did not innovate, but the extent to which the First Class was asked to devote its time and energies to this task is arresting. In his maiden political speech on behalf of the Institut, Laplace underlined the alliance between learning and government when he remarked that "the great number of reports prepared by the three Classes, and for the most part requested by the government, prove . . . the usefulness of the rapprochement of power and enlightenment."[72] The volume of correspondence with government officials and the number of reports drafted by the First Class on matters of only peripheral connection with the advancement of science is staggering.[73] This kind of activity was an obvious way of filling the vacuum left by the reduced function of the First Class in the scientific enterprise. But a new dimension was added, borrowed from Condorcet's conception of intellectuals as civil servants and reinforced by the experience of the war years. By the Constitution of 1795 and through the policies of the Directorate, the academies had become nationalized and intimately drawn into the governmental apparatus. Science and its promotion, recognized as an instrument of national propaganda

71. Aucoc, *Lois,* pp. 62–63. Two such reports were in fact drawn up in 1807 by Delambre and Cuvier and published in 1808.
72. Laplace, p. 4.
73. Crosland, *Society of Arcueil,* pp. 153–154. A look at the index of *Institut, Reg* I and II, under "Ministre," corroborates this impression.

and prosperity, had become a political imperative for the new regime. Moreover, the scientific elite had proven to be a useful, loyal, and docile asset for national survival. The Republic at war had in fact learned that it needed science. Now it was ready to use it to further other national goals.

One of the most celebrated examples of science's service to France and, by virtue of the country's missionary zeal outside its borders, to humanity as a whole, was the standardization of weights and measures. In their reports to the legislature, spokesmen for the Institut constantly reminded the nation of the potential importance of this reform and of the risks devoted geodesists ran during the Terror to complete their task.[74] We know that early in the Revolution the scientific elite had exploited this project for the advancement of scientific goals far removed from national utility.[75] In 1798, it was the government's turn to capitalize on scientists' exploits to further its policies. Talleyrand, the Minister of Foreign Affairs, gladly acceded to the Institut's request to convoke an international congress in Paris to verify and approve the choice of standards. Its initial purpose undoubtedly was a genuine scientific *raison d'être*, and in fact the Congress modified French plans for defining the unit of weight.[76] Beyond this, the meeting was seized upon by the government as an opportunity to further its *Kulturpolitik*. Purportedly a French-sponsored international conclave for the benefit of mankind, it was in fact restricted to delegates from states under French rule or in its sphere of cultural dominance.[77] In spite of their scientific importance and size, Great Britain, Prussia, and Russia were not invited to send delegates, readily understandable in view of the state of war between France and the Allies. It was nevertheless implied that the conclave demonstrated France's altruistic and genuine concern for civilization as a whole. The propagandistic uses made

74. Bitaubé, pp. 4–5; Bigourdan, *Système Métrique*, pp. 160–166. References to the importance of this French gift to humanity are quoted in Bigourdan, *Système Métrique*, pp. 160, 169.
75. Biot, *Essai*, p. 36; Favre, ch. IV.
76. The Congress is discussed by Crosland, *Isis*, LX, 226–231, and the problems of standardization by Birembaut, *RHS*, XII, 49–50.
77. The identification of the idea of civilizing humanity with the pursuit of national goals is discussed by Curtius, ch. I.

of the event were in keeping with French foreign policy under the Directorate, and pointed toward the increasingly imperialistic tone that followed Bonaparte's *coup d'état* in 1799.

General Bonaparte inherited the practice of using science and its practitioners for national purposes from executive committees of the Convention.[78] During the Italian campaigns of 1796, three distinguished academicians, Berthollet, Monge, and Thouin, were assigned by the government to the army to select scientific and artistic objects worth sending back to Paris, which everyone assumed was to become the world's cultural capital. Two years later, after Bonaparte's election to the First Class, he took a large corps of learned men with his expeditionary force to Egypt.[79] It was his plan to establish an "Institut d'Egypte" which, like the Institut National, would at the same time enrich the world's storehouse of knowledge and soften the impact of France's military aims by forcing world attention onto its cultural merit. However one assesses his motives, Bonaparte was following a path marked out by the Directorate, but with characteristically more enthusiasm and efficiency than that body could muster. Under his regime, all national activities, including the cultural assets the Institut represented, were placed at the government's disposal.

Napoleon Bonaparte carried this new *Kulturpolitik* to its logical conclusion. Men of science and learning were enticed into government service by the prospect of continued government support of their profession and the bestowal of political honors.[80] Laplace, after he served briefly as Minister of Interior, was appointed to the Senate and later became one of its leading officers. Berthollet, Bougainville, Chaptal, Darcet, Daubenton, Fourcroy, Lacepède, Lagrange, and Monge, among others, were also given positions as senators. Close to half of the Institut was named to the new Legion of Honor shortly after its creation in 1802, and Lacepède was designated its principal officer.

78. Aubry, ch. X.
79. Aubry, ch. XII; Laissus, *Revue de Synthèse*.
80. As early as 1796, the government was reducing its support of private associations like the Lycée des Arts in favor of the Institut. On 15 nivôse IV [5 Jan. 1796], an official of the Ministry of the Interior wrote that "the government must devote its preferential treatment to public institutions founded by the Constitution and the legislative body": AN, F^{17}1143, pièce 68. I owe this reference to Professor L. Pearce Williams.

The Napoleonic era propelled France's scientific elite into the limelight of national prominence. Thenceforth, they were to constitute part of the new "aristocracy," the *notables*.[81]

As many have shown in detail, Napoleon patronized science and its devotees on a scale heretofore unknown.[82] He donated money to set up special prizes to be administered by the Institut; personally welcomed Volta, the Italian genius of electrical science, when he came to Paris to demonstrate his new device for manufacturing current electricity; favored the application of science to agriculture; and encouraged the creation of industrial fairs to promote technological innovation.[83] Through his generous efforts to support science, Napoleon also firmly settled the trend toward the subordination of the learned society to state ends. He patronized science, but demanded unflinching loyalty to the nation and to his own political views. The relatively high degree of personal and institutional independence, which had been one of the cherished traditions in the Old Regime, was significantly reduced as Napoleon's autocratic regime progressed.[84] He required cultural institutions to follow his leadership. The instances of his personal interference with the Institut's activities in the name of national interest illustrate the extent to which he conceived of the learned society as an organ of state. Lalande was ordered to cease preaching atheism when it embarrassed Napoleon's relations with the Church.[85] The Ecole Polytechnique was stripped of most of its research budget and experimental laboratories when Napoleon decided to militarize it, in the face of contrary advice from scientific associates.[86] When the Emperor wrote in 1807 that the Institut existed to carry out his wishes, and therefore was not entitled to oppose tasks assigned to it, he was merely affirming a state of affairs that had already been in operation for a number of years.[87]

The most notorious of the incursions of political consid-

81. Reinhard, pp. 20–26, 32–35; Tudesq, pp. 456–478.
82. Barral; Maras; Crosland, *Society of Arcueil*, ch. I.
83. Chaptal, pp. 278–293.
84. For a detailed and perceptive discussion of Napoleon's relations with intellectuals in general, see Moravia, ch. V.
85. Aulard, *Etudes*, ser. 4, pp. 313–315.
86. Raymond-Latour, II, 70–74; Crosland, *Society of Arcueil*, pp. 204–205.
87. Maras, p. 56; Crosland, *Society of Arcueil*, p. 42.

erations into the life of the Institut came in 1803 when, by
Napoleon's direct orders, a structural reorganization was decreed for it. It is now generally agreed that he demanded this
change to cover the suppression of the Second Class of Moral
and Political Sciences, where his *idéologue* critics were
lodged.[88] Chaptal, who prepared the new statutes for the Institut, seized the opportunity to insert a few important structural changes desired by the scientists, chief among which was
to provide the Classes with greater independence from each
other.[89] While temporarily beneficial to the daily operations of
the scientists in the Institut, these changes did nothing to reverse
the trend that had turned the learned society into a storehouse
of knowledge, a distributor of honors, the official voice of learning in the nation, and, above all, the direct servant of the governing elite. By supporting the Institut and reorganizing it by
decree from above, Napoleon completed the process of raising
the learned society to its position as the official representative
of French culture, at the same time accelerating the process of
making it subservient to the state.

For science as well as arts and letters, this bureaucratization
of learning was a convenient way of insuring loyalty and
placing the Republic of Letters at the beck and call of the state.
The new Institut embodied fully the changes that had been
brewing in the academic world since the last decades of the Old
Regime. Considered as the nerve center of science, the Academy now had passed through its full life cycle, from birth to
fossilization. At the same time, it had been transmuted from an
informal, private club of scientific enthusiasts with little renown
to the most honored and prestigious voice of science on the
Continent.

88. La Revellière-Lépeaux, II, 463–481; Simon, ch. XIV; Stein, ch. VII; Moravia, pp. 564–569.
89. Chaptal. pp. 59, 94. The text of the new regulations and the speech that
introduced it to the legislature are in Aucoc, *Lois*, pp. 67–77. For details on
the personnel involved in the transformation of the Institut, see Dehérain,
pp. 372–373.

Conclusion

OUR STORY of the Academy's transit through more than a century of history has illustrated the special position of scientific institutions in a world of change. On the one hand, in the Old Regime, the Academy acted as a passive mediator between the Republic of Science and the larger community, guaranteeing through its openness that the interests of all would be accommodated. On the other hand, the Academy assumed the active task of forging and conveying through several generations of scientists the values essential to their community, creating thereby a tradition for French science that has undergone few fundamental changes since the Revolution. Chief among these values was the coupling of doctrinal freedom on scientific questions with their rigorous evaluation by professional peers. Because these principles of action emerged with the development of science as an autonomous discipline throughout the Western world, they were shared by other communities outside France. Yet the form the Academy assumed to transmit them was defined by the nature of French culture and politics. The preference for elitist forms of governance and state sponsorship, though temporarily abandoned during the height of the Revolution, persisted and was in fact consolidated by the social atmosphere of Napoleonic times.

These special characteristics of the Academy offer substantial help in understanding the high quality of scientific activity in France during the eighteenth century, and its relative decline thereafter. Initially, great advances were made because no country was so well suited as France to aid in professionalizing an activity that had previously been in the hands of enthusiasts and amateurs. At first the need was for a general

scientific group, strongly supported by a paternalistic administration and accepted by a society organized in corporate units.
Serious tensions arose when the political crisis of the Revolution, which rejected the aristocratic organization of learning,
coincided with an expansion of science that demanded disci-
pline-oriented groupings. The reorganization of French science, dictated by special historical circumstances during the
Revolution, provided the nineteenth century with a peculiar
heritage.

On the one hand, new or refurbished specialized institutions captured the leading edge of research, while on the other
the general scientific society assumed the public relations of
science. Progress was carried on in one type of institutional
framework; its official voice was sounded by another body.
When the Imperial University was created in 1808, still another separate and autonomous unit of scientific activity
emerged.[1] Coordination was in practice assured by overlapping
membership in these bodies, but this practice tended to place
the interests of individuals over those of their disciplines. In
contrast to the organization of science in the Anglo-Saxon
world where the Associations for the Advancement of Science
function as federations of scientific professions, the Institut
continued to act as if its constituency had not significantly
changed since the eighteenth century. Conservatism in the Institut was thus encouraged not only by the age of its membership, but more critically by the assumptions about the life of
science its structure implied.

The conservative bent of the Institut was reinforced by its
alliance with governmental bureaucracy. In Napoleonic times,
the conscious blurring of lines separating administration and
culture insured the subservience of learning to national goals,
and promoted the bestowal of national honors through the Institut.[2] Its members were given dress uniforms on which to pin
their medals of merit, provided with a handsome building

1. Gilpin's characterization of the French scientific scene as one of "institutional proliferation and intellectual fragmentation" (p. 81) applies here.
2. Roddier, p. 54, sees this blurring as responsible for the decline of the vitality
of the academic system in the nineteenth century. My assessment also backs
the views of Vaughan, pp. 74–85, on the French educational system.

topped by a now-famous *coupole* on the Seine, and granted the rank of notables in public gatherings.[3] This recognition was in some measure a replica of the ceremony enjoyed by the academies under the Old Regime. But these new trappings served to mask the separation between the official symbol and the actual locus of scientific progress. Thus, while France's espousal of the concept of *culture générale* embodied in the whole Institut minimized the differences between scientific and literary cultures, it also permitted a serious, undetected rift to develop between the government of science and its citizenry.[4] Not until after the 1930's, when the Centre National de la Recherche Scientifique and the various ministerial councils and committees for scientific research were set up, did the scientific community possess an alternative to the conservative Institut for expressing its collective voice.[5]

Another serious consequence ensued from the legacy of the academic system. In contrast to the experience of other nations in the nineteenth century, science in France was rarely carried out as a "private enterprise." In the face of the dominance of publicly sponsored institutions, the impact of voluntary associations of scientists and research in private industries was inconsequential.[6] Under Napoleonic rule favorable to the scientific enterprise, this paternalism was a temporary asset. It was otherwise under less benevolent regimes, particularly when the debate about public instruction turned on its religious or secular character. Moreover, the peculiarly French tradition of the centralization and bureaucratization of science in public institutions kept the directors of the enterprise from responding quickly and adequately to the needs of an ever-growing set of scientific communities. What French science needed most for

3. The costume (shown as figure 4) dates from 1801; the move to the Collège Mazarin was decreed in 1805. See Maindron, *Académie,* pp. 71–76, 244. This consecration of merit has often been seen as a legitimate and useful function, even by critics. See Cap, II, 284; Sénard.
4. Curtius; Dufresne, p. 308; Guerlac, *Modern France,* pp. 81–82; and Sas each have noted, in different contexts, that the French word "savant" has a broader cultural meaning than its English counterpart "scientist."
5. Guerlac, *Modern France,* pp. 97–105; Gilpin, pp. 158–159, 188–198.
6. This is related to the characteristically French suspicion of small, private groupings, discussed by Rose. The Société d'Arcueil, an important though short-lived counterexample, is in my opinion atypical.

its continued vitality in the nineteenth century—support, freedom, autonomy, and interspecific competition—was largely denied because of its elitist posture and its bureaucratization.

The history of the Academy, followed through a series of opposing political regimes, also offers evidence of the close connection existing between ostensibly irrelevant cultural factors and the development of scientific institutions. We have seen the organization of scientific activity pass through characteristic phases closely related to their political contexts.[7] At the same time, it is evident that in the long run the advancement of science was less dependent upon any particular form of institutional organization than upon one that protected its autonomy. Most professional scientists of the eighteenth century were able to pursue their work despite political change, provided they received a modicum of financial support and intellectual independence. Institutions of science like the Academy became critical for professionals both as the official representatives of corporate interests and as buffers against societal pressures. Hence the title "membre de l'Institut" meant more than the formal recognition of intellectual merit. It was also taken as an official license to pursue a career of scientific research that did not require constant justification. In a symbolic fashion, the academies and the Institut legitimized the place of intellectuals in modern French society, as distinct from their utilitarian roles as teachers or applied scientists. If France has developed pride in fostering the life of the mind, it is in no small measure due to its support of the Institut and the public espousal of its principles.

The historical development discussed in this study may also be usefully reviewed by reference to a sociological model that has promising application for other attempts to understand the evolution of scientific institutions. For that reason, it is important to sketch out the model explicitly and identify the now familiar parts of our story with the elements of the theory.

In the Parsonian theory of action, the scientific institution may be considered as a social unit.[8] It possesses a set of activities, clearly defined functions, and a concrete organization. While

7. Hahn, *XII Congrès International.*
8. Parsons, pp. 337–338; Parsons and Shils; Smelser, chs. II, III, XV.

initially the role and institutional form of the unit are defined by a complex set of antecedent cultural traditions and historical circumstances, its very existence in a state of equilibrium for a period of time leads to the establishment of expectations, modes of behavior, and the adoption of values that are peculiar to the unit itself. These are strongly reinforced when the institution is provided with a measure of autonomy that includes self-perpetuation. As the parameters of the social system change, the unit seeks to maintain its original equilibrium by adjusting its internal dynamics. But when the tensions caused by changing conditions prove too strenuous, the institution inevitably disintegrates. Its activities and functions, however, may be taken over by specially designed or modified units that have a logical and sequential relationship to the defunct institution. In the instance of science, change normally follows the pattern observed in all growth activities, through a process of functional differentiation that accompanies the classical "division of labor."[9]

In our story, the unit is the Academy which, during the early part of the eighteenth century, evolved its own traditions in response to the dual needs of the scientific community and Old Regime society.[10] As demands from both of these social units increased and diverged, there were signs of attempts to adjust the Academy's practices without betraying its own value system. But when the conflicting requirements of science and society could not be resolved in the Academy, it was dissolved. In the period following its demise the functions once assumed by the Academy were taken over by newly emerging groups, that is, by voluntary associations, research centers, educational institutions, and government commissions. While there were short-term dislocations caused by the revolutionary circumstances, the process of social change of the units designed to foster scientific advancement proceeded along lines predictable from the Parsonian model. The age of general academies was giving way to the more complex and specialized age of higher education and research institutes.

9. Clark.
10. This period illustrates the "high degree of structural congruence" between the culture and the social unit required for stability. See Parsons, p. 46.

The peculiar condition in France was the successful attempt to restore a unit whose major scientific functions had been assumed by other bodies.[11] Overlapping membership in the younger institutions and the Institut for a time masked the inevitable fact that the latter's creative role was on the wane. By a process of elimination, its only significant functions became those of recording scientific achievements and serving as a symbol for science. For all the apparent vitality and the great pomp that has always surrounded it, the Institut, considered as an academy with a new name, became an obsolete institution. Its new-found function, confirmed by Napoleon's cultural policies, has been to serve the nation by making periodic reports summarizing the state of the art, and to record scientific achievement and merit through its elections, publications, and awards of prizes. As an arm of the state, the Institut remained for over a century the only organized unit that spoke on behalf of the scientific community and through which national policies were transmitted. Today it is a glorious relic of the past, more akin to a Hall of Fame than an Olympic stadium. Time and the very nature of the growth of science, which the Academy had so succesfully stimulated, were its undoing. Age, wisdom, and ceremony now prevail where once youth, creativity, and debate reigned supreme.

11. There is need for a detailed investigation of Frenchmen's strong commitment to old institutional forms, which far exceeds the normal conservative attachment to value systems. This penchant for the "classical" values of order and structure may well be the counterweight necessary to offset France's frequent and radical political changes.

APPENDIXES

Appendix I

A Bibliographical Note

THE HISTORY of the Academy was the subject of three important books late in the nineteenth century, each of which can still be read with profit today. The works of Maury, *Académie des Sciences* (1864), and Bertrand (1869) are the product of two well-informed and perceptive academicians, the first focusing on the annals of the society's achievements represented through its publications, and the second dealing with the life and habits of the institution itself. Useful as they are, both are seriously marred as historical literature by the absence of references so that a number of their claims cannot be substantiated with documents. The third book, Maindron, *L'Académie* (1888), is by contrast thoroughly reliable. Written by one of the Academy's finest archivists, it does, however, read more like a source book than a balanced or critical study. Later studies by Gauja, Plantefol, and the handsome catalogues of tricentenary exhibits add details without altering the composite assessment derived from Maury, Bertrand, Maindron, and the statutes of the Academy (published by Aucoc, *Lois*, in 1889).

Monographic study on the Academy has been most intensive for its origins and its demise. There are pioneering studies by Bigourdan (*Premières Sociétés*), King, H. Brown (*Scientific Organizations*), and George, which have been conveniently summarized in Taton's succinct *Origines*. Articles by Schiller and McKeon and the current publication of the Oldenburg *Correspondence* nevertheless demonstrate that the subject is far from closed. No attention, for example, has been paid to Hirschfield's suggestion about the connections between Jansenists and the founding of the Academy. Little help for assessing the Colbertian academic system as a whole can be expected from historians of culture who, despite the availability of printed sources for several academies (Académie Française, *Registres*; Lemonnier, *Procès-Verbaux*; Montaiglon), have made no comparably detailed studies of sister institutions. A new work on Colbert's attitude toward learning is needed to supersede Clément, V, xxx-c.

Work now in progress by Olmsted and Chapin on astronomy, Costabel on physics, and Laissus and Roger on the life sciences promises to shed more light on the deliberately obscure and secret proceedings of the Academy in the seventeenth century. A collaborative project to publish the first volumes of the Registres (AdS, Reg) announced by Taton in X *Congrès International*, pp. 283–286, has unfortunately been deferred for lack of funds, but it is likely to yield further information even if published in fragments. A major desideratum for this period is the preparation of an icono-bibliographical study of the Academy's early publications, including the accounts of meetings given in the *Journal des Savants*.

For the early eighteenth century, the most suggestive new approach has been Jacquot's use of contemporary scientific correspondence. In Paris, for example, a careful use of the papers of Jean-Paul Bignon (see Appendix II) would vastly enrich our limited knowledge. More attention should also be given to the archives on publishing and censorship discussed by Birn, Estivals, Furet, and Pottinger. Another potential source of enlightenment would be a study of the *Descriptions des Arts et Métiers*, and the *Machines et Inventions* published by the Academy. Proust (*Diderot et l'Encyclopédie*), Cole and Watts, and Birembaut ("Quelques Réflexions") provide excellent starting grounds. Recent archival exploration by Daumas, Doyon and Liaigre, Parker, and Payen suggest there exists a wealth of untapped sources for technological history.

For the remainder of the Academy's history, the richest new sources, which I plan to publish *in extenso*, are Lalande's annotated copy of the Academy's internal regulations (Lalande "Collection") and Lavoisier's treasury minute-book (JT). The former was originally called to my attention by Suzanne Delorme and the latter consulted when it was in the private collection of Denis I. Duveen, in New York. Both documents have repeatedly made evident the complexity of the organization and the intensity with which the life of the institution was debated within the membership. They have also served to establish the inadequacy of the records set down in the Academy's official proceedings (AdS, Reg), a point which has been verified almost every time they are compared with independent eyewitness accounts. Such accounts include: BN, MS. n.a.f. 9186; Croÿ, II, 54–55; Sanders, II, 96–99; Birembaut ("L'Académie Royale"); and J. E. Smith, I, 131–132.

Records of the Academy's sessions in AdS, Reg should be compared to the original drafts (*plumitifs*) made by the secretary (for example in AdS, Lav 2) and can often be supplemented by documents kept in AdS, dossiers des séances, and the occasional minutes kept by individuals, such as those of Duhamel du Monceau and Fougeroux de Bondaroy in AN, 127AP4 dossier 2; AdS, Lav 873; and BS, MS. Box 132. Attendance records were kept in AdS, feuilles de présence, to calculate the *jeton* money to which certain academicians were entitled. The Academy's financial records are also a rich source of information. These are to be found in bound registers in AdS; AdS, Lav 864–869 and 1034–1209; and in AN, F^{17} 1218 and 2499. A good deal of material useful for the internal history of the Academy should have turned up in the private papers of its officers (see Appendix II), though, with the exception of Condorcet's papers, they have not been used much in this book. It may well be that in 1793 some of the Academy's papers were taken for safekeeping by the caretaker Lucas or Condorcet's private secretary, Etienne Cardot, but not returned at the end of the Revolution. There is also evidence that specific items disappeared at the time Maindron and Grimaux were working in the society's archives.

Additional information about the Academy, its collections, library, and activities is scattered in Parisian depositories: BN, MS. n.a.f. 5133–5153; Bibliothèque de l'Arsenal, MS. 4624 and 7464, fols. 43–52; BI, MS. 1064–1065, 1385–1387, 1826–1827, 1962, 1986; and AN, AA 34 (1037), AA 63 (137), F^{17}1081 dossier 3, 1032 dossier 3, and 1336. There are a few unexploited documents on academic reform plans in AdS, Lav 86, 409, 932; AN, F^{17}1383 dossier 4; and at Cornell University. For the most part, holdings in AN related to the government ($O^1$611 and 666–688, and F^{17} 1094–1097) have been exploited by Guillaume in *CIPCN* or are disappointingly meager. Scattered and generally untouched information about technology is in AN, F^{12}, inventoried by Bonnassieux and Lelong and at the Conservatoire des Arts et Métiers.

The Academy's last years have been studied from a variety of viewpoints by Baker, Birembaut, Chapin, Fayet, McKie, Scheler, Smeaton, and Taton, as well as in the longer unpublished theses of Chapin and Billingsley. The modern writings of Guerlac, *Scientific Monthly*, LXXX, 93–101, and in *Critical Problems*, pp. 317–320; of Gillispie in *Critical Problems*, pp. 255–289, and *Behavioral Science*, IV, 67–73; and my own unpublished thesis are reviewed by Chapin, *French Historical Studies*, V, 371–374. They have stimulated this study by offering explicit and alternate explanations for the Academy's demise. My present views owe a great deal to them, but have also been markedly affected by three other factors. Studies on a variety of aspects of eighteenth-century science in France in *Enseignement et Diffusion* underscored the centrality of the Academy's role. Olivier-Martin's book provided the suggestion for considering the institution principally as another of the Old Regime's corporate structures. Finally, a consideration of the work of Baker, Bouillier, Grosclaude, and Manuel, supplemented by the documents in *CIPAL*, convinced me of the strategic and ideological importance of Condorcet's attempt to extend academic ideals and practices to all parts of learning, thereby institutionalizing its supremacy. This way of looking at the Academy was given additional support by finding a continuous line of disappointed scientific amateurs criticizing this conception long before it emerged distinctly as part of Condorcet's plan. In my analysis, Guerlac's preference for a Rousseauistic source of these attacks and Gillispie's emphasis on Diderot are subsumed under the larger heading of anticorporate sentiments. Such an interpretation is rendered all the more plausible when the antagonistic critics are set against the academic reformers and the conservative academicians. My study of the reform debates of 1789 in *RHS*, XVIII, 15–28, and especially of Cassini's "Mes Annales," which I also intend to publish, document the extent to which the Academy was thus faced with an internal as well as an external crisis in revolutionary times.

The least well explored aspect of academic competition and opposition is the history of other official institutions and voluntary associations. Among the former, the other academies and royal societies, the Collège Royal, and the Jardin du Roi are best known. Reference to printed sources on them is given in the footnotes. Certain important manuscript sources that have not been used require special mention. For the Royal Society of Medicine, there are dozens of uncatalogued boxes in the library of the Académie de Médecine. The same repository also houses better catalogued papers of the Academy of Surgery. These can be supplemented by BS, MS. box 128, and AN, F^{17}1318–1319. The Académie Royale de Marine has its proceedings deposited at the Service Historique de la Marine. For the Collège Royal, there are important archives still in the hands of the institution today, and the revealing diary of Lalande in Paris, Bibliothèque Victor-Cousin, MS. 99. Many of the Observatory's archives are now deposited in its library, accessible through the published catalogue by Bigourdan. Other items may be found in AN, F^{17}1064–1065, 1290 dossier 2, 1331B dossier 3, 1707 dossier 1, and F^{17A}1012 dossier 1; and BHVP, MS. 808, fol. 551.

For all these investigations, the most informative starting point is Tourneux, *Bibliographie*, III, 643–680. For activities during the Revolution, methodical use of the tables of Sigismond Lacroix; Gerbaux and Schmidt; *AP*; *CIPAL*; and *CIPN* should be undertaken first, to save time and effort. It may be helpful, nonetheless, to indicate briefly for each organization the sources I found most useful, even at the risk of duplicating some of their indications.

Musée and Lycée Républicain: BHVP, MS. 772, 808–809, 919–920; AN, F^{17}1331B.

Lycée des Arts: Paris, Ecole des Ponts et Chaussées, MS. 2072 dossier 9; AN, F^{17}1004B (580), 1004C (604), 1331B dossier 3, 1143 dossier 3.

Société Philomatique: BMHN, MS. 647; BS, MS. boxes 123–132; AN, F^{17}1309 dossier 5; correspondence between Riche, Silvestre, and Brongniart in BMHN, MS. 1967 and 1989.

Société Philotechnique: BS, MS. 2067–2069.

Société Libre d'Emulation: AN, T 160; F^{17}1148.

Société des Inventions et Découvertes: BN, MS. fr. 8045, fol. 9.

Société du Point Central des Arts et Métiers: Archives du Conservatoire des Arts et Métiers, MS. 10–391; AN, F^{12}995, F^{17}1014 (45) and 1097 dossier 3; London, Wellcome Historical Medical Library, Molard folder.

Bureau de Consultation des Arts et Métiers: Archives du Conservatoire des Arts et Métiers (uncatalogued papers); BHVP, MS. 798; AN, F^{17}1136–1138; BS, MS. box 128.

Société d'Economie Rurale: BMHN, MS. 464; BS, MS. boxes 123, 128, 133.

Société Linnéenne: Paris, Bibliothèque Mazarine, MS. 4441.

Société d'Histoire Naturelle: BMHN, MS. 298–299, 464; BS, MS. boxes 123, 128–129; AN, AA 54 (1509), F^{17}1003.

Société d'Agriculture: BS, MS. boxes 128, 132.

Société Médicale d'Emulation: BS, MS. box 128.

Institut (Société) Libre des Sciences, Belles-Lettres et Arts: BS, MS. box 133; AN, F^{17}3038.

Société en Faveur des Savants et Hommes de Lettres: AN, 27AP 16.

Ecole Polytechnique: AN, F^{17}1139; Bibliothèque de l'Ecole Polytechnique, "Registre Contenant les Déliberations du Conseil de l'Ecole Centrale des Travaux Publics" and "Procès-Verbaux des Séances du Conseil de l'Ecole Polytechnique."

The story of the fate of academicians during the Revolution is best pursued through biographical studies like those indicated in Appendix II for Bailly, Cassini IV, Condorcet, Fourcroy, Guyton, Lavoisier, and Monge. In addition to the manuscript sources indicated there, I have found a surprisingly large quantity of scattered unpublished material outside Paris. Chief among these and least well known are the letters exchanged between Parisians and foreigners. I have consulted the following libraries outside France with great profit:

Boston: Boston Public Library, Chamberlain Collection.

Brussels: Académie Royale de Belgique, Collection Stassart.

Geneva: Bibliothèque Publique et Universitaire, Cabinet des Manuscrits; Conservatoire de Botanique.

Haarlem: Hollandsche Maatschappij von Wetenschappen, Van Marum correspondence.

London: British Museum, Banks correspondence.

Linnean Society, J. E. Smith correspondence (catalogue by Dawson).

Royal Society, Blagden correspondence.

Wellcome Historical Medical Library, autograph collection.

New York: Columbia University, David Eugene Smith Collection.

Philadelphia: American Philosophical Society.

Pennsylvania Historical Society, Dreer and Gratz Collections.

In France, there is a seemingly endless number of autograph collections outside Paris, of which I have only consulted a fraction. The following municipal libraries' manuscript holdings proved useful: Amiens; Angers (MS. 615–632 and Grille papers); Avignon; Clermont-Ferrand (Collections Bertrand and De Chazelles); Laon; Lille; Mantes (Collection Clerc de Landresse); Nantes (Collection Labouchère); Orléans (MS. 1421–1423); Rouen (Collections De Blosseville, Girardin, and Duputel). There is obviously need for a general catalogue of scientific correspondence in the eighteenth century that would call attention to this rich but scattered mine of information.

The study of the science of the revolutionary period was supplemented by the study of the Lakanal papers in AN, AB XIX 333, in BHVP, MS. 815, and at Tulane University, New

Orleans; the Brongniart papers in BMHN; the Ginguené papers in BN, MS. n.a.f. 9192–9193; and the Daunou papers in BN, MS. n.a.f. 21880–21933. For printed materials, aside from the rich collections at BN and at AN, AD VIII, I fruitfully consulted BI, BHVP, BM (including the pamphlets catalogued by Fortescue), the A. D. White Collection at Cornell University, and the abundant but uncatalogued pamphlet collection at the University of California, Berkeley. By far the most important single source of printed materials was in *CIPAL* and *CIPCN*, masterfully edited by Guillaume, and recently indexed. Most of the newspapers listed were consulted at AN (ADXX^A), BN, BHVP, and Cornell and Harvard Universities.

For the post-Thermidorean period, the secondary literature is incomplete and unreliable. I found myself in disagreement with the shallow or distorted treatments of the beginnings of the Institut by Lakanal, Taillandier, and Marin. Instead of emphasizing the role of Lakanal, Daunou, or Grégoire in its creation, I followed some of the implications of Moravia's study of the *idéologues*, though I consider Boissy d'Anglas and Fourcroy to have been more instrumental than any other author suggests. For the government's role in cultural affairs, and particularly Bonaparte's conception, I have taken a position somewhere between Williams' anti-Napoleonic assessment (*Isis*, XLVII, 369–382) and Maras's and Crosland's too ready acceptance of his benevolence. It is clear that the subject deserves a lengthier and more balanced picture.

A final word must be said about the way in which my study is indebted to the work of philosophically oriented historians of science and to sociologists. Until recent times, secondary literature on science during the Revolution has been concerned with biography or framed in terms of classical problems of political or intellectual history: change and continuity, Voltaire vs. Rousseau, monarchy and republicanism, Girondins vs. Jacobins. Initially, as I read the sources through the eyes of Biot, Taine, Despois, Pouchet, and Guillaume, my preoccupations were the same. My professional interest in the history of science eventually made me realize the importance of the forgotten dimension in the story, the development of science as a profession. Only when this other perspective was considered seriously did the relationships of the political and cultural movements to science begin to have a direct bearing. To perceive this dimension of the story, the writings of Ben-David, Foucault, Gusdorf, and a group of Parsonians have been critically significant.

Appendix II

Bibliographical Data
on Academicians

*D*URING the period under study, 424 men were elected or appointed to the Academy, not counting its corresponding and nonresident members. If honorary and foreign members are also excluded, the number of "working" academicians is reduced to 315. This was the group for which bibliographical information was systematically sought. The search yielded data for 224 of them.

A serious effort was made to exclude articles or books that have been superseded by more recent titles. Special attention was paid to lesser-known figures for whom information is scarce, so that the length of an entry or even its presence is not necessarily a measure of the scientist's relative importance either in his discipline or in the Academy's affairs. A less intensive but equally fruitful search for manuscripts by or about the academicians was also instituted and its results have been noted at the end of each entry.

One type of biographical source has been excluded from our lists because of easy accessibility: the official *éloges*, prepared for reading at the public sessions and printed in the *Histoire* section of the society's publications. They constitute one of the most valued sources of information about academicians and can be conveniently consulted in the collected works of the perpetual secretaries of the Academy. A list of these *éloges* and their authors is given for each academician in the important Institut de France, *Index Biographique des Membres et Correspondants de l'Académie des Sciences du 22 Décembre 1666 au 15 Décembre 1967* (Paris, 1968). For those wishing further information, the biographical dossiers in AdS, the *DBF*, and the forthcoming volumes of the *Dictionary of Scientific Biography* should also be consulted.

A list of collected *éloges* is given below, followed by the bibliographical data for "working" academicians, ordered alphabetically. Following this series, there is a select list of biographies of honorary and nonresident academicians, with special attention to titles containing information about science and the Academy. These are also arranged alphabetically according to the academician.

A. *Collected Works*

Arago, Dominique-François, *Œuvres*. Paris, 1854–1862, I–III.

Condorcet, Marie-Jean-Antoine-Nicolas Caritat de, *Œuvres*. Paris, 1847–1849, II–III.

Cuvier, Jean-Léopold-Nicolas-Frédéric, *Recueil des Eloges Historiques*, new ed. Paris, 1861, 3 vols.

Fontenelle, Bernard le Bovier de, *Œuvres*. Paris, 1742, V–VI.

Fouchy, Jean-Paul Grandjean de, *Eloges des Académiciens de l'Académie Royale des Sciences Morts depuis l'An 1744*. Paris, 1761.

Louis, Antoine, *Eloges Lus dans les Séances Publiques de l'Académie Royale de Chirurgie de 1750 à 1792*. Paris, 1859.

Mairan, Jean-Jacques Dortous de, *Eloges des Académiciens de l'Académie Royale des Sciences Morts dans les Années 1741, 1742 et 1743*. Paris, 1747.

Pariset, Etienne, *Histoire des Membres de l'Académie Royale de Médecine*. Paris, 1850, 2 vols.

Vicq d'Azyr, Félix, *Œuvres*. Paris, an XIII [1805], I–III.

Information about the contents of these volumes and the location of other *éloges* is in Maindron, *Académie*, pp. 324–333.

B. *"Working" Academicians*

ADANSON, Michel (1727–1806)

Adanson: The Bicentennial of Michel Adanson's "Famille des Plantes." Pittsburgh, 1963–1964. Articles by J. P. Nicolas, F. A.

Stafleu, W. D. Margadant, G. Duprat, P. H. A. Sneath, T. Monod, and R. de Vilmorin.

Chevalier, Auguste, *Michel Adanson*. Paris, 1934.

Guedès, Michel, "La Méthode Taxonomique d'Adanson," *RHS*, XX (1967) 361–386.

Nicolas, Jean-Paul, *Adanson et les Encyclopédistes*. Paris, 1965.

Scheler, Lucien, "Antoine-Laurent Lavoisier et Michel Adanson, Rédacteurs de Programmes de Prix à l'Académie des Sciences," *RHS*, XIV (1961) 257–284.

MSS. in BMHN; BI, MS. 3185; and Pittsburgh, Pennsylvania, Hunt Botanical Library, described in *Adanson*, pp. 265–368.

ALBERT, Charles d' (1686–1751)

MSS. in Paris, Archives de la Marine, C⁷3, dossier Albert.

ALEMBERT, Jean le Rond d' (1717–1783)

Bertrand, Joseph L. F., *D'Alembert*. Paris, 1889.

Briggs, J. Morton, Jr., "D'Alembert: Philosophy and Mechanics in the Eighteenth Century," *University of Colorado Studies in History* (1964), no. 3, 38–56.

Butts, Robert E., "Rationalism in Modern Science: D'Alembert and the 'Esprit Simpliste,' " *Bucknell Review*, VIII (1959) 127–139.

Grimsley, Ronald, *Jean d'Alembert, 1717–1783*. Oxford, 1963.

Hankins, Thomas L., *Jean d'Alembert, Science and the Enlightenment*. Oxford, 1970.

Laissus, Yves, "Une Lettre Inédite de D'Alembert," *RHS*, VII (1954) 1–5.

Ley, Hermann, "Sur l'Importance de D'Alembert," *Pensée* (1952), no. 44, 49–57, (1953), no. 46, 39–50.

Mayer, Jean, "D'Alembert et l'Académie des Sciences," *Literature and Science*. Oxford, 1955, pp. 202–205.

Muller, Maurice, *Essai sur la Philosophie de Jean d'Alembert*. Paris, 1926.

Pappas, John N., "Rousseau and D'Alembert," *Publications of the Modern Language Association of America*, LXXV (1960) 46–60.

——, *Voltaire and D'Alembert*. Blooomington, 1962.

Sagnac, Philippe, "Les Conflits de la Science et de la Religion au XVIIIè Siècle: D'Alembert et Buffon," *La Révolution Française*, LXXVIII (1925) 5–15.

Vollgraff, Johann Adrian, "Christiaan Huygens et Jean le Rond d'Alembert," *Janus*, XX (1915) 269–313.

MSS. in BI, MS. 1786–1793, 2466–2474.

ANGIVILLER, Charles-Claude de Flahaut de La Billarderie, comte d' (1730–1809)

Mémoires de Charles Claude Flahaut Comte de La Billarderie d'Angiviller; Notes sur les Mémoires de Marmontel, ed. Louis Bobé. Copenhagen, 1933.

Silvestre de Sacy, Jacques, *Le Comte d'Angiviller, Dernier Directeur Général des Batiments du Roi*. Paris, 1953.

ANVILLE, Jean-Baptiste Bourguignon d' (1697–1782)

Barbié du Boccage, Jean-Denis, *Notice des Ouvrages de M. d'Anville*. Paris, 1802.

Du Bus, Charles, "Les Collections D'Anville à la Bibliothèque Nationale," Comité des Travaux Historiques et Scientifiques, *Bulletin de la Section de Géographie*, XLI (1926) 93–145.

Mascart, Jean, "La Correspondance de D'Anville," *Ciel et Terre*, XXXIII (1912) 8–13.

ARCY, Patrick d' (1725–1779)

MSS. in AN, $O^1$592, 126–128.

AUZOUT, Adrien (1622–1691)

McKeon, Robert, *Etablissement de l'Astronomie de Précision et Œuvre d'Adrien Auzout*. Mimeographed thesis, Ecole Pratique des Hautes Etudes, Paris, 1965.

——, "Le Récit d'Auzout au Sujet des Expériences sur le Vide," *XI Congrès International*, III, 355–363.

BAILLY, Jean-Sylvain (1736–1793)

Brucker, Gene A., *Jean-Sylvain Bailly, Revolutionary Mayor of Paris*. Urbana, 1950.

Hahn, Roger, "Quelques Nouveaux Documents sur Jean-Sylvain Bailly," *RHS*, VIII (1955) 338–353.

Smith, Edwin Burrows, "Jean-Sylvain Bailly—Astronomer, Mystic, Revolutionary—1736–1793," *Transactions APS*, XLIV (1954) 427–538.

BARTHEZ, Paul-Joseph (1734–1806)

Barthez, Antoine-Charles-Ernest, *Sur le Vitalisme de Barthez*. Paris, 1864.

Baumes, Jean-Baptiste-Théodore, *Eloge de Paul-Joseph Barthez*. Montpellier, 1807.

Euzière, Jules, "Les Maîtres de l'Ecole de Médecine de Montpellier et la Philosophie," *Monspeliensis Hippocrates*, VII (1964) 13–22.

Lordat, Jacques, *Exposition de la Doctrine Médicale de P. J. Barthez*. Paris, 1818.

Pagès, Paul, "Défense et Illustration du Vitalisme Montpelliérain," *Monspeliensis Hippocrates*, V (1962) 19–23.

BAUMÉ, Antoine (1728–1804)
Cadet de Gassicourt, Charles L., *Eloge de Baumé*. Brussels, an XIV [1805].
Davy, René, *L'Apothicaire Antoine Baumé (1728–1804)*. Cahors, 1955.
Dorveaux, Paul, "Apothicaires Membres de l'Académie Royale des Sciences. XII. Antoine Baumé," *Revue d'Histoire de la Pharmacie*, XXIV (1936) 345–353.

BAYEN, Pierre (1725–1798)
Blaessinger, Edmond, *Quelques Grandes Figures de la Pharmacie Militaire*. Paris, 1948, pp. 49–69.
Cap, Paul-Antoine, "Notice Biographique de Pierre Bayen," *Gazette Médicale de Paris*, ser. 3, XX (1865) 1–13.
Dorveaux, Paul, "Une Lettre de Bayen Pendant la Terreur," *Revue d'Histoire de la Pharmacie*, XXVI (1938) 240–244.

BÉLIDOR, Bernard Forest de (1693–1761)
Du Bois, Lucien, *Un Précurseur: Bélidor*. Lausanne, 1931.
MSS. in Paris and Vincennes, Archives de la Guerre.

BERTHOLLET, Claude-Louis (1748–1822)
Crosland, Maurice, *The Society of Arcueil*. London, 1967, p. 480.
Daumas, Maurice, *Les Savants d'Arcueil et la Science du XIXè Siècle*. Paris, 1954.
Jomard, Edme-François, *Notice sur la Vie et les Ouvrages de C. L. Berthollet*. Annecy, 1844.
Partington, James Riddick, *A History of Chemistry*. London, 1962, III, 496–516.
Rocchietta, Sergio, and Domenico Cavanna, "Claudio Berthollet (1748–1822), Gloria Italiana e Francese," *Collection des Travaux de l'Académie Internationale d'Histoire des Sciences*, XII (1962) 109–122.
Revue Savoisienne (1948) 114–144.
Articles by Tapponier and Aubry.

BERTIN, Exupère-Joseph (1712–1781)
Le Nécrologe des Hommes Célèbres de France, XVII (1782) 137–151.

BÉZOUT, Etienne (1730–1783)
Duveen, Denis I., and Roger Hahn, "Laplace's Succession to Bézout's Post of 'Examinateur des Elèves de l'Artillerie,' " *Isis*, XLVIII (1957) 416–427.

Vinot, Joseph, *Bézout. Sa Vie et ses Œuvres.* Nemours, 1883.

BIOT, Jean-Baptiste (1774–1862)

Crosland, Maurice, *The Society of Arcueil.* London, 1967, p. 481.

Lefort, Francisque, *Notice sur la Vie et les Travaux de Biot.* Paris, 1867.

Martius, Carl F. P. von, *Zum Gedächtniss an Jean-Baptiste Biot.* Munich, 1862.

Picard, Emile, *La Vie et l'Œuvre de J. B. Biot.* Paris, 1927.

MSS. in BI, MS. 4895–4896.

BLONDEL, Nicolas-François (1618–1686)

Mauclaire, Placide, and C. Vigoureux, *Nicolas-François de Blondel. Ingénieur et Architecte du Roi, 1618–1686.* Laon, 1938.

BONAPARTE, Napoléon (1769–1821)

Barral, Georges, *Histoire des Sciences sous Napoléon Bonaparte.* Paris, 1889.

Crosland, Maurice, *The Society of Arcueil. A View of French Science at the Time of Napoleon I.* London, 1967.

Lacour-Gayet, Georges, *Bonaparte, Membre de l'Institut.* Paris, 1921.

Maindron, Ernest, *L'Académie des Sciences.* Paris, 1888, pp. 201–310.

Maras, Raymond J., "Napoleon: Patron of Science," *The Historian*, XXI (1958) 46–62.

BORDA, Jean-Charles de (1733–1799)

Mascart, Jean, *La Vie et les Travaux du Chevalier Jean-Charles de Borda.* Lyons, 1919.

Société de Borda. *Bicentenaire de la Naissance du Chevalier Jean-Charles de Borda.* Dax, 1933.

BORDENAVE, Toussaint (1728–1782)

MSS. in BI, MS. 4508–4511.

BORELLY, alias BOREL, Jacques (?–1689)

Chabbert, Pierre, "Pierre Borel (1620?–1671)," *RHS*, XXI (1968) 303–343.

BORY, Gabriel (1720–1801)

Chapin, Seymour, "Les Associés Libres de l'Académie Royale des Sciences. Un Projet Inédit pour la Modification de leurs Statuts (1788)," *RHS*, XVIII (1965) 7–13.

MSS. in BN, MS. fr. 6349–6350.

BOSSUT, Charles (1730–1814)

Doublet, Edouard, "L'Abbé Bossut," *Bulletin des Sciences Mathématiques*, ser. 2, XXXVIII (1914) part I, 93–96, 121–125, 158–159, 186–190, 220–224.

Hahn, Roger, "The Chair of Hydrodynamics in Paris, 1775–1791: A Creation of Turgot," *X Congrès International*, pp. 751–754.

BOUGAINVILLE, Louis-Antoine (1729–1811)

Martin-Allanic, Jean-Etienne, *Bougainville, Navigateur et les Découvertes de son Temps*. Paris, 1964.

BOUGUER, Pierre (1698–1758)

Lamontagne, Roland, *La Vie et l'Œuvre de Pierre Bouguer*. Paris, 1964.

Maheu, Gilles, "Bibliographie de Pierre Bouguer (1698–1758); Lettres de Bouguer à Euler," *RHS*, XIX (1966) 193–224.

Morère, Jean-Edouard, "La Photométrie: Les Sources de *L'Essai d'Optique sur la Gradation de la Lumière* de Pierre Bouguer, 1729," *RHS*, XVIII (1965) 337–384.

BOULDUC, Gilles-François (1675–1742)

Dorveaux, Paul, "Apothicaires Membres de l'Académie Royale des Sciences. IV. Gilles François Boulduc," *Revue d'Histoire de la Pharmacie*, XIX (1931) 113–117.

———, "Les Boulduc, Apothicaires de la Princesse Palatine," *Revue d'Histoire de la Pharmacie*, XXI (1933) 110–111.

BOULDUC, Simon (1652–1729)

Dorveaux, Paul, "Apothicaires Membres de l'Académie Royale des Sciences. III. Simon Boulduc," *Revue d'Histoire de la Pharmacie*, XVIII (1930), 5–15.

BOURDELIN, Claude I (1621–1699)

Dorveaux, Paul, "Les Grands Pharmaciens Apothicaires Membres de l'Académie Royale des Sciences. I. Claude Bourdelin," *Bulletin de la Société d'Histoire de la Pharmacie*, XVII (1929) 289–297.

MSS. in BN, MS. n.a.f. 5133–5149.

BOUVARD, Alexis (1767–1843)

Forni, Jules. *Bouvard*. Chambéry, 1888.

Gautier, Alfred, and Adolphe Quételet, "Notice sur Alexis Bouvard," *Annuaire de l'Académie Royale des Sciences et Belles-Lettres de Bruxelles*, X (1844) 108–132.

Philippe, Charles, "Notice sur l'Astronome Bouvard," *Revue Savoisienne*, XXXIV (1893) 152–168, 214–225, 285–301.

BOUVARD, Michel-Philippe (1717–1787)

Delaunay, Paul, *Le Monde Médical Parisien au Dix-Huitième Siècle*. Paris, 1906, pp. 431–440.

Guénet, Antoine-Jean-Baptiste-Maclou, *Eloge Historique de Michel Philippe Bouvart*. Paris, 1787.

BRISSON, Mathurin-Jacques (1723–1806)

Birembaut, Arthur, "Les Liens de Famille entre Réaumur et Brisson, son Dernier Elève," *RHS*, XI (1958) 167–169.

Merland, Constant, in *Biographies Vendéennes*, II, 1–47. Nantes, 1883.

See BI, MS. 2041, fol. 90.

BROUSSONET, Pierre-Marie-Auguste (1761–1807)

Candolle, Augustin Pyramus de, *Eloge Historique de M. Auguste Broussonnet*. Montpellier, 1809.

Thiébaut de Berneaud, Arsenne, *Eloge de Broussonnet*. Paris, 1824.

MSS. in BI, 3186.

BUACHE, Philippe (1700–1773)

Doublet, Edouard, "Une Famille d'Astronomes et de Géographes," *Revue de Géographie Commerciale de Bordeaux*, ser. 3, XVIII (1934) 1–42.

Drapeyron, Ludovic, "Les Deux Buache," *Revue de Géographie*, XXI (1897) 6–16.

Vuacheux, Ferdinand, "Un Mémoire et une Lettre du Géographe Philippe Buache (1763)," *Bulletin de Géographie Historique et Descriptive* (1901) 269–274.

MSS. in Paris, Archives de la Marine, 2 JJ 17–19; BI, MS. 2302–2326.

BUACHE de la NEUVILLE, Jean-Nicolas (1741–1825)

For Doublet and Drapeyron, see BUACHE, Philippe.

MSS. in BI, MS. 2302–2326.

BUCQUET, Jean-Baptiste-Marie (1746–1780)

McDonald, Eric, "The Collaboration of Bucquet and Lavoisier," *Ambix*, XIII (1966) 74–83.

———, *Jean-Baptiste-Michel Bucquet (1746–1780)*. University of London, M.Sc. thesis, 1965.

BUFFON, Georges-Louis Leclerc (1707–1788)

Bertin, Léon, et al., *Buffon*. Paris, 1953.

Falls, William F., *Buffon et l'Agrandissement du Jardin du Roi*. Paris, 1933.

François, Yves, "Notes pour l'Histoire du Jardin des Plantes, sur Quelques Projets d'Aménagement du Jardin du Roi au Temps de Buffon," *Bulletin du Muséum National d'Histoire Naturelle*, XXII (1950) 675–681.

Hanks, Lesley, *Buffon avant l'Histoire Naturelle*. Paris, 1966.

Piveteau, Jean, ed., *Œuvres Philosophiques de Buffon*. Paris, 1954.

Roger, Jacques, "Buffon, les Epoques de la Nature," *Mémoires du Muséum National d'Histoire Naturelle*, n.s., ser. C, X (1962).

Sagnac, see ALEMBERT.

Starobinski, Jean, "Rousseau et Buffon," *Gesnerus*, XXI (1964) 83–94.

Weil, Françoise, "La Correspondance Buffon-Cramer," *RHS*, XIV (1961) 97–136.

MSS. listed in Piveteau, ed., *Œuvres*, pp. 513–570.

BURLET, Claude (1664–1731)

Hazon, Jacques-Albert, *Notice des Hommes les Plus Célèbres de la Faculté de Médecine*. Paris, 1788, p. 175.

CADET de GASSICOURT, Louis-Claude (1731–1799)

Berman, Alex, "The Cadet Circle: Representatives of an Era of French Pharmacy," *Bulletin of the History of Medicine*, XL (1966) 101–111.

Dorveaux, Paul, "Apothicaires Membres de l'Académie Royale des Sciences. X. Louis-Claude Cadet, dit Cadet de Gassicourt," *Revue d'Histoire de la Pharmacie*, XXII (1934) 385–397, XXIII (1935) 1–13.

Toraude, Léon G., *Etude Anecdotique, Historique et Critique sur les Cadet (1695–1900)*. Paris, 1902.

CAMUS, Charles-Etienne-Louis (1699–1768)

L'Huillier, Jean-Baptiste-Théophile, "Essai Biographique sur le Mathématicien Camus," *Almanach Historique de Seine-et-Marne pour 1863*. Meaux, 1863.

CARCAVI, Pierre de (1603–1684)

Henry, Charles, "Pierre Carcavy, Intermédiaire de Fermat, de Pascal et de Huygens, Bibliothécaire de Colbert et du Roi, Directeur de l'Académie des Sciences," *Bullettino di Bibliografia e di Storia delle Scienze Matematiche e Fisiche*, XVII (1884) 317–391.

CARNOT, Lazare-Nicolas-Marguerite (1753–1823)

Anthouard, R., "Aperçus sur la Recherche Scientifique en France sous la Révolution et l'Empire," *Thalès*, IV (1937–1939) 186–198.

Reinhard, Marcel, *Le Grand Carnot*. Paris, 1950–1952, 2 vols.

CASSINI, Giovanni Domenico (Cassini I) (1625–1712)

Aiton, Eric John, "The Vortex Theory of Planetary Motions," *AS*, XIII (1957) 249–264, XIV (1958) 132–147.

Cassini, Jean-Dominique (Cassini IV), *Mémoires pour Servir à l'Histoire des Sciences*. Paris, 1810, pp. 255–324.

Derenzini, Tullio, "Alcune Lettere di Giovanni Alfonso Borelli a Gian Domenico Cassini," *Physis*, II (1960) 235–241.

Gallois, Lucien L. J., "L'Académie des Sciences et les Origines de la Carte de Cassini," *Annales de Géographie*, XVIII (1909) 193–204, 289–310.

Laissus, Joseph, "A Propos de Jean Dominique Cassini," *90è Congrès des Sociétés Savantes*, III, 9–16.

Wailly, Jacques de, "Comment Devinrent Picards des Gentils-hommes Italiens et d'un Trompe l'Oeil et d'une Lettre sur lesquels s'Acheva la Glorieuse Carrière de la Maison Cassini," *Société d'Emulation Historique et Littéraire d'Abbeville* (1964) 399–441.

CASSINI, Jean-Dominique (Cassini IV) (1748–1845)

Devic, Jean-François Schlisteur, *Histoire de la Vie et des Travaux Scientifiques et Littéraires de J. D. Cassini IV*. Clermont, 1851.

Wailly, see CASSINI I.

MSS. in Paris, Observatoire; Clermont (Oise), Bibliothèque Municipale.

CASSINI de THURY, César-François (Cassini III) (1714–1784)

Berthaut, Henri M. A., *La Carte de France 1750–1898. Essai Historique*. Paris, 1898, I, 1–65.

Bigourdan, Guillaume, *L'Astronomie à Béziers. La Querelle Cassini-Lalande*. Paris, 1927.

Drapeyron, Ludovic, "Enquête à Instituer sur l'Exécution de la Grande Carte Topographique de France de Cassini de Thury," *Revue de Géographie*, XXXVIII (1896) 1–16.

———, "La Vie et les Travaux Géographiques de Cassini de Thury Auteur de la Première Carte Topographique de la France," *Revue de Géographie*, XXXIX (1896) 241–251.

CELS, Jacques-Philippe-Martin (1740–1806)

Silvestre, Augustin-François, *Discours Prononcé à Montrouge*. Paris, 1806.

MSS. in BI, MS. 3184.

CHABERT, Joseph-Bernard (1724–1805)

Lamontagne, Robert, *Chabert de Cogolin et l'Expédition de Louisbourg*. Montreal, 1964.

MSS. in Paris, Archives de la Marine, 2JJ 20–48; AN, $F^{17}1081$ (2).

CHAPPE D'AUTEROCHE, Jean-Baptiste (1722–1769)

Armitage, Angus, "Chappe d'Auteroche: A Pathfinder for Astronomy," *AS*, X (1954) 277–293.

Chaptal, Jean-Antoine (1756–1832)
Pigeire, Jean, *La Vie et l'Œuvre de Chaptal (1756–1832)*. Paris, 1932.
Tresse, René, "J. A. Chaptal et l'Enseignement Technique de 1800 à 1819," *RHS*, X (1957) 167–174.
"Lettres de J. A. Chaptal à son Fils," *Revue du Gévaudan*, n.s., V (1959) 68–73.
Mes Souvenirs sur Napoléon. ed. A. Chaptal. Paris, 1893.
Charas, Moyse (1619–1698)
Cap, Paul-Antoine, *Etudes Biographiques*, I, 117–129. Paris, 1857.
Dorveaux, Paul, "Apothicaires Membres de l'Académie Royale des Sciences. II. Moyse Charas," *Bulletin de la Société d'Histoire de la Pharmacie*, XVII (1929) 329–340, 377–390.
Phisalix, Marie, "Les Vipères au Jardin du Roy et à l'Académie des Sciences," *Bulletin de la Société Zoologique de France*, LXII (1937) 5–21.
———, "Moyse Charas et les Vipères du Jardin du Roi," *Bulletin de la Montagne Ste. Geneviève*, VIII (1920–1938) 149–159.
Charles, Jacques (le Géomètre) (?–1791)
Ragut, Camille, "Charles," *Statistique du Département de Saône-et-Loire*, I, 360. Mâcon, 1838.
Charles, Jacques-Alexandre-César (1746–1823)
Champeix, Robert, *Savants Méconnus*. Paris, 1966, pp. 1–52.
MSS. in BI, 2104, and 2038 fols. 134–152.
Chicoyneau, François (1672–1752)
Coste, Ulysse, *F. Chicoyneau et la Peste de 1720*. Montpellier, 1880.
Chirac, Pierre (1648–1732)
Dulieu, Louis, "Pierre Chirac, sa Vie, ses Ecrits, ses Idées," *Montpellier Médical*, ser. 3, LI (1957) 767–786.
Hamy, Ernest-Théodore, "Chirac et la Salle des Squelettes du Jardin du Roi (1731)," *Bulletin du Muséum d'Histoire Naturelle*, XIII (1907) 102–104.
Chomel, Pierre-Jean-Baptiste (1671–1740)
Les Chomel, Médecins (1639–1853) et leur Famille. Paris, 1901.
Clairaut, Alexis-Claude (1713–1765)
Brunet, Pierre, *La Vie et l'Œuvre de Clairaut (1713–1765)*. Paris, 1952.
Condorcet, Marie-Jean-Antoine-Nicolas Caritat de (1743–1794)
Baker, Keith, "Les Débuts de Condorcet au Secrétariat de l'Académie Royale des Sciences (1773–1776)," *RHS*, XX (1967) 229–280.

————, "Scientism, Elitism and Liberalism; The Case of Condorcet," *Studies on Voltaire and the Eighteenth Century*, LV (1967) 129–165.

Cahen, Léon, *Condorcet et la Révolution Française*. Paris, 1904.

Cento, Alberto, *Condorcet e l'Idea di Progresso*. Florence, 1956.

Delsaux, Hélène, *Condorcet Journaliste, 1790–1794*. Paris, 1931.

Granger, Gilles-Gaston, *La Mathématique Sociale du Marquis de Condorcet*. Paris, 1956.

Pelseneer, Jean, "Lettres Inédites de Condorcet," *Osiris*, X (1952) 322–327.

Robinet, Jean-François-Eugène, *Condorcet, sa Vie, son Œuvre, 1743–1794*. Paris, 1893.

Taton, René, "Condorcet et Sylvestre-François Lacroix," *RHS*, XII (1959) 127–158, 243–262.

MSS. in BI, MS. 848–885.

CORNETTE, Claude-Melchior (1744–1794)

Desormonts, André, *Claude-Melchior Cornette, Apothicaire, Chimiste*. Paris, 1933.

MSS. in Versailles, Archives Départementales.

COULOMB, Charles-Augustin (1736–1806)

Bauer, Edmond, *L'Electromagnétisme, Hier et Aujourd'hui*. Paris, 1949, pp. 213–235.

Gillmor, Charles Stewart, "Charles-Augustin Coulomb," unpublished Ph.D. dissertation, Princeton University, 1968.

COURTIVRON, Gaspard (1715–1785)

Brunet, Pierre, "Gaspard de Courtivron," *Mémoires de l'Académie des Sciences, Arts et Belles-Lettres de Dijon* (1927–1931) 115–134.

COUSIN, Jacques-Antoine-Joseph (1739–1800)

Poisson, Charles, *Les Fournisseurs aux Armées sous la Révolution Française*. Paris, 1932.

MSS. in BI, MS. 2041, fols. 77–80.

CUVIER, Jean-Léopold-Nicolas-Frédéric, alias Georges (1769–1832)

Coleman, William, *Georges Cuvier, Zoologist. A Study of Evolution Theory*. Cambridge, Mass., 1964.

Daudin, Henri, *Cuvier et Lamarck. Les Classes Zoologiques et l'Idée de Série Animale, 1790–1830*. Paris, 1926, 2 vols.

Dehérain, Henri, *Catalogue des Manuscrits du Fonds Cuvier*. Hendaye, 1922.

Hamy, see LAMARCK.

Roule, Louis, *Cuvier et la Science de la Nature*. Paris, 1929.

Lettres de Georges Cuvier à C. M. Pfaff sur l'Histoire Naturelle, la Politique et la Littérature 1788–1792, tr. Louis Marchant. Paris, 1858.

MSS. in BI, MS. 3001–3347.

DAGELET, alias D'AGELET, Joseph Le Paute (1751–1788)
"Centenaire de la Mort de Lapérouse," *Bulletin de la Société de Géographie*, ser. 7, IX (1888) 293–302.

DARCET, alias D'ARCET, Jean (1725–1801)
Cuzacq, René, *Un Savant Chalossais: Le Chimiste Jean Darcet (1724–1801) et sa Famille*. Mont de Marsan, 1955.

Dizé, Michel, *Précis Historique sur la Vie et les Travaux de J. D'Arcet*. Paris, an X [1802].

DAUBENTON, alias D'AUBENTON, Louis-Jean-Marie (1716–1800)
Guillaume, James, *Etudes Révolutionnaires*, I, 178–199. Paris, 1908.

Roule, Louis, *Daubenton et l'Exploitation de la Nature*. Paris, 1925.

MSS. in BI, MS. 3176.

DELAMBRE, Jean-Baptiste-Joseph (1749–1822)
Caulle, Joseph, "Delambre. Sa Participation à la Détermination du Mètre," *Recueil des Publications de la Société Havraise d'Etudes Diverses*, CIII (1936) 141–157.

Desboves, Adolphe, *Delambre et Ampère*. Amiens, 1881.

Warmé, Vulfran, *Eloge Historique de M. Delambre*. Amiens, 1824.

MSS. in BI, MS. 2041–2042.

DELISLE, Guillaume (1675–1726)
Doublet, see BUACHE.

MSS. in Paris, Chambre des Députés; BI, MS. 2302–2326.

DELISLE, Joseph-Nicolas (1688–1768)
Isnard, Albert, "Joseph-Nicolas Delisle, sa Biographie et sa Collection de Cartes Géographiques à la Bibliothèque Nationale," Comité des Travaux Historiques et Scientifiques. *Bulletin de la Section de Géographie*, XXX (1915) 34–164.

Woolf, Harry, *The Transits of Venus*. Princeton, 1959, pp. 23–70.

MSS. in Paris, Archives de la Marine, 2 JJ 50–91, and as indicated in Woolf, pp. 215–216.

DEPARCIEUX, alias DE PARCIEUX, Antoine (1703–1768)
Durand, Ernest, *Notice Biographique sur Antoine Deparcieux (1703–1763)*. Alais, 1904.

DES BILLETTES, Gilles-Filleau (1634–1720)

Gachet, Henri, "Monsieur Des Billettes, Historien du Papier," *Papier Geschichte*, VII (1957) 49–50.

DES ESSARTZ, Jean-Charles (1729–1811)
Socard, Emile, *Biographies des Personnages de Troyes*. Troyes, 1882, p. 131.

DESFONTAINES, René Louiche (1750–1833)
Chevalier, Auguste, *La Vie et l'Œuvre de René Desfontaines*. Paris, 1939.
Laissus, Joseph, "La Succession de Le Monnier au Jardin du Roi. Antoine-Laurent de Jussieu et René Louiche-Desfontaines," *91è Congrès des Sociétés Savantes*, I, 137–152.
Laissus, Yves, "Lettres Inédites de René Desfontaines à Louis-Guillaume Le Monnier," *91è Congrès des Sociétés Savantes*, I, 153–169.

DESLANDES, André-François Boureau (1690–1757)
Carr, John L., "Deslandes and the Encyclopédie," *French Studies*, XVI (1962) 154–160.
———, "The Life and Works of André-François Boureau-Deslandes," unpublished Ph.D. dissertation, University of Glasgow, 1954.
Laurent, Charles, "Monsieur Deslandes," *Bulletin de la Société Archéologique du Finistère*, XC (1964) 134–275.

DESMAREST, Nicolas (1725–1815)
Lacroix, Alfred, *Figure des Savants*, I, 7–18. Paris, 1932.
Silvestre, Augustin-François, *Notices Biographiques sur MM. Journu-Auber, Cotte, Allaire, Desmarest et Tenon*. Paris, 1816.
MSS. in BI, MS. 3199.

DEYEUX, Nicolas (1745–1837)
Dorveaux, Paul, "Nicolas Deyeux," *Bulletin de la Société d'Histoire de la Pharmacie*, II (1921) 318–322.

DIETRICH, Philippe-Frédéric (1748–1793)
Kugler, Ernst, *Philipp Friedrich von Dietrich. Ein Beitrag zur Geschichte der Vulkanologie*. Munich, 1896.
Mathiez, Albert, "Frédéric Dietrich d'après des Documents Inédits," *Annales Révolutionnaires*, XII (1920) 389–408, 471–499.
Ramon, Gabriel G., *Frédéric de Dietrich, Premier Maire de Strasbourg sous la Révolution Française*. Nancy, 1919.

DIONIS DU SÉJOUR, Achille-Pierre (1734–1794)
Lalande, Joseph-Jérôme le François de, "Eloge de Duséjour," *Connaissance des Temps, an VII*, pp. 312–317.

DODART, Denis (1634–1707)

Nomdedeu, Georges, *Les Médecins de Port-Royal.* Bordeaux, 1931, pp. 34–36.

Puech-Milhau, M. L., "An Interview on Canada with La Salle in 1678," *Canadian Historical Review*, XVIII (1937) 163–177. MSS. in BN, MS. n.a.f. 5149.

DOLOMIEU, Dieudonné-Sylvain-Guy-Tancrède, alias Déodat, Gratet de (1750–1801)

Lacroix, Alfred, *Déodat Dolomieu.* Paris, 1921, 2 vols.

DUFAY, alias DU FAY, Charles-François de Cisternay (1698–1739)

Brunet, Pierre, "L'Œuvre Scientifique de Charles-François du Fay (1698–1739)," *Petrus Nonius*, III (1940).

Cohen, I. Bernard, "Guericke and Dufay," *AS*, VII (1951) 207–209.

"Lettres de Dufay à Réaumur," *Correspondance Historique et Archéologique*, V (1898) 306–309.

DU HAMEL, Jean-Baptiste (1623–1706)

Tolmer, Léon, "J. B. Du Hamel," *Bocage Normand* (1939) 58–72.

Vialard, Augustin, *J. B. Du Hamel, Prêtre de l'Oratoire, Chancelier de l'Eglise de Bayeux.* Paris, 1884.

DUHAMEL, Jean-Pierre-François Guillot (1730–1816)

Lacroix, Alfred, *Figure des Savants*, I, 19–23. Paris, 1932.

DUHAMEL DU MONCEAU, Henri-Louis (1700–1782)

Bourde, André J., *Agronomie et Agronomes en France au XVIIIè Siècle.* Paris, 1967, pp. 253–424.

Chinard, Gilbert, "Recently Acquired Botanical Documents," *Library Bulletin. Studies of Historical Documents in the Library of the APS* (1957) 508–522.

Ewan, Jacob, "Fougeroux de Bondaroy (1732–1789) and his Projected Revision of Duhamel du Monceau's 'Traité' (1755) on Trees and Shrubs," *Proceedings APS*, CIII (1959) 807–818.

Lamontagne, Roland, "Rapport sur le Traité des Arbres et Arbustes. . . de Duhamel du Monceau," *RHS*, XVI (1963) 221–225.

MSS. in AN, 127AP, in APS, and in Harvard University.

DU VERNEY, alias DUVERNEY, Joseph-Guichard (1648–1730)

Jeudy, René-Louis, *Les Principaux Anatomistes Français du XVIIè Siècle.* Paris, 1941.

Renaud, Jean, *Les Communautés de Maîtres Chirurgiens avant la Révolution de 1789 en Forez.* St. Etienne, 1946, pp. 107–115.

Viry, Octave de, *Etudes Historiques et Médicales sur le Forez. Les du Verney.* Lyons, 1869.

FERREIN, Antoine (1693–1769)
Barritault, Georges, *L'Anatomie en France au XVIIIè Siècle.*
Angers, 1940, pp. 65–68.

FLEURIEU, Charles-Pierre Claret de (1738–1810)
Chasseriau, Frédéric, *Notice sur M. le Comte de Fleurieu.* Paris,
1856.

Doublet, Edouard, "Le Centenaire de M. Fleurieu," *Bulletin de la
Société de Géographie Commerciale de Bordeaux,* XXXIII
(1910) 257–269, 291–301.

FONTENELLE, Bernard le Bovier de (1657–1757)
Birembaut, Arthur, Pierre Costabel, and Suzanne Delorme, "La
Correspondance Leibniz-Fontenelle et les Relations de Leibniz
avec l'Académie Royale des Sciences en 1700–1701," *RHS,*
XIX (1966) 115–132.

Carré, J. R., *La Philosophie de Fontenelle.* Paris, 1932.

Counillon, Jean-François, *Fontenelle, Ecrivain, Savant, Philosophe.* Fécamp, 1959.

Delorme, Suzanne, "La Vie Scientifique à l'Epoque de Fontenelle
d'après les 'Eloges des Savants'," *Archeion,* XIX (1937) 217–
235.

Marsak, Leonard M., "Bernard de Fontenelle: The Idea of Science in the French Enlightenment," *Transactions APS,* XLIV
(1959) 1–64.

Fontenelle, sa Vie et son Œuvre, 1657–1757. Paris, 1961.
Articles by A. Adam, A. Couder, S. Delorme, A. Robinet, and
J. Rostand.

FOUGEROUX de BONDAROY, Auguste-Denis (1732–1789)
Chinard and Ewan, see DUHAMEL DU MONCEAU

Scheler, Lucien, *Lavoisier et la Révolution Française. II. Le Journal de Fougeroux de Bondaroy.* Paris, 1960.

FOURCROY, Antoine-François (1755–1809)
Duveen, Denis I., "Lavoisier Writes to Fourcroy from Prison,"
Notes and Records of the Royal Society, XIII (1958) 59–60.

Hahn, Roger, "Fourcroy, Advocate of Lavoisier?" *AIHS,* XII
(1959) 285–288.

Kersaint, Georges, "Antoine-François de Fourcroy (1755–1809).
Sa Vie et son Œuvre," *Mémoires du Muséum National d'Histoire Naturelle,* n.s., ser. D, II (1966) 1–296.

———, "L'Usine de Vauquelin et Fourcroy," *Revue d'Histoire
de la Pharmacie,* XLVII (1959) 25–30.

———, "Sur la Fabrique de Fourcroy, Vauquelin, Deserres,"

Revue Générale des Sciences Pures et Appliquées, LXVII (1960) 93–101.

Scheler, Lucien, "A propos d'une Lettre de Fourcroy à Lavoisier du 3 Septembre 1793," *RHS,* XV (1962) 43–50.

Smeaton, William A., "The Early Years of the Lycée and the Lycée des Arts. A Chapter in the Lives of A. L. Lavoisier and A. F. de Fourcroy," *AS,* XI (1955) 257–267, 309–319.

———, *Fourcroy, Chemist and Revolutionary.* Cambridge, 1962.

GAMACHES, Etienne-Simon de (1672–1756)

Pizzurosso, Arnaldo, "Etienne-Simon de Gamaches e il suo 'Sistema': Lo Stile e la *Science du Cœur," Annali della Scuola Normale Superiore di Pisa. Lettere, Storia e Filosofia,* XXXI (1962) 1–32.

GAYANT, Louis (?–1673)

Hoff, Hebel E., and Roger Guillemin, "The First Experiments on Transfusion in France," *Journal of the History of Medicine,* XVIII (1963) 103–124.

GEOFFROY, Claude-François (1729–1753)

Dorveaux, Paul, "Apothicaires Membres de l'Académie Royale des Sciences. VIII. Claude-François Geoffroy," *Revue d'Histoire de la Pharmacie,* XX (1932) 122–126.

GEOFFROY, Claude-Joseph (1685–1752)

Dorveaux, Paul, "Apothicaires Membres de l'Académie Royale des Sciences. VII. Claude-Joseph Geoffroy," *Revue d'Histoire de la Pharmacie,* XX (1932) 113–122.

Lamy, Edouard, "Deux Conchyliologistes Français du XVIIIè Siècle: Les Geoffroy Oncle et Neveu," *Journal de Conchyliologie,* LXXIII (1929) 129–132.

Partington, James Riddick, *A History of Chemistry,* III, 55–57. London, 1962.

GEOFFROY, Etienne-François (1672–1731)

Cohen, I. Bernard, "Isaac Newton, Hans Sloane and the Académie Royale des Sciences," *Mélanges Alexandre Koyré,* I, 61–116. Paris, 1964.

Dorveaux, Paul, "Apothicaires Membres de l'Académie Royale des Sciences. V. Etienne-François Geoffroy," *Revue d'Histoire de la Pharmacie,* XIX (1931) 118–126.

Lamy, see GEOFFROY, C.-J.

Partington, James Riddick, *A History of Chemistry,* III, 49–55. London, 1962.

GILBERT, François-Hilaire (1757–1800)

Delafourchardière, Alphonse, *François-Hilaire Gilbert. Sa Vie, sa Correspondance.* Châtelleraut, 1843.

Railliet, Alcide-Louis-Joseph, "Gilbert," *Bulletin de la Société Centrale de Médecine Vétérinaire,* LVIII (1904) 696–736.

GILLET de LAUMONT, François-Pierre-Nicolas (1747–1834)

Birembaut, Arthur, in *Enseignement et Diffusion,* pp. 403–404.

GROSSE, Jean, alias Johann GROSS (?–1744)

Dorveaux, Paul, "Jean Grosse, Médecin Allemand, et l'Invention de l'Ether Sulfurique," *Bulletin de la Société d'Histoire de la Pharmacie,* XVII (1929) 182–186.

GUA de MALVES, Jean-Paul de (1712–1786)

Montbas, Hugues de, "Quelques Encyclopédistes Oubliés," *Revue des Travaux de l'Académie des Sciences Morales et Politiques,* ser. 4, CV (1952) 32–40.

Sauerbeck, Paul, "Einleitung in die Analytische Geometrie der Höheren Algebraischen Kurven nach den Methoden von Jean Paul de Gua de Malves," *Abhandlungen zur Geschichte der Mathematischen Wissenschaften,* XV (1902).

GUETTARD, Jean-Etienne (1715–1786)

Lamontagne, Roland, "La Participation Canadienne à l'Œuvre Minéralogique de Guettard," *RHS,* XVIII (1965) 385–388.

Rappaport, Rhoda, "Lavoisier's Geologic Activities, 1763–1792," *Isis,* LVIII (1968) 375–384.

Soland, Aimé de, *Etude sur Guettard.* Angers, 1873.

MSS. in BMHN, MS. 1996, 2184–2194.

GUGLIELMINI, Domenico (1655–1710)

Colonne, F. J. M., *Doménique Guglielmini (1655–1710).* Paris, 1929.

Hooykaas, Reijer, "Domenico Guglielmini et le Développement de la Cristallographie," *Atti della Fondazione Giorgio Ronchi,* VIII (1953) 13–28.

Lombardini, Elia, *Dell'Origine e del Progresso della Scienza Idraulica; Osservazioni Storico-Critiche Concernenti i Lavori di . . . Domenico Guglielmini,* 3rd ed. Milan, 1872.

GUYTON de MORVEAU, Louis-Bernard (1737–1816)

Bouchard, Georges, *Guyton-Morveau, Chimiste et Conventionnel, 1737–1816.* Paris, 1938.

Smeaton, William A., "The Contribution of P. J. Macquer, T. O. Bergman and L. B. Guyton de Morveau to the Reform of Chemical Nomenclature," *AS,* X (1954) 97–106.

———, "L. B. Guyton de Morveau (1737–1816)," *Ambix,* VI (1957), 18–34.

———, "Guyton de Morveau and Chemical Affinity," *Ambix*, XI (1963), 55–64.

———, "Guyton de Morveau and the Phlogiston Theory," *Mélanges Alexandre Koyré*, I, 522–540. Paris, 1964.

———, "Guyton de Morveau's Course of Chemistry in the Dijon Academy," *Ambix*, IX (1961) 53–69.

———, "The Portable Chemical Laboratories of Guyton de Morveau, Cronstedt and Göttling," *Ambix*, XIII (1966) 84–91.

HALLÉ, Jean-Noël (1754–1822)

Estournet, O., *La Famille des Hallé*. Paris, 1905, pp. 167–170.

Malato, Marco T., "Hallé, Igienista Francese del Secolo XVIII," *Collana di Pagine di Storia della Medicina* (1958), no. 4, 22–34.

HAÜY, René-Just (1743–1822)

Amoros, J. L., "Notas sobre la Historia de la Cristalografía. I. La Controversia Haüy-Mitscherlich," *Boletín de la Sociedad Española de Historia Natural. Sección Geológica*, LVII (1959) 5–30.

Hooykaas, Reijer, "Les Débuts de la Théorie Cristallographique de R. J. Haüy," *RHS*, VIII (1955) 319–337.

Lacroix, Alfred, "La Vie et l'Œuvre de l'Abbé René-Just Haüy," *Bulletin de la Société Française de Minéralogie*, LXVII (1944) 15–226.

Moulin, H., "Lhomond et Haüy, Professeurs au Collège du Cardinal Lemoine et Amis Intimes 1772–1822," *Mémoires de l'Académie Nationale des Sciences, Arts et Belles-Lettres de Caen* (1884) 424–438.

MSS. in BMHN, MS. 811–813, 1398–1403.

HELLOT, Jean (1685–1766)

Beer, John J., "Eighteenth–Century Theory on the Process of Dyeing," *Isis*, LI (1960) 21–30.

Birembaut, Arthur, and Guy Thuillier, "Une Source Inédite: Les Cahiers du Chimiste Jean Hellot (1685–1766)," *Annales. Economies. Sociétés Civilisations*, XXI (1966) 357–364.

HELVÉTIUS, Jean-Claude-Adrien (1685–1755)

Levy, Edmond, "Le Testament d'Helvétius, Premier Médecin de la Reine Marie Leczinska," *Revue de l'Histoire de Versailles et de Seine-et-Oise*, XXVII (1926) 219–220.

HOMBERG, Wilhelm, alias Guillaume (1652–1715)

Cap, Paul-Antoine, *Etudes Biographiques*, II, 214–232. Paris, 1864.

Huard, Pierre, "L'Autopsie de Guillaume Homberg," *89è Congrès des Sociétés Savantes*, pp. 155–156.

Partington, James Riddick, *A History of Chemistry*, III, 42–47. London, 1962.

HUNAULD, alias HUNAUD, François-Joseph (1701–1742)

Hamy, Ernest-Théodore, "Un Intérieur du Savant Parisien Anatomiste et Médecin au Milieu du XVIIIè Siècle; Scellés et Inventaire Chez l'Académicien Hunauld [1742]," *Bulletin de la Société Française d'Histoire de la Médecine*, V (1906) 199–209.

HUYGENS, alias HUYGHENS de ZUYLICHEM, Christiaan (1629–1695)

Brugmans, Henri L., *Le Séjour de Christian Huygens à Paris*. Paris, 1935.

Sergescu, Pierre, "La Littérature Mathématique dans la Première Période (1665–1705) du 'Journal des Savants'," *AIHS*, I (1947) 60–99.

Vollgraff, see ALEMBERT.

HUZARD, Jean-Baptiste (1755–1838)

Bouchard-Huzard, Louis, *Notice Biographique sur J. B. Huzard*. Paris, 1839.

Silvestre, Augustin-François, *Notice Biographique sur M. J.-B. Huzard*. Paris, 1838.

MSS. in BI, MS. 1961–1977.

ISNARD, Antoine-Tristan Danty d' (1663–1743)

Hamy, Ernest-Théodore, "Un Manuscrit de Danty d'Isnard à la Bibliothèque d'Arras," *Bulletin du Muséum d'Histoire Naturelle*, VIII (1902) 293–294.

JARS, Antoine-Gabriel (1732–1769)

Chevalier, Jean, *Le Creusot, Berceau de la Grande Industrie Française*, new ed. Paris, 1946.

JAUGEON, Jacques (1646–1724)

Morison, Stanley, *Four Centuries of Fine Printing*. New York, 1949, pp. 36–37.

MSS. in BI, MS. 2741; BN, MS. fr. 9157–9158, 9181–9183.

JEAURAT, Edme-Sébastien (1724–1803)

Faddegon, J. M., and Boizard de Guise, "L'Astéréomètre de Jeaurat," *Société Astronomique de France*, L (1936) 553–559.

MSS. in BI, MS. 2041.

JUSSIEU, Antoine de (1686–1758)

Lacroix, Alfred, *Figure des Savants*, IV, 109–124. Paris, 1938.

MSS. in BMHN.

JUSSIEU, Antoine-Laurent de (1748–1836)

Bréard, Charles, "Lettres Inédites d'Antoine-Laurent de Jussieu," *La Normandie Littéraire, Archéologique, Historique, etc.,* VIII (1893) 161–169.

Brongniart, Adolphe, *Notice Historique sur Antoine Laurent de Jussieu.* Paris, 1837.

Courtney, C. P., "Antoine-Laurent de Jussieu, Collaborateur de l'Abbé Raynal. Documents Inédits," *Revue d'Histoire Littéraire,* LXIII (1963) 217–227.

Lacroix, Alfred, *Figure des Savants,* IV, 141–158. Paris, 1938.

Laissus, Joseph, "Antoine-Laurent de Jussieu 'L'Aimable Professeur'," *89è Congrès des Sociétés Savantes,* pp. 27–39.

Laissus, see DESFONTAINES.

MSS. in BMHN.

JUSSIEU, Bernard de (1699–1777)

Lacroix, Alfred, *Figure des Savants,* IV, 125–140. Paris, 1938.

MSS. in BMHN.

JUSSIEU, Joseph de (1704–1779)

Lacroix, Alfred, *Figure des Savants,* IV, 159–173. Paris, 1938.

Laissus, Joseph, "Note sur les Manuscrits de Joseph de Jussieu (1704–1779) Conservés à la Bibliothèque Centrale du Muséum d'Histoire Naturelle," *89è Congrès des Sociétés Savantes,* pp. 9–16.

MSS. in BMHN.

LA BILLARDIÈRE, Jacques-Julien Houtou de (1755–1834)

Chevalier, Auguste, "J. J. Houtou La Billardière (1755–1834). Sa Vie, son Œuvre," *Revue Internationale de la Botanique Appliquée,* XXXIII (1953) 97–202.

LA CAILLE, Nicolas-Louis de (1713–1762)

Armitage, Angus, "The Astronomical Work of Nicolas-Louis de la Caille," *AS,* XII (1956) 163–191.

Boquet, F., "Le Bicentenaire de Lacaille," *L'Astronomie,* XXVII (1913) 457–473.

LACEPÈDE, Bernard-Germain-Etienne de la Ville-sur-Illon (1756–1825)

Roule, Louis, *Lacepède et la Sociologie Humanitaire selon la Nature.* Paris, 1932.

———, "La Vie et l'Œuvre de Lacepède," *Mémoires de la Société Zoologique de France,* XXVII (1917) 139–237.

MSS. in BI, MS. 3209.

LA CHAMBRE, Marin Cureau de (1596–1669)

Diamond, Solomon, "Marin Cureau de La Chambre (1594–

1669)," *Journal of the History of the Behavioral Sciences*, IV (1968) 40–54.

Doranlo, Robert, *La Médecine au XVIIè Siècle. Marin Cureau de la Chambre*. Paris, 1939.

Hamy, Ernest-Théodore, "Les Débuts de l'Anthropologie et de l'Anatomie Humaine au Jardin des Plantes. M. Cureau de La Chambre et P. Dionis," *L'Anthropologie*, V (1894) 257–275.

Savelli, Roberto, "Un Opticien Méconnu: Le Sieur de La Chambre," *VIII Congrès International*, I, 275–281.

LA CONDAMINE, Charles-Marie de (1701–1774)

Conlon, Pierre M., "La Condamine the Inquisitive," *Studies on Voltaire and the Eighteenth Century*, LV (1967) 361–393.

Jones, Tom B., *The Figure of the Earth*. Lawrence, 1967.

Le Sueur, Achille, "La Condamine, d'après ses Papiers Inédits," *Mémoires de l'Académie des Sciences, des Lettres et des Arts d'Amiens*, LVI (1909) 1–80.

Revue Générale du Caoutchouc, XIII (1936).

LACROIX, Sylvestre-François (1765–1843)

Taton, René, "Laplace et Sylvestre-François Lacroix," *RHS*, VI (1953) 350–360.

———, "L'Ecole Polytechnique et le Renouveau de la Géometrie Analytique," *Mélanges Alexandre Koyré*, I, 552–564. Paris, 1964.

Taton, see CONDORCET.

MSS. in BI, MS. 2396–2403.

LA GALISSONNIÈRE, Roland-Michel Barrin (1693–1756)

Lamontagne, Roland, *La Galissonière et le Canada*. Montreal, 1962.

LAGRANGE, Joseph-Louis de (1736–1813)

Burzio, Filipo, *Lagrange*. Turin, 1945.

Loria, Gino, "Saggio di una Bibliografia Lagrangiana," *Isis*, XL (1949) 112–117.

Sarton, George, "Lagrange's Personality (1736–1813)," *Proceedings APS*, LXXXVIII (1949) 456–496.

———, René Taton, and Guy Beaujouan, "Documents Nouveaux Concernant Lagrange," *RHS*, III (1950) 110–132.

MSS. in BI, MS. 886–916, and Paris, Bibliothèque de l'Ecole des Ponts et Chaussées, MS. 2704.

LA HIRE, Gabriel-Philippe de (1677–1719)

MSS. in Paris, Observatoire.

LA HIRE, Philippe de (1640–1718)

Sergescu, see HUYGENS.

Taton, René, "La Première Œuvre Géometrique de Philippe de La Hire," *RHS*, VI (1953) 93–111.

LALANDE, Joseph-Jérôme Lefrançois de (1732–1807)
Amiable, Louis, *Le Franc-Maçon Jérôme Lalande*. Paris, 1889.
Bluche, Joseph, "Jérôme Lalande," *Annales de la Société d'Emulation et d'Agriculture de l'Ain*, XXXVII (1904) 5–34.
Denizet, "Lalande et l'Art de l'Ingénieur," *Annales de la Société d'Emulation et d'Agriculture de l'Ain*, XXXVIII (1905) 232–261.
Doublet, Edouard, "L'Astronome Lalande et la Géographie," *Revue de Géographie Commerciale*, LXIV (1920) 132–146, 165–170.
Marchand, Charles E. H., "Jérôme Lalande et l'Astronomie au XVIIIè Siècle," *Annales de la Société d'Emulation et d'Agriculture de l'Ain*, XL (1907) 82–145, XLI (1908) 313–417 (also published as a book, Bourg, 1909).
Monod-Cassidy, Hélène, "Un Astronome Philosophe, Jérôme de Lalande," *Studies on Voltaire and the Eighteenth Century*, LVI (1967) 907–930.
Pavlova, Galina E., "J. J. Lalande and the Petersburg Academy of Sciences," *X Congrès International*, pp. 743–746.
———, *Lalande 1732–1807*. Leningrad, 1967 (in Russian).
MSS. in BN, MS. fr. 12271–12275.

LAMARCK, Jean-Baptiste-Pierre-Antoine de Monet (1744–1829)
Carozzi, Albert, "Lamarck's Theory of the Earth: *Hydréologie*," *Isis*, LV (1964) 293–307.
Daudin, see CUVIER.
Gillispie, Charles C., "The Formation of Lamarck's Evolutionary Theory," *AIHS*, IX (1956) 323–338.
———, "Lamarck and Darwin in the History of Science," *Forerunners of Darwin*. Baltimore, 1959, pp. 265–291.
Grassé, Pierre-Paul, "La Biologie, Texte Inédit de Lamarck," *Revue Scientifique*, LXXXII (1944) 267–276.
Hamy, Ernest-Théodore, *Les Débuts de Lamarck, Suivis de Recherches sur Adanson, Jussieu, Pallas, Geoffroy St. Hilaire, Georges Cuvier, etc*. Paris, 1909.
Landrieu, Marcel, *Lamarck*. Paris, 1909.
Poljakov, I. M., "The Manuscripts of Lamarck," *Russian Review of Biology*, XLVIII (1959) 289–296.
Packard, Alpheus Spring, *Lamarck*. New York, 1901.
Roule, Louis, *Lamarck et l'Interprétation de la Nature*. Paris, 1927.

MSS. in AN, 23 AP; BI, MS. 3156; and Harvard University, Museum of Comparative Zoology.

LA PEYRONIE, François Gigot de (1678–1747)
Briot, Pierre-François, *De l'Influence de La Peyronie sur le Lustre et les Progrès de la Chirurgie Française.* Besançon, 1820.
Estor, Henri, "Madame d'Issert et le Testament de La Peyronie," *Monspeliensis Hippocrates,* X (1961) no. 36, 21–28.

LAPLACE, Pierre-Simon (1749–1827)
Andoyer, Henri, *L'Œuvre Scientifique de Laplace.* Paris, 1922.
Crosland, Maurice, *The Society of Arcueil.* London, 1967, pp. 482–483.
Duveen, Denis I., and Roger Hahn, "Deux Lettres de Laplace à Lavoisier," *RHS,* XI (1958) 337–342.
———, and ———, "Laplace's Succession to Bézout's Post of Examinateur des Elèves de l'Artillerie," *Isis,* XLVIII (1957) 416–427.
Finn, Bernard, "Laplace and the Speed of Sound," *Isis,* LV (1964) 7–19.
Hahn, Roger, *Laplace as a Newtonian Scientist.* Los Angeles, 1967.
———, "Laplace's First Formulation of Scientific Determinism in 1773," *XI Congrès International,* II, 167–171.
———, "Laplace's Religious Views," *AIHS,* VIII (1955) 38–40.
Laissus, Yves, "Deux Lettres de Laplace," *RHS,* XIV (1961) 285–296.
Simon, Georges-Abel-René, "Laplace. Ses Origines Familiales et ses Premiers Débuts," *Normannia,* X (1937) 477–496.
———, "Les Origines de Laplace; sa Généalogie; ses Études," *Biometrika,* XXI (1929) 217–230.
Taton, see LACROIX.
Vuillemin, Jules, "Sur la Généralisation de l'Estimation de la Force chez Laplace," *Thalès,* IX (1958) 61–75.
MSS. in BI, MS. 2242.

LASSONE, Joseph-François-Marie de (1717–1788)
MSS. in Paris, Académie de Médecine, MS. box 109.

LASSUS, Pierre (1741–1807)
Sue, Pierre, *Eloge Historique de Lassus.* Paris, 1808.

LAVOISIER, Antoine-Laurent de (1743–1794)
Baker, Keith M., and William A. Smeaton, "The Origins and Authorship of the Educational Proposals Published in 1793 by the 'Bureau de Consultation des Arts et Métiers' and Generally Ascribed to Lavoisier," *AS,* XXI (1965) 33–46.

Birembaut, Arthur, "A propos des Biographies de Lavoisier," *VII Congrès International,* pp. 216–220.

————, "A propos d'une Publication Récente sur Lavoisier et le Lycée des Arts," *RHS,* XI (1958) 267–273.

————, "Quelques Aspects de la Personalité de Lavoisier," *Actes du 72è Congrès de l'Association Française pour l'Avancement des Sciences,* pp. 539–542.

Carozzi, Albert V., "Lavoisier's Fundamental Contribution to Stratigraphy," *Ohio Journal of Science,* LXV (1965) 71–85.

Daumas, Maurice, *Lavoisier. Théoricien et Expérimentateur.* Paris, 1955.

Duveen, Denis I., "Antoine-Laurent Lavoisier and the French Revolution," *Journal of Chemical Education,* XXXI (1954), 60–65, XXXIV (1957) 502–503, XXXV (1958) 233–234.

————, "Madame Lavoisier," *Chymia,* IV (1953) 13–29.

————, *Supplement to A Bibliography of the Works of Antoine-Laurent Lavoisier.* London, 1965.

————, and Herbert S. Klickstein, "Antoine-Laurent Lavoisier's Contributions to Medicine and Public Health," *Bulletin of the History of Medicine,* XXIX (1955) 164–179.

————, and ————, "Benjamin Franklin and Antoine-Laurent Lavoisier," *AS,* XI (1955) 103–128, 271–302, XIII (1957) 30–46.

————, and ————, *A Bibliography of the Works of Antoine-Laurent Lavoisier.* London, 1954.

Duveen, see FOURCROY.

Duveen and Hahn, see LAPLACE.

Gillispie, Charles C., "Notice Biographique de Lavoisier par Mme Lavoisier," *RHS,* IX (1956) 52–61.

Gough, Jerry B., "Lavoisier's Early Career in Science," *British Journal for the History of Science,* IV (1968) 52–57.

Grimaux, Edouard, *Lavoisier, 1743–1794,* 3d ed. Paris, 1899.

Guerlac, Henry, "A Note on Lavoisier's Scientific Education," *Isis,* XLVII (1956) 211–216.

————, "Lavoisier and his Biographers," *Isis,* XLV (1954) 51–62.

————, *Lavoisier, the Crucial Year.* Ithaca, 1961.

Hahn, see FOURCROY.

Hollister, S. C., *Antoine-Laurent Lavoisier. An Exhibition.* Ithaca, 1963.

McDonald, see BUCQUET.

McKie, Douglas, *Antoine Lavoisier, the Father of Modern Chemistry*. Philadelphia, 1935.

Rappaport, see GUETTARD.

Scheler, Lucien, "Antoine-Laurent Lavoisier et le *Journal d'Histoire Naturelle*," *RHS*, XIV (1961) 1–9.

————, *Lavoisier et la Révolution Française*. Paris, 1956–1960, 2 vols.

Scheler, see ADANSON, FOUGEROUX de BONDAROY, and FOURCROY.

Smeaton, William A., "L'Avant-Coureur, the Journal in which some of Lavoisier's Earliest Research Was Reported," *AS*, XIII (1957) 219–234.

————, "Lavoisier's Membership of the Assembly of Representatives of the Commune of Paris, 1789–1790," *AS*, XIII (1957) 235–248.

————, "Lavoisier's Membership of the Société Royale d'Agriculture and the Comité d'Agriculture," *AS*, XII (1956) 267–277.

————, "Lavoisier's Membership of the Société Royale de Médecine," *AS*, XII (1956) 228–244.

————, "New Light on Lavoisier: The Research of the Last Ten Years," *History of Science*, II (1963) 51–69.

Smeaton, see FOURCROY.

Storrs, F. C., "Lavoisier's Technical Reports," *AS*, XXII (1966) 251–275, XXIV (1968) 179–197.

Vergnaud, Marguerite, "Science et Progrès d'après Lavoisier," *Cahiers Internationaux de Sociologie*, XV (1953) 174–186.

————, "Un Savant pendant la Révolution," *Cahiers Internationaux de Sociologie*, XVII (1954) 123–139.

MSS. in AdS, Lav; Paris, Conservatoire des Arts et Métiers; Clermont-Ferrand, Bibliothèque Municipale; and Cornell University, described in Hollister, above.

LE FÈBVRE, Jean (1650–1706)

Tissot, Amédée, *Etude sur Jean Le Fevre, Ouvrier, Tisserand, Astronome*. Paris, 1872.

LE GENDRE, Adrien-Marie (1752–1833)

Beaumont, Léonce-Elie de, *Eloge Historique de A.-M. Legendre*. Paris, 1861.

Hellman, C. Doris, "Legendre and the French Reform of Weights and Measures," *Osiris*, I (1936) 314–340.

Maurice, Frédéric, "Mémoire sur les Travaux et les Ecrits de M. Legendre," *Bibliothèque Universelle, Sciences et Arts*, LII (1833) 45–82

LE GENTIL de la GALAISIÈRE, Guillaume-Joseph-Hyacinthe-Jean-Baptiste (1725–1792)

Cassini, Jean-Dominique (Cassini IV), *Mémoires pour Servir à l'Histoire des Sciences*. Paris, 1810, pp. 358–372.

Lacroix, Alfred, *Figure des Savants*, III, 169–176. Paris, 1938.

LEIBNIZ, alias LEIBNITZ, Gottfried Wilhelm (1646–1716)

Barber, William Henry, *Leibniz in France from Arnauld to Voltaire*. Oxford, 1955.

Belaval, Yvon, *Leibniz Critique de Descartes*. Paris, 1960.

Birembaut, Costabel, and Delorme, see FONTENELLE.

Cohen, I. Bernard, "Leibniz on Elliptical Orbits—as Seen in his Correspondence with the Académie Royale des Sciences in 1700," *Journal of the History of Medicine*, XVII (1962) 72–82.

Hofmann, Josef Ehrenfried, *Die Entwicklungsgeschichte der Leibnizschen Mathematik während des Aufenthaltes in Paris (1672–1676)*. Munich, 1949.

Robinet, André, *Malebranche et Leibniz; Relations Personelles*. Paris, 1955.

Sergescu, see HUYGENS.

LELIÈVRE, Claude-Hugues (1752–1835)

Bonnard, "Notice sur M. Lelièvre," *Annales des Mines*, ser. 4, VII (1845) 506–522.

LÉMERY, Nicolas (1645–1715)

Cap, Paul-Antoine, *Etudes Biographiques*, I, 180–226. Paris, 1857.

Dorveaux, Paul, "Apothicaires Membres de l'Académie Royale des Sciences. VI. Nicolas Lémery," *Revue d'Histoire de la Pharmacie*, XIX (1931) 208–219.

Partington, James Riddick, *A History of Chemistry*, III, 28–41. London, 1962.

LE MONNIER, Louis-Guillaume (1717–1799)

Challan, Antoine-Didier-Jean-Baptiste, *Essai Historique sur la Vie du Citoyen Louis-Guillaume Lemonnier*. Versailles, an VIII [1800].

Laissus, see DESFONTAINES.

Mercet, Suzanne, "La Maison des Italiens au Grand Montreuil," *Revue de l'Histoire de Versailles et de Seine-et-Oise*, XXVIII (1926) 168–190, 245–263.

Robida, Michel, *Les Bourgeois de Paris. Trois Siècles de Chronique Familiale, de 1675 à nos Jours*. Paris, 1955.

MSS. in BMHN, MS. 357, 1995, and others; Paris, Bibliothèque de l'Ecole des Ponts et Chaussées, MS. 2704; BI, MS. 3178.

LE MONNIER, Pierre (1675–1757)

Robida, see LE MONNIER, Louis.

MSS. in BI, MS. 3178.

LE MONNIER, Pierre-Charles (1715–1799)

Robida, see LE MONNIER, Louis.

L'HÉRITIER de BRUTELLE, Charles-Louis (1746–1800)

Stafleu, Frans Antonie, "L'Héritier de Brutelle: The Man and his Works," in L'Héritier, *Sertum Anglicum 1788*. Pittsburgh, 1963, pp. xiii–xliii.

MSS. in Geneva, Conservatoire Botanique, MS. 79A; BI, MS. 3180; and the references in Stafleu, pp. xlii–xliii.

L'HOSPITAL, Guillaume-François-Antoine de (1661–1704)

Costabel, Pierre, "Une Lettre Inédite du Marquis de L'Hôpital sur la Résolution de l'Equation du Troisième Degré," *RHS*, XVIII (1965) 29–43.

Sergescu, see HUYGENS.

Spiess, Otto, "Une Section de l'Œuvre des Mathématiciens Bernoulli," *AIHS*, II (1948) 354–362.

LIEUTAUD, Joseph (1703–1780)

Bariéty, Maurice, "Lieutaud et la Méthode Anatoclinique," *La Presse Médicale*, LV, 1 (1947) 138.

Chavernac, J. F. E., *Le Botaniste Garidel et son Neveu Lieutaud*. Marseille, 1877.

Laservolle, Pierre de, *Eloge Historique de M. Lieutaud*. Paris, 1781.

LITTRE, Alexis (1654–1725)

Jeudy, René-Louis, *Les Principaux Anatomistes Français du XVIIè Siècle*. Paris, 1941.

MACQUER, Pierre-Joseph (1718–1784)

Coleby, Leslie J. M., *The Chemical Studies of P. J. Macquer*. London, 1938.

Neville, R. G., "Macquer and the First Chemical Dictionary, 1766," *Journal of Chemical Education*, XLIII (1966) 486–490.

Smeaton, William A., 'Macquer's Course of Chemistry at the Jardin du Roi," *X Congrès International*, pp. 847–849.

Smeaton, see GUYTON de MORVEAU.

MSS. in BN, MS. fr. 9127–9135 (see catalogue by Bondois, BN, MS. n.a.f. 13260) fr. 12305–12306; BMHN, MS. 283.

MAGNOL, Pierre (1638–1715)

Dulieu, Louis, "Les Magnol," *RHS*, XII (1959) 209–244.

MAIRAN, Jean-Jacques Dortous de (1678–1771)

Duboul, J., "Dortous de Mairan: Etude sur sa Vie et ses Travaux," *Actes de l'Académie Impériale des Sciences, Belles-Lettres et Arts de Bordeaux*, ser. 3, XXIV (1862) 163–197.

"Lettres Inédites de Mairan à Bouillet," *Bulletin de la Société Archéologique, Scientifique et Littéraire de Béziers*, ser. 2, II (1860) 20–23.

Méditations Métaphysiques et Correspondance de N. Malebranche avec J.-J. Dortous de Mairan, sur des Sujets de Métaphysique. Paris, 1841.

MALOUIN, Paul-Jacques (1701–1777)

Des Essartz, Jean-Charles, *Eloge de Malouin*. Paris, 1778.

MARALDI, Giacomo Filippo (1665–1729)

Cassini, Jean-Dominique (Cassini IV), *Mémoires pour Servir à l'Histoire des Sciences*. Paris, 1810, pp. 325–347.

MSS. in Paris, Observatoire.

MARCHANT, Jean (?–1738)

Hus, H., "Jean Marchant, an Eighteenth Century Mutationist," *American Naturalist*, XLV (1911) 491–506.

MARIOTTE, Edme (1620–1684)

Brunet, Pierre, "La Méthodologie de Mariotte," *AIHS*, I (1947) 26–59.

Bugler, G., "Un Précurseur de la Biologie Expérimentale: Edme Mariotte," *RHS*, III (1950) 242–250.

Costabel, Pierre, "Le Paradoxe de Mariotte," *AIHS*, II (1949) 864–881.

————, "Mariotte et le Phénomène Elastique," *84è Congrès des Sociétés Savantes*, pp. 67–69.

Pelseneer, Jean, "Petite Contribution à la Connaissance de Mariotte," *Isis*, XLII (1951) 299–301.

Rochot, Bernard, "Roberval, Mariotte et la Logique," *AIHS*, VI (1953) 38–43.

MAUPERTUIS, Pierre-Louis Moreau de (1698–1759)

Bachelard, Suzanne, *Les Polémiques Concernant le Principe de Moindre Action au XVIIIè Siècle*. Paris, 1961.

Brown, Harcourt, "Maupertuis 'Philosophe': Enlightenment and the Berlin Academy," *Studies on Voltaire and the Eighteenth Century*, XXIV (1963) 255–269.

Brunet, Pierre, *Maupertuis. Etude Biographique*. Paris, 1929.

Dufrenoy, Marie-Louise, "Maupertuis et le Progrès Scientifique," *Studies on Voltaire and the Eighteenth Century*, XXV (1963) 519–587.

Glass, Bentley, "Maupertuis, Pioneer of Genetics and Evolution," *Forerunners of Darwin*. Baltimore, 1959, pp. 51–83.

Jones, see LA CONDAMINE.
Nordmann, Claude J., "L'Expédition de Maupertuis et Celsius en Laponie," *Cahiers d'Histoire Mondiale*, X (1966) 74–97.
Schumaker, John A., "Pierre de Maupertuis and the History of Comets," *Scripta Mathematica*, XXIII (1957) 97–108.
Speziali, Pierre, "Une Lettre Inédite de Maupertuis à Gabriel Cramer," *Archives des Sciences* (Geneva), VI (1953) 89–93.
MÉCHAIN, Pierre-François-André (1744–1804)
Dougados, "Lettres de l'Astronome Méchain à M. Rolland," *Mémoires de la Société des Arts et des Sciences de Carcassonne*, II (1856) 74–130.
"Lettres Inédites de Méchain Adressées à J. Pons-Bernard (1785–1788)," *Les Archives de Trans en Provence*, XI (1938) 337, XII (1939) 55–61, 106–110, 197–202.
MSS. in Laon, Bibliothèque Municipale; Paris, Observatoire.
MERY, Jean (1645–1722)
Franklin, Kenneth James, "Jean Méry (1645–1722) and his Ideas on the Foetal Blood Flow," *Annals of Science*, V (1945) 203–228.
Herpin, J. C., "Notice Historique sur la Vie et les Travaux de Jean Mery," *Comptes-Rendus des Travaux de la Société du Berry à Paris*, XI (1863) 434–474.
MESSIER, Charles-Joseph (1730–1817)
Flammarion, Gabrielle Camille, "Bicentenaire de la Naissance de Messier," *Bulletin de la Société Astronomique de France*, XLIV (1930) 249–259.
Floquet, Gaston, "L'Astronome Messier," *Mémoires de l'Académie Stanislas*, ser. 5, XIX (1901–1902) xxiii–lxvii.
Gingerich, Owen, "Messier and his Catalogue," *Sky and Telescope*, XII (1953) 255–257.
MEUSNIER de la PLACE, Jean-Baptiste-Marie-Charles (1754–1793)
Darboux, Gaston, ed., *Mémoires et Travaux de Meusnier Relatif à l'Aérostation, Précédés d'une Notice Historique.* Paris, 1910.
MOLIÈRES, Joseph Privat de (1676–1742)
Aiton, see CASSINI I.
MONGE, Gaspard (1746–1818)
Aubry, Paul V., *Monge. Le Savant Ami de Napoléon Bonaparte 1746–1818.* Paris, 1954.
Laissus, Yves, "Gaspard Monge et l'Expédition d'Egypte (1798–1799)," *Revue de Synthèse*, ser. 3, LXXXI (1960) 309–336.
Launay, Louis de, *Monge, Fondateur de l'Ecole Polytechnique.* Paris, 1933.

Taton, René, "La Première Note Mathématique de Gaspard Monge (Juin 1769)," *RHS*, XIX (1966) 143–149.

———, "L'Ecole Polytechnique et le Renouveau de la Géométrie Analytique," *Mélanges Alexandre Koyré*, I, 552–564. Paris, 1964.

———, *L'Œuvre Scientifique de Monge*. Paris, 1951.

MONTALEMBERT, Marc-René (1714–1800)

Gaudin, R., "Marc-René, Marquis de Montalembert," *Bulletins et Mémoires de la Société Archéologique et Historique de la Charente* (1938) 49–127.

Lloyd, Ernest Marsh, *Vauban, Montalembert, Carnot: Engineering Studies*. London, 1887, pp. 98–154.

MONTIGNY, Etienne Mignot de (1714–1782)

Bourde la Rogerie, H., "Les Voyageurs en Bretagne. Le Voyage de Mignot de Montigny en Bretagne, en 1752," *Mémoires de la Société d'Histoire et d'Archéologie de Bretagne*, VI (1925) 225–301.

MSS. in Paris, Bibliothèque Mazarine, MS. 2840, 3596–3597.

MORAND, Sauveur-François (1697–1773)

Elaut, L., "L'Œuvre Anatomique de Sauveur-François Morand (1697–1773)," *Histoire de la Médecine* (1963) numéro spécial 4, 19–27.

MSS. in Paris, Académie de Médecine, MS. 66; BI, MS. 1797.

MORIN de SAINT-VICTOR, Louis (1635–1715)

Delaunay, Paul, *Vieux Médecins Sarthois*. Paris, 1906, pp. 86–106.

MORIN de TOULON (?–1707)

Warner, Marjorie, "The Morins," *National Horticultural Magazine* (1954) 167–176.

NOLLET, Jean-Antoine (1700–1770)

Torlais, Jean, *L'Abbé Nollet, un Physicien au Siècle des Lumières*. Paris, 1954.

OLIVIER, Guillaume-Antoine (1756–1814)

Olivier, Ernest, *G. A. Olivier, sa Vie, ses Travaux, ses Voyages*. Moulins, 1880.

Théodoridès, Jean, "Jean-Guillaume Bruguière (1749–1798) et Guillaume Olivier (1756–1814), Médecins Naturalistes et Voyageurs," *86è Congrès des Sociétés Savantes*, pp. 173–183.

OZANAM, Jacques (1640–1718)

Sergescu, see HUYGENS.

PALISOT de BEAUVOIS, Ambroise-Marie-François-Joseph (1752–1820)

Silvestre, Augustin-François, *Notice Biographique sur M. Palisot.* Paris, 1820.

Thiébaut de Berneaud, Arsenne, *Eloge Historique d'Ambroise-Marie-François-Joseph Palisot de Beauvois.* Paris, 1821.

MSS. in Belgium, Bibliothèque de Mons; BI, MS. 3200.

PARMENTIER, Antoine-Augustin (1737–1813)

Balland, Antoine, *La Chimie Alimentaire dans l'Œuvre de Parmentier.* Paris, 1902.

Blaessinger, Edmond, *Quelques Grandes Figures de la Pharmacie Militaire.* Paris, 1948, pp. 71–118.

Bouvet, Marcel, "Quelques Documents Inédits sur Parmentier," *Revue d'Histoire de la Pharmacie,* XLVIII (1960) 419–423, XLIX (1961) 1–7.

Mouchon, Emile, *Notice Historique sur Antoine Parmentier.* Lyons, 1843.

Seince, Franck-François-Charles, *Parmentier et l'Hygiène Alimentaire.* Bordeaux, 1935.

Silvestre, Augustin-François, *Notice Biographique sur M. Antoine-Augustin Parmentier.* Paris, 1815.

PECQUET, Jean (1622–1674)

Bentata, Yves, *Un Médecin Anatomiste du XVIIè Siècle, Jean Pecquet.* Paris, 1932.

Gillis, Paul, "Un Anatomiste Français: Jean Pecquet," *Revue Internationale de l'Enseignement,* LXXVII (1923) 271–284 and 352–358, LXXVIII (1924) 35–42.

Jeudy, René-Louis, *Les Principaux Anatomistes Français du XVIIè Siècle.* Paris, 1941.

Lucq, Jean, *Jean Pecquet, 1622–1674.* Paris, 1925.

PELLETAN, Philippe-Jean (1747–1829)

Genty, Maurice, "Philippe-Jean Pelletan (1747-1829)," *Les Biographies Médicales,* VIII, no. 8 (1934).

PELLETIER, Bertrand (1761–1797)

Blaessinger, Edmond, *Quelques Grandes Figures de la Pharmacie Militaire.* Paris, 1948, pp. 119–144.

Dorveaux, Paul, "Apothicaires Membres de l'Académie Royale des Sciences. XIII. Bertrand Pelletier," *Revue d'Histoire de la Pharmacie,* XXV (1937) 5–24.

Valette, Guillaume, "Sur Deux Miniatures Représentant Bertrand Pelletier et Hilaire Marie Rouelle, Trouvées dans les Documents du Secrétariat de la Faculté de Pharmacie de Paris," *Revue d'Histoire de la Pharmacie,* L (1962) 217–218.

PÉRIER, Jacques-Constantin (1742–1818)

Bouchary, Jean, *L'Eau à Paris à la Fin du 18è Siècle*. Paris, 1946.

Payen, Jacques, *Recherches sur les Frères Périer et l'Introduction en France de la Machine à Vapeur de Watt*. Mimeographed thesis, Paris, Ecole des Hautes Etudes, VIè section. Paris, 1966.

PERRAULT, Claude (1613–1688)

Barchillon, Jacques, "Les Frères Perrault à travers la Correspondance et les Œuvres de Christian Huygens," *XVIIè Siècle* (1962), no. 56, 19–36.

Colombe, Jean, "Portraits d'Ancêtres. III: Claude Perrault," *Hippocrate*, XVI (1949) 1–47.

Hallays, André, *Les Perrault*. Paris, 1926.

Lebovits, Joseph, *Claude Perrault, Physiologiste*. Paris, 1931.

Schiller, Netty, "L'Iconographie de Claude Perrault (1613–1688)," *91è Congrès des Sociétés Savantes*, I, 215–234.

Tenenti, Alberto, "Claude Perrault et la Pensée Scientifique Française dans la Seconde Moitié du XVIIè Siècle," *Hommage à Lucien Febvre. Eventail de l'Histoire Vivante*, II, 303–316. Paris, 1953.

PERRONET, Jean-Rodolphe (1708–1794)

Cheguillaume, H., "Perronet, Ingénieur de la Généralité d'Alençon, 1737–1747," *Bulletin de la Société Historique et Archéologique de l'Orne*, X (1891) 40–117.

Debauve, Alphonse, *Les Travaux Publics et les Ingénieurs des Ponts et Chaussées depuis le 17è Siècle*. Paris, 1893.

Lesage, Pierre-Charles, *Notice pour Servir à l'Eloge de M. Perronet*. Paris, an XIII [1805].

MSS. in Paris, Bibliothèque de l'Ecole des Ponts et Chaussées, MS. 336.

PETIT, Antoine (1722–1794)

Cuissard, Charles, "Notice sur Antoine Petit d'Orléans (1722–1794)," *Mémoires de la Société d'Agriculture, Sciences, Belles-Lettres et Arts d'Orléans*, ser. 5, I (1901) 127–164.

PETIT, alias POURFOUR du PETIT, François (1664–1741)

Hamy, Ernest-Théodore, "Petit contre Petit; une Pétition de Pourfour du Petit pour Obtenir la Place d'Académicien Pensionnaire (1725)," *France Médicale*, LIII (1906) 217–219.

Zehnder, E., *François Pourfour du Petit (1664–1741) und seine Experimentelle Forschung über das Nervensystem*. Zurich, 1968.

PETIT, Jean-Louis (1674–1750)

Hamy, see PETIT, François.

Louis, Antoine, *Eloge de M. Petit*. Paris, 1750.

Mondor, Henri, *Anatomistes et Chirurgiens*. Paris, 1949, pp. 165–216.

PICARD, Jean (1620–1682)

Armitage, Angus, "Jean Picard and his Circle," *Endeavor*, XIII (1954) 17–21.

PINEL, Philippe (1745–1826)

Grange, Kathleen M., "Pinel and Eighteenth-Century Psychiatry," *Bulletin of the History of Medicine*, XXXV (1961) 442–453.

Lechler, Walter Helmut, *Philippe Pinel. Seine Familie, seine Jugend- und Studienjahre, 1745–1778*. Munich, 1960.

Semelaigne, René, *Philippe Pinel et son Œuvre*. Paris, 1888.

Woods, Evelyn A., and Eric T. Carlson, "The Psychiatry of Philippe Pinel," *Bulletin of the History of Medicine*, XXXV (1961) 14–25.

PINGRÉ, Alexandre-Guy (1711–1796)

Armitage, Angus, "The Pilgrimage of Pingré (1711–1796). An Astronomer-Monk of Eighteenth-Century France," *AS*, IX (1953) 47–63.

Lacroix, Alfred. *Figure des Savants*, III, 177–184. Paris, 1938.

PITOT, Henri (1695–1771)

Humbert, Pierre, "La Vie et l'Œuvre d'Henri Pitot," *Bulletin de la Société Languedocienne de Géographie*, ser. 2, XXIV (1953) 167–198.

———, "L'Œuvre Mathématique d'Henri Pitot," *RHS*, VI (1953) 322–328.

POISSONNIER, Pierre-Isaac (1720–1798)

Sue, Pierre, *Eloge de Pierre-Isaac Poissonnier*. Paris, an VII [1805].

Vallery-Radot, Pierre, "La Vie Ardente de Pierre-Isaac Poissonnier," *Bulletin de la Société Française d'Histoire de la Médecine*, XXXII (1938) 44–54.

PORTAL, Antoine (1742–1832)

Barritault, Georges, *L'Anatomie en France au XVIIIè Siècle*. Angers, 1940, pp. 73–83.

Hamy, Ernest-Théodore, "Lettres de Noblesse en Faveur du Sieur Portal et de sa Postérité," *Bulletin de la Société Française d'Histoire de la Médecine*, V (1906) 361–365.

POUPART, François (1661–1709)

Delaunay, Paul, "F. Poupart," *Bulletin de la Société Française d'Histoire de la Médecine*, III (1904) 323–342.

PRONY, Gaspard-Clair-François-Marie Riche (1755–1839)

Dupin, Charles, *Eloge de M. le Baron de Prony*. Paris, 1840.

Walckenaer, "La Vie de Prony," *Bulletin de la Société d'Encouragement pour l'Industrie Nationale*, CXXXIX (1940) 68–98, 166.

QUESNAY, François (1694–1774)

Doyon and Liaigre, see VAUCANSON

Huard, Pierre, "Le Bicentenaire du Tableau Economique de François Quesnay (1758)," *Histoire de la Médecine*, numéro spécial (1958) 155–169.

Schelle, Gustav, *Le Docteur Quesnay*. Paris, 1907.

François Quesnay et la Physiocratie. Paris, 1958.

Articles by J. Benard, J. Conan, L. Einaudi, J. Hecht, A. Kubota, A. Landry, J. Molinier, L. Salleron, J. J. Spengler, and H. Woog.

RAMOND de CARBONNIÈRES, Louis-François-Elisabeth (1755–1827)

Cornet, Lucien, *Ramond pendant la Tourmente Révolutionnaire. Notes Chronologiques 1789–1795*. Pau, 1955.

Girdlestone, Cuthbert, *Louis-François Ramond (1755–1827). Sa Vie, son Œuvre Littéraire et Politique*. Paris, 1968.

Reboul, Jacques, *Un Grand Précurseur des Romantiques. Ramond (1755–1827)*. Nice, 1910.

MSS. in BI, MS. 3154.

RÉAUMUR, René-Antoine Ferchault de (1683–1757)

Birembaut, see BRISSON

Caullery, Maurice, *Les Papiers Laissés par de Réaumur*. Paris, 1929.

Torlais, Jean, *Un Esprit Encyclopédique en dehors de l'Encyclopédie. Réaumur*, rev. ed. Paris, 1961.

La Vie et l'Œuvre de Réaumur (1683–1757). Paris, 1962.

Articles by A. Birembaut, A. Davy de Virville, P. P. Grassé, R. Mortier, J. Rostand, P. Speziali, R. Taton, and J. Torlais.

RÉGIS, Pierre-Sylvain (1632–1707)

Watson, Richard A., "A Note on the Probabilistic Physics of Pierre-Sylvain Régis, 1632–1707," *AIHS*, XVII (1964) 33–36.

RENEAUME, alias RENEAULME de la GARANNE, Michel-Louys (1675–1739)

Dupré, Alexandre, "Essais Biographiques sur Quelques Médecins Blésois," *Mémoires de la Société des Sciences et Lettres de Blois*, V (1856) 483–488.

RICHARD, Louis-Claude-Marie (1754–1821)

Bertemes, G., "Correspondance de Linné Père et Fils avec André Thouin . . . et Notices Biographiques sur Linné, Thouin, les

Richard, Dombey et Commerson," *Bulletin de la Société d'Histoire Naturelle des Ardennes*, XXX (1935) 39–120.

Lamy, Edouard, "Note sur une Collection Conchyliologique du Commencement du XIXè Siècle," *Bulletin du Muséum d'Histoire Naturelle*, XXI (1915) 101–104.

Landrin, Armand, "Correspondance Inédite de Linné avec Claude Richard et Antoine Richard," *Mémoires de la Société des Sciences Naturelles de Seine-et-Oise*, VIII (1862–1863) 73–120.

MSS. in AN, E 350; BI, MS. 3180.

RICHER, Jean (1630–1696)

Olmsted, John W., "The Scientific Expedition of Jean Richer to Cayenne, 1672–1673," *Isis*, XXXIV (1942) 117–128.

———, "The Voyage of Jean Richer to Arcadia in 1670: A Study on the Relations of Science and Navigation under Colbert," *Proceedings APS*, CIV (1960) 612–634.

ROBERVAL, Gilles Personne, alias Personier, de (1602–1675)

Auger, Léon, *Un Savant Méconnu: Gilles Personne de Roberval (1602–1675)*. Paris, 1962.

Rochot, see MARIOTTE.

ROCHON, Alexis-Marie de (1741–1817)

Lacroix, Alfred, *Figure des Savants*, IV, 15–24. Paris, 1938.

Levot, Prosper-Jean, *Biographie Bretonne*, II, 739–746. Paris, 1857.

RØMER, Ole alias Olaüs or Olof (1644–1710)

Cohen, I. Bernard, *Roemer and the First Determination of the Velocity of Light*. New York, 1944.

Pihl, Mogens, *Ole Rømers Videnskabelige Liv*. Copenhagen, 1944.

Strömgren, Elis, *Ole Rømer som Astronom*. Copenhagen, 1944.

ROLLE, Michel (1652–1719)

Sergescu, Pierre, "Un Episod din Batalia pentru Triumful Calculului Diferential: Polemica Rolle-Saurin, 1702–1705," *Essais Scientifiques*. Timişoara, 1944.

ROUELLE, Guillaume-François (1703–1770)

Dorveaux, Paul, "Apothicaires Membres de l'Académie Royale des Sciences. IX. Guillaume-François Rouelle," *Revue d'Histoire de la Pharmacie*, XXI (1933), 169–186.

Rappaport, Rhoda, "G. F. Rouelle: An Eighteenth-Century Chemist and Teacher," *Chymia*, VI (1960) 68–101.

———, "Rouelle and Stahl—The Phlogistic Revolution in France," *Chymia*, VII (1961) 73–102.

Secrétan, Claude, *Un Aspect de la Chimie Pré-Lavoisienne, le Cours de G. F. Rouelle.* Lausanne, 1943.

SABATIER, Raphaël-Bienvenu (1732–1811)
Percy, Pierre-François, *Eloge Historique de M. Sabatier.* Paris, 1812.

SAGE, Balthazar-Georges (1740–1824)
Dorveaux, Paul, "Apothicaires Membres de l'Académie Royale des Sciences. XI. Balthazar-Georges Sage," *Revue d'Histoire de la Pharmacie,* XXIII (1935) 152–166, XXIV (1936) 216–232.

SAURIN, Joseph (1655–1737)
Sergescu, see ROLLE.

SAUVEUR, Joseph (1653–1716)
Auger, Léon, *Un Fondateur de l'Acoustique: Joseph Sauveur.* Paris, 1956.

SENAC, Jean-Baptiste (1693–1770)
Degris, Georges, *Etude sur Sénac, Premier Médecin de Louis XV (1693–1770).* Paris, 1901.

TAUVRY, Daniel (1669–1701)
Delaunay, Paul, *Vieux Médecins Mayennais.* Paris, 1903, pp. 45–57.

TENON, Jacques-René (1724–1816)
Huard, Pierre, "Les Papiers de Tenon," *XIXth International Congress for the History of Medicine,* pp. 567–572.
Laignel-Lavastine, Maxime, "Mémoires de Tenon sur les Hôpitaux de Paris," *Histoire de la Médecine,* I (1951) 38–44.
Silvestre, see DESMAREST.

TERRASSON, Jean-Baptiste (1670–1750)
Thomasseau de Cursay, Jean-Marie-Joseph, *Mémoires sur les Sçavans de la Famille de Terrasson.* Paris, 1761.

TESSIER, Henri-Alexandre (1741–1837)
Menault, Ernest, *Souvenirs de Beauce.* Etampes, 1858, pp. 67–112.
Silvestre, Augustin-François, *Notice Biographique sur Feu M. Tessier.* Paris, 1839.
Inauguration du Monument Elevé en l'Honneur de Tessier à Angerville. Versailles, 1876.
MSS. in Paris, Académie de Médecine, MS. box 120.

THÉVENOT, Melchisédech (1620–1692)
Lenox, James, *The Voyages of Thévenot.* New York, 1879.
McKeon, Robert M., "Une Lettre de Melchisédech Thévenot sur les Débuts de l'Académie Royale des Sciences," *RHS,* XVIII (1965) 1–6.

Nordstrom, Johan, "Swammerdamiana," *Lychnos* (1954–1955) 21–65.

Thouin, André (1747–1824)

Bertemes, see Richard.

Guillaumin, André, "André Thouin et l'Enrichissement des Collections de Plantes Vivantes au Muséum aux Dépens des Jardins de la Liste Civile des Emigrés et Condamnés," *Bulletin du Muséum d'Histoire Naturelle*, ser. 2, XVI (1944) 483–490.

————, and V. Chaudun, "La Collection des Modèles Réduits d'Instruments Agricoles et Horticoles du Muséum à propos d'une Lettre Inédite de A. Thouin," *Bulletin du Muséum d'Histoire Naturelle*, ser. 2, XVI (1944) 137–141.

Letouzey, Yvonne, "Les Arbres de la Liberté en l'An II, avec un Texte Inédit d'André Thouin (1747–1824)," *Revue Forestière Française*, XI (1961) 685–692.

Silvestre, Augustin-François, *Notice Biographique sur M. André Thouin*. Paris, 1825.

Thiébaut de Berneaud, Arsenne, *Eloge Historique de André Thouin*. Paris, 1825.

MSS. in BMHN, MS. 308–316, 1971–1989, 1997–1998; BI, MS. 3208.

Tillet, Mathieu (1714–1791)

Wehnelt, B., "Mathieu Tillet. Tilletia," *Beiträge zur Geschichte der Phytopathologie*, III (1937) 45–146.

Tournefort, Joseph Pitton de (1656–1708)

Heim, Roger, ed., *Tournefort*. Paris, 1957.

Articles by G. Becker, H. Bianchi, C. Carrière, J. P. Coste, P. J. Dufert, R. Dughi, G. Duprat, H. Gaussen, P. Guiral, H. Humbert, J. F. Leroy, F. Meyer, J. Motte, and G. Ranson.

Tressan, Louis-Elisabeth de la Vergne (1705–1783)

Jacoubet, Henri, *Le Comte de Tressan et les Origines du Genre Troubadour*. Paris, 1923.

Tressan, Henri A. G., *Souvenirs du Comte de Tressan*. Versailles, 1897.

Tschirnhaus, Ehrenfried Walther (1651–1708)

Winter, Emil, *E. W. von Tschirnhaus (1651–1708), ein Leben im Dienste des Akademiegedankens*. Berlin, 1959.

————, ed., *E. W. von Tschirnhaus und die Frühaufklärung in Mittel- und Osteuropa*. Berlin, 1960.

Turgot, Etienne-François (1721–1788)

Lacroix, Alfred, *Figure des Savants*, III, 61–69. Paris, 1938.

Lamy, Edouard, "Un Collectionneur Naturaliste du XVIIIè

Siècle: Le Chevalier Turgot," *Bulletin du Muséum National d'Histoire Naturelle*, ser. 2, II (1930) 657–659.

VAILLANT, Sébastien (1669–1722)

Elaut, L., "Een Parabotanisch Gedicht op het Levenswerk van Sebastien Vaillant," *Biologisch Jaarboek*, XXXI (1963) 313–318.

Ramsbottom, John, "S. Vaillant's 'Discours sur la Structure des Fleurs' 1718," *Journal of the Society for the Bibliography of Natural History*, IV (1963) 194–196.

VANDERMONDE, Alexandre-Théophile (1735–1796)

Birembaut, Arthur, "Précisions sur la Biographie du Mathématicien Vandermonde et de sa Famille," *72è Congrès de l'Association Française pour l'Avancement des Sciences*. Paris, 1953, pp. 530–533.

Lebesgue, Henri, "L'Œuvre Mathématique de Vandermonde," *L'Enseignement Mathématique*, ser. 2, I (1956) 203–223.

VARIGNON, Pierre (1654–1722)

Costabel, Pierre, *Pierre Varignon (1654–1722) et la Diffusion en France du Calcul Différentiel et Intégral*. Paris, 1966.

———, "Varignon, Lamy et le Parallélogramme des Forces," *AIHS*, XIX (1966) 103–124.

Fleckenstein, Joachim Otto, "Pierre Varignon und die Mathematischen Wissenschaften im Zeitalter des Cartesianismus," *AIHS*, II (1948) 76–138.

VAUCANSON, Jacques de (1709–1782)

Doyon, André, and Lucien Liaigre, *Jacques Vaucanson, Mécanicien de Génie*. Paris, 1967.

Maingot, Eliane, "Histoire Anecdotique d'un Méconnu Célèbre: Vaucanson," *Aesculape*, XLVII (1964), no. 5, 3–46.

VAUQUELIN, Baptiste-Nicolas-Louis (1763–1829)

Bonnemain, Henri, "Vauquelin et la Société Philomatique," *Revue d'Histoire de la Pharmacie*, LII (1964) 149–154.

Bouvet, Maurice, "Vauquelin Fut-il Membre de l'Académie Royale des Sciences?" *Revue d'Histoire de la Pharmacie*, XLIII (1955) 66–70.

———, and Georges Kersaint, "A la Gloire de Nicolas Vauquelin," *Revue d'Histoire de la Pharmacie*, LI (1963) 17–25.

Delaunay, Albert, "Louis-Nicolas Vauquelin," *Histoire de la Médecine*, XII (1963), no. 6, 2–15.

Dorveaux, Paul, "Vauquelin Fut-il Membre de l'Académie Royale des Sciences?" *Revue d'Histoire de la Pharmacie*, XX (1932) 57–60.

Kersaint, Georges, "Sur une Correspondance Inédite de Nicolas-Louis Vauquelin (1763–1829)," *Bulletin de la Société Chimique et Biologique de France* (1958) 603–619.

Kersaint, see FOURCROY.

Partington, James Riddick, *A History of Chemistry*, III, 551–557. London, 1962.

Revue d'Histoire de la Pharmacie, LI (1963), no. 177.
Issue devoted entirely to Vauquelin.

VENTENAT, Etienne-Pierre (1757–1808)
MSS. in BI, MS. 3188.

VICQ D'AZYR, Félix (1748–1794)
Cabanis, Pierre-Jean-Georges, "Eloge de Vicq-d'Azyr," *Œuvres Complètes*, V, 179–216. Paris, 1825.

Dubois, Emile-Frédéric, "Recherches Historiques sur les Dernières Années de Louis et Vicq d'Azyr," *Gazette Médicale de Paris*, ser. 3, XXI (1866) 641–650, 655–667, 669–680.

Dufresne, A. J. L., *Notes sur la Vie et les Œuvres de Vicq d'Azyr (1748–1794)*. Bordeaux, 1906.

Moreau, Jacques-Louis, *Eloge de Félix Vicq d'Azyr*. Paris, an VI [1798].

MSS. in Paris, Académie de Médecine, MS. boxes 106, 109; BI, MS. 3154; BMHN, MS. 1452–1459.

VIEUSSENS, Raymond (1641–1715)
Dulieu, Louis, "Raymond Vieussens," *Monspeliensis Hippocrates*, X (1967), no. 35, 8–26.

Jeudy, René-Louis, *Les Principaux Anatomistes Français du XVIIè Siècle*. Paris, 1941.

Sachs, B., "Raymond de Vieussens, Noted Neuroanatomist and Physician of the XVII Century," *Proceedings of the Charaka Club*, III (1910) 99–105.

WINSLOW, Jacques-Bénigne (1669–1760)
Barritault, Georges, *L'Anatomie en France au XVIIIè Siècle*. Angers, 1940, pp. 43–63.

Snorrason, Egill, *Anatomen J. B. Winsløw 1669–1760*. Copenhagen, 1969.

Autobiographie, ed. Vilhelm Maar. Paris, 1912.

MSS. in Paris, Bibliothèque Mazarine, MS. 1167.

C. *Other Academicians*

AIGUILLON, Armand-Louis Duplessis de Richelieu (1683–1750)
Locard, Arnould, "Correspondance Inédite entre le Comte

d'Agenois, Duc d'Aiguillon, le Comte de Seignelay et le Cardinal de Polignac sur la Divisibilité de la Matière," *Mémoires de l'Académie des Sciences, Belles-Lettres et Arts de Lyon,* Sciences, ser. 2, XXVI (1833–1834) 279–326. (Reprinted in *Revue de l'Agenais,* XI (1884) 193–209, 303–329.)

BERTIN, Henri-Léonard-Jean-Baptiste (1720–1792)

Antoine, Michel, *Le Secrétariat d'Etat de Bertin 1763–1780.* Paris, 1948.

Bourde, André J., *Agronomie et Agronomes en France au XVIIIè Siècle.* Paris, 1967, pp. 1079–1289.

Dronne, Michel, *Bertin et l'Elevage Français au XVIIIè Siècle.* Paris, 1966.

BIGNON, Jean-Paul (1662–1743)

Amsler, M., "Une Lettre de Leibniz," *Gesnerus,* XIX (1962) 87–88.

Bonnardet, E., "L'Abbé Bignon," *L'Oratoire de France* (1937) 46–50.

Jacquot, Jean, "Sir Hans Sloane and French Men of Science," *Notes and Records of the Royal Society of London,* X (1952) 85–98.

MSS. in BN, MS. fr. 22225–22236.

BOCHART DE SARON, Jean-Baptiste-Gaspard (1730–1794)

Cassini, Jean-Dominique, *Mémoires pour Servir à l'Histoire des Sciences.* Paris, 1810, pp. 373–391.

Montjoie, Félix-Louis-Christophe, *Eloge Historique de Jean-Baptiste-Gaspard Bochart de Saron.* Paris, [1800].

CHAULNES, Michel-Ferdinand d'Albert d'Ailly (1714–1769)

Mahler, P., "Un Savant au XVIIIè Siècle: Le Duc de Chaulnes," *Bulletin de la Société Philomatique de Paris,* ser. 10, I (1909) 72–76.

CHAUSSIER, François (1746–1828)

Romontet, Ernest, *François Chaussier (1746–1828). Sa Vie, son Œuvre.* Montbrison, 1921.

COQUEBERT DE MONTBRET, Charles-Etienne (1755–1831)

Silvestre, Augustin-François, *Notice Biographique sur M. C. E. Baron Coquebert de Montbret.* Paris, 1832.

FAGON, Guy-Crescent (1638–1718)

Beguin, Marie Colette, *Monsieur Fagon Médecin du Roy.* Mimeographed thesis. Bordeaux, 1958.

Fréguel, Jean-Baptiste, *Guy-Crescent Fagon, Premier Médecin de Louis XIV.* Bordeaux, 1923.

Grozieux de Laguérenne, Augustin-Marie-François-Jean, *Guy-*

Crescent Fagon, Archiatre de Louis XIV, Surintendant du Jardin Royal des Plantes (1638–1718). Paris, 1930.

LA ROCHEFOUCAULD D'ENVILLE, Louis-Alexandre (1743–1792)
Hahn, Roger, "La Réforme des Statuts de l'Académie Royale des Sciences en 1789," *RHS*, XVIII (1965) 15–28.
Jovy, Ernest, "La Correspondance du Duc de La Rochefoucauld d'Enville et de Georges Louis Le Sage Conservée à la Bibliothèque de Genève," *Bulletin du Bibliophile et du Bibliothécaire* (1917) 461–490, (1918) 33–55 and 135–151.

LATREILLE, Pierre-André (1762–1833)
Nussac, Louis de, *Les Débuts d'un Savant Naturaliste. Le Prince d'Entomologie: P. A. Latreille à Brive, de 1762 à 1798*. Paris, 1906.
Ross, Herbert H., "The Early History of Sawfly Study in North America," *Transactions of the American Entomological Society*, LXXXV (1959) 315–321.

LE PELETIER, alias LE PELLETIER DES FORTS, Michel-Robert (1675–1740)
Ross, see LATREILLE.

LOMÉNIE DE BRIENNE, Etienne-Charles de (1727–1794)
Bonnet, Edmond, "Le Cabinet d'Histoire Naturelle des Frères Loménie de Brienne," *Bulletin du Muséum d'Histoire Naturelle*, XII (1916) 278–287.

LOWENDAL, alias LOWENDAHL, Ulric-Frédéric Woldemar (1700–1755)
Sinéty, André-Louis-Woldemar-Alphée, *Vie du Maréchal de Lowendal*. Paris, 1867, 2 vols.

LUYNES, Paul d'Albert (1703–1788)
Blondel, "Notice Historique sur le Cardinal de Luynes, Archevêque de Sens 1703–1788," *Bulletin des Sciences Historiques et Naturelles de l'Yonne*, L (1896) supplément, 99–106.
Fournerat, C., "Recherches sur les Personnes qui dans le Département de l'Yonne se Sont Occupées de l'Astronomie et des Sciences qui s'y Rapportent," *Bulletin des Sciences Historiques et Naturelles de l'Yonne*, XVI (1862) Sciences, 47–55.

MALEBRANCHE, Nicolas (1638–1715)
Robinet, André, "La Vocation Académicienne de Malebranche," *RHS*, XII (1959) 1–18.
Robinet, see LEIBNIZ.
Rodis-Lewis, Geneviève, *Nicolas Malebranche*. Paris, 1963.
Malebranche, l'Homme et l'Œuvre, ed. Centre International de Synthèse. Paris, 1967.

Malebranche, Nicolas, *Œuvres Complètes*. Paris, 1958–1967, 20 vols.

Extensive bibliography, XX, 339–441.

MALESHERBES, Chrétien-Guillaume de Lamoignon de (1721–1794)

François, Marcel, "Rousseau et Malesherbes: Nouvelles Datations de Quelques Lettres sur la Botanique," *Studi Francesi*, II (1958) 15–23.

Grosclaude, Pierre, *Malesherbes, Témoin et Interprète de son Temps*. Paris, 1961.

PERCY, Pierre-François (1754–1825)

Bourdin, Ernest, *Percy*. Besançon, 1906.

POLIGNAC, Melchior de (1661–1741)

Fusil, Casimir Alexandre, *L'Anti-Lucrèce du Cardinal de Polignac*. Paris, 1917.

Locard, see AIGUILLON.

Patry, Edouard, *L'Anti-Lucrèce du Cardinal de Polignac*. Auch, 1872.

RENAU, alias RENAU D'ELISSAGARAY, Bernard (1652–1719)

Bagneux, Jean de, *Renau d'Elissagaray 1652–1719*. Bayonne, 1932.

Costabel, Pierre, "Une Leçon Magistrale de Jean I Bernoulli," *Hommage à Gaston Bachelard*. Paris, 1957, pp. 83–92.

TALLARD, Camille d'Hostun (1652–1728)

Bry, Joseph, "Le Duc de Tallard et son Duché," *Annales de la Faculté de Droit d'Aix*, n.s., XL (1947) 5–25.

TORCY, Jean-Baptiste Colbert (1665–1746)

Duffo, François-Albert, *J. B. Colbert Marquis de Torcy*. Paris, 1934.

Marchand, Jean, "Abrégé de la Vie de M. le Marquis de Torcy Escrite par Mme la Marquise d'Ancezune, sa Fille," *Revue d'Histoire Diplomatique*, XLVI (1932) 310–343, XLVII (1933) 51–76 and 188–214.

TRUCHET, Jean, alias Le Père Sébastien (1657–1729)

Léry, Edmond, "Le P. Sébastien Truchet. Ses Travaux à Versailles et à Marly," *Revue de l'Histoire de Versailles et de Seine-et-Oise*, XXXI (1929) 220–241.

MSS. in AN, M 803, 849–851, 855; BS, MS. 1025.

TRUDAINE, Daniel-Charles (1703–1769)

Delorme, Suzanne, "Une Famille de Grands Commis de l'Etat, Amis des Sciences au XVIIIè Siècle: Les Trudaine," *RHS*, III (1950) 101–109.

TRUDAINE DE MONTIGNY, Jean-Charles-Philibert (1733–1777)

Delorme, see TRUDAINE.

VAUBAN, Sébastien le Prestre (1633–1707)

    Esmonin, Edmond, "Quelques Données Inédites sur Vauban et les Premiers Recensements," *Population*, IX (1954) 507–512.

    Guerlac, Henry, "Vauban: The Impact of Science on War," *Makers of Modern Strategy*, ed. Edward M. Earle. Princeton, 1943, pp. 26–48.

    Guttin, Jacques, *Vauban et le Corps des Ingénieurs Militaires.* Paris, 1957.

Appendix III

Bibliography

L'Abréviateur Universel. Paris, 1792–an V[1797].

Académie des Sciences, *Index Biographique des Membres et Correspondants de l'Académie des Sciences*. Paris, 1968.

Académie Française, *Les Registres de l'Académie Française, 1672–1793*. Paris, 1895–1906, 4 vols.

Ackerknecht, Erwin H., *Medicine at the Paris Hospitals 1794–1848*. Baltimore, 1967.

Actes de la Société d'Histoire Naturelle de Paris, ed. A. L. Millin. Paris, 1792.

Adams, Percy G., *Travelers and Travel Liars 1660–1800*. Berkeley, 1962.

Affiches, Annonces et Avis Divers, ou Journal Général de France. Paris, 1777–1795.

Aguillon, Louis, "L'École des Mines de Paris: Notice Historique," *Annales des Mines*, ser. 8, XV (1889) 433–686.

Alembert, Jean le Rond d', "Essai sur la Société des Gens de Lettres et des Grands," *Œuvres Philosophiques, Historiques et Littéraires*, Vol. III, 19–104. Paris, an XIII [1805].

———, *Œuvres et Correspondances Inédites de d'Alembert*, ed. Charles Henry. Paris, 1887.

———, *Œuvres Philosophiques, Historiques et Littéraires*. Paris, an XIII–1805, 18 vols.

———, *Preliminary Discourse to the Encyclopedia of Diderot*, tr. Richard N. Schwab and Walter E. Rex. Indianapolis, 1963.

———, "Réflexions sur l'Etat Présent de la République des Lettres pour l'Article Gens de Lettres," *Œuvres et Correspondances Inédites de d'Alembert*, ed. Charles Henry, pp. 67–79. Paris, 1887.

Almanach Royal. Paris, 1700–1803.
    After 1793 the title changed to *Almanach National*.

L'Amateur d'Autographes. Paris, 1 September 1865.

A Messieurs les Electeurs du Département de Paris de l'Année 1791. [Paris, 6 September 1791.]
Pamphlet in BM, 936.f.9.

Amiable, Louis, *Le Franc-Maçon Jérôme Lalande.* Paris, 1889.

————, "Les Origines Maçonniques du Musée de Paris et du Lycée," *La Révolution Française*, XXXI (1896) 484–500.

————, *Une Loge Maçonnique d'avant 1789. La R*** L*** Les Neuf Sœurs.* Paris, 1897.

L'Ami du Roi, des Français, de l'Ordre et surtout de la Vérité. Paris, 1790–1792.

Annales de Chimie. Paris, 1789–an VIII [1800].

Annales Politiques, Civiles et Littéraires du Dix-Huitième Siècle, ed. Simon N. H. Linguet. London [etc.], 1777–1792, 19 vols.

Arbelet, Paul, "Prieur de la Côte-d'Or. Ministre des Munitions," *Revue Bleue*, LVI (1918) 14–19, 76–80, 108–112.

Archives Parlementaires de 1789 à 1860. Recueil Complet des Débats Législatifs et Politiques des Chambres Françaises, ed. Jérôme Mavidal and Emile Laurent. Paris, 1862–1969. (cited in notes as *AP.*)
Only série 1, vols. 1–87, covering 1787–1794, was used.

Arnault, Antoine-Vincent, *Souvenirs d'un Sexagénaire.* Paris, 1833, 4 vols.

Artz, Frederick B., *The Development of Technical Education in France, 1500–1850.* Cambridge, Mass., 1966.

Atkinson, Geoffroy, *The Extraordinary Voyage in French Literature before 1700.* New York, 1920.

————, *The Extraordinary Voyage in French Literature from 1700 to 1720.* Paris, 1922.

[Aubert de Vitry, François-Jean-Philibert], *J. J. Rousseau à l'Assemblée Nationale.* Paris, 1789.

Aubry, Paul V., *Monge. Le Savant Ami de Napoléon Bonaparte 1746–1818.* Paris, 1954.

Aucoc, Léon, *L'Institut de France et les Anciennes Académies.* Paris, 1889.

————, ed., *L'Institut de France. Lois, Statuts et Règlements Concernant les Anciennes Académies et l'Institut de 1635 à 1889.* Paris, 1889. (cited in notes as Aucoc, *Lois*)

Audrein, Yves-Marie, *Mémoire sur l'Education Nationale Française Suivi d'un Projet de Décret Présenté à l'Assemblée Nationale le 11 Décembre 1791.* Paris, 1791.

Aufrère, Léon, *Soulavie et son Secret, un Conflit entre l'Actualisme et le Créationisme.* Paris, 1952.

Auguis, Pierre-René, *Eloge Historique de L. A.* [sic] *Millin.* Paris, 1819.

Aulanier, Christiane, *Histoire du Palais et du Musée du Louvre. La Salle des Caryatides.* Paris, 1957.

Aulard, François-Victor-Alphonse, "Napoléon et l'Athée Lalande," *Etudes et Leçons sur la Révolution Française,* ser. 4, pp. 303–316. Paris, 1908.

———, *Recueil des Actes du Comité de Salut Public.* Paris, 1889–1951, 28 vols., and Tables, 3 vols.

———, *La Société des Jacobins.* Paris, 1889–1897, 6 vols.

———, "Talleyrand–Périgord," *Dictionnaire de Pédagogie et Instruction Primaire,* part I, 2853–2856. Paris, 1882–1887.

Bachaumont, Louis Petit de, *Mémoires Secrets pour Servir à l'Histoire de la République des Lettres en France.* London, 1780–1789, 36 vols. in -18°.

Bacon, Francis, *La Nouvelle Atlantide de François Bacon, Chancelier d'Angleterre, Traduite en François, et Continuée: Avec des Reflexions sur l'Institution et les Occupations des Academies Françoise, des Sciences, et des Inscriptions par M. R.* [Gilles–Bernard Raguet]. Paris, 1702.

Bacquié, Franc, "Les Inspecteurs des Manufactures sous l'Ancien Régime, 1669–1792," *Mémoires et Documents pour Servir à l'Histoire du Commerce et de l'Industrie en France,* XI, ed. J. Hayem. Paris, 1927.

Bailly, Jean-Sylvain, *Mémoires d'un Témoin de la Révolution.* Paris, 1804, 3 vols.

———, *Recueil de Pièces Intéressantes sur les Arts, les Sciences et la Littérature. Paris,* 1810.

Baker, Keith M., "Les Débuts de Condorcet au Secrétariat de l'Académie Royale des Sciences (1773–1776)," *Revue d'Histoire des Sciences,* XX (1967) 229–280.

———, "The Early History of the Term 'Social Science'," *Annals of Science,* XX (1964) 211–226.

———, "Scientism, Elitism and Liberalism: The Case of Condorcet," *Studies on Voltaire and the Eighteenth Century,* LV (1967) 129–165.

———, "An Unpublished Essay of Condorcet on Technical Methods of Classification," *Annals of Science,* XVIII (1962) 99–123.

———, and William A. Smeaton, "The Origins and Authorship of the Educational Proposals Published in 1793 by the Bureau de

Consultation des Arts et Métiers and Generally Ascribed to Lavoisier," *Annals of Science*, XXI (1965) 33–46.

Ballot, Charles, "Procès-Verbaux du Bureau de Consultation des Arts et Métiers," *Bulletin d'Histoire Economique de la Révolution* (1913) 15–160.

Barnes, Annie, *Jean Le Clerc, 1657–1736 et la République des Lettres*. Paris, 1938.

Barral, Georges, *Histoire des Sciences sous Napoléon*. Paris, 1889.

Barrière, Pierre, *L'Académie de Bordeaux, Centre de Culture Internationale au XVIIIè Siècle (1712–1792)*. Bordeaux, 1951.

————, *La Vie Intellectuelle en France du XVIè Siècle à l'Epoque Contemporaine*. Paris, 1961.

Batteux, Charles, *Les Beaux Arts Réduits à un Même Principe*. Paris, 1747.

Baumes, Jean-Baptiste-Théodore, *Discours sur la Dignité et les Avantages des Réunions Académiques*. Montpellier, an X [1802].

————, *Discours sur la Nécessité des Sciences dans une Nation Libre*. Montpellier, an III [1795].

Bellier de la Chavignerie, Emile, "Les Artistes Français du XVIIIè Siècle Oubliés ou Dédaignés," *Revue Universelle des Arts*, XIX (1864) 203–224, 249–252.

Ben-David, Joseph, "Scientific Growth: A Sociological View," *Minerva*, II (1964) 455–476.

————, "The Scientific Role: The Conditions of its Establishment in Europe," *Minerva*, IV (1965) 15–54.

Benveniste, E., "Civilisation. Contribution à l'Histoire du Mot," *Hommage à Lucien Febvre, Eventail de l'Histoire Vivante*, I, 47–54. Paris, 1953.

Benzing, Josef, *Die Buchdrucker des 16. und 17. Jahrhunderts*. Wiesbaden, 1963.

Bergasse du Petit-Thouars, Aristide, *Un Défenseur des Principes Traditionnels sous la Révolution. Nicolas Bergasse*. Paris, 1910.

Bergman, Torbern, *Torbern Bergman's Foreign Correspondance*, ed. Göte Carlid and Johan Nordstrom. Stockholm, 1965.

Bernal, John D., "Les Rapports Scientifiques entre la Grande-Bretagne et la France au XVIIIe Siècle," *Revue d'Histoire des Sciences*, IX (1956) 289–300.

Bernardin, Edith, *Jean-Marie Roland et le Ministère de l'Intérieur (1792–1793)*. Paris, 1964.

Berthelot, Marcelin, "Notice sur les Origines et sur l'Histoire de la Société Philomatique," *Mémoires Publiés par la Société Philo-*

matique à l'Occasion du Centenaire de sa Fondation, 1788–1888, pp. i–xvii. Paris, 1888.

Bertholon, Pierre, *Des Avantages que la Physique et les Arts qui en Dépendent Peuvent Tirer des Globes Aérostatiques*. Montpellier, 1784.

Berthoud, Ferdinand, *Essai sur l'Horlogerie*. Paris, 1763, 2 vols.

Bertrand, Joseph, *L'Académie des Sciences et les Académiciens de 1666 à 1793*. Paris, 1869.

Bigourdan, Guillaume, "Le Bureau des Longitudes. Son Histoire et ses Travaux de l'Origine (1795) à ce Jour," Bureau des Longitudes, *Annuaire* (1928) A1–A72, (1929) C1–C92, (1930) A1–A110, (1931) A1–A145, (1932) A1–A117, (1933) A1–A91.

———, "Inventaire Général et Sommaire des Manuscrits de la Bibliothèque de l'Observatoire de Paris," *Annales de l'Observatoire de Paris, Mémoires*, XXI (1895) F1–F60.

———, *Les Premières Sociétés Savantes de Paris et les Origines de l'Académie des Sciences*. Paris, 1919.

———, *Le Système Métrique des Poids et Mesures*. Paris, 1901.

Billingsley, Susan Vaughan, "L'Académie Royale des Sciences, 1785–1793: An Interpretation," unpublished Ph.D. dissertation, University of California, Berkeley, 1958.

Bioche, Charles, "Les Savants et le Gouvernement au XVIIè et XVIIIè Siècles," *Revue des Deux Mondes*, CVII (1937) 174–189.

Biographie Universelle, Ancienne et Moderne, ed. Jean F. Michaud and Louis G. Michaud. Paris, n.d., 45 vols.

Biollay, Léon, *Etudes Economiques sur le XVIIIè Siècle. Le Pacte de Famine. L'Administration du Commerce*. Paris, 1885.

Biot, Jean-Baptiste, *Essai sur l'Histoire Générale des Sciences pendant la Révolution Française*. Paris, an XI [1803].

———, *Mélanges Scientifiques et Littéraires*. Paris, 1858, 3 vols.

Bird, C. Wesley, "Une Source de la Chaumière Indienne," *Revue d'Histoire Littéraire de la France*, XLV (1938) 520–526.

Birembaut, Arthur, "L'Académie Royale des Sciences en 1780 Vue par l'Astronome Suédois Lexell (1740–1784)," *Revue d'Histoire des Sciences*, X (1957) 148–166.

———, "Les Deux Déterminations de l'Unité de Masse du Système Métrique," *Revue d'Histoire des Sciences*, XII (1959) 25–54.

———, "Les Ecoles Gratuites de Dessin," *Enseignement et Diffusion*, pp. 441–476.

———, "L'Enseignement de la Minéralogie et des Techniques Minières," *Enseignement et Diffusion*, pp. 365–418.

————, "Une Lettre Inédite de Marat à Roume," *Annales Historiques de la Révolution Française*, XXIX (1967) 395–399.

————, "Précisions sur la Biographie du Mathématicien Vandermonde et de sa Famille," *72è Session de l'Association Française pour l'Avancement des Sciences*, pp. 530–533. Paris, 1953.

————, "A propos d'une Publication Récente sur Lavoisier et le Lycée des Arts," *Revue d'Histoire des Sciences*, XI (1958) 267–273.

————, "Quelques Réflexions sur les Problèmes Posés par la Conservation et la Consultation des Archives Techniques Françaises," *Archives Internationales d'Histoire des Sciences*, XIX (1966) 21–102.

————, "Sur les Lettres du Physicien Magellan Conservées aux Archives Nationales," *Revue d'Histoire des Sciences*, IX (1956) 150–161.

————, and Guy Thuillier, "Une Source Inédite: Les Cahiers du Chimiste Jean Hellot (1685–1766)," *Annales*, XXI (1966) 357–364.

Birn, Raymond, "Le Journal des Savants sous l'Ancien Régime," *Journal des Savants* (1965) 15–35.

[Bitaubé, Paul-Jérémie], *Discours du Président de l'Institut Prononcé à la Barre du Conseil des Anciens, le Deuxième Jour Complémentaire [an VI]*. [Paris], an VII [1799].

Blake, John, "Scientific Institutions since the Renaissance: Their Role in Medical Research," *Proceedings of the American Philosophical Society*, CI (1957) 31–62.

Blanvillain, J. François C., *Le Pariséum ou Tableau Actuel de Paris*, 2nd ed. Paris, 1807–1808.

Bloch, Camille, ed., *Procès-Verbaux du Comité des Finances de l'Assemblée Constituante*. Rennes, 1922.

Boatner, Charlotte H., "Certain Unpublished Letters from French Scientists of the Revolutionary Period Taken from the Files of Joseph Lakanal," *Osiris*, I (1936) 173–183.

Bodley, John Edward Courtenay, *Cardinal Manning. The Decay of Idealism in France. The Institute of France*. London, 1912.

Boisseau, Jean, "Les Membres Parisiens de l'Académie Royale de Chirurgie," unpublished thesis, Rennes, Faculté de Médecine, 1962, 2 vols.

Boissy d'Anglas, François–Antoine, *Quelques Idées sur les Arts, sur la Nécessité de les Encourager, sur les Institutions qui Peuvent en Assurer le Perfectionnement, et sur Divers Etablissemens Nécessaires à l'Enseignement Public*. [Paris, 1795.]

Bondois, Paul M., "Le Privilège Exclusif au XVIIIè Siècle," *Revue d'Histoire Economique et Sociale*, XXI (1933) 140–189.

Bonnassieux, Pierre, and Eugène Lelong, eds., "Conseil de Commerce et Bureau du Commerce 1700–1791," *Inventaire Analytique des Procès-Verbaux*. Paris, 1900.

Bonnet, Edmond, "Histoire Anecdotique des Herborisations Parisiennes Précédée de Quelques Recherches sur l'Enseignement de la Botanique Médicale à Paris," *Bulletin des Sciences Pharmacologiques*, X (1904) 94–111.

Bouchard, Georges, *Guyton-Morveau, Chimiste et Conventionnel (1737–1816)*. Paris, 1938.

———, *Prieur de la Côte-d'Or*. Paris, 1946.

Bouchard, Marcel, *L'Académie de Dijon et le Premier Discours de Rousseau*. Paris, 1950.

Bouchary, Jean, *L'Eau à Paris à la Fin du XVIIIè Siècle. La Compagnie des Eaux de Paris et l'Enterprise de l'Yvette*. Paris, 1946.

Bouillier, Francisque, "Divers Projets de Réorganisation des Anciennes Académies," *Séances et Travaux de l'Académie des Sciences Morales et Politiques*, CXV (1881) 636–673.

———, "Une Parfaite Académie d'après Bacon et Leibniz," *Revue des Deux Mondes*, XXIX (1878) 673–697.

———, "La Révolution et les Académies," *Le Correspondant*, n.s. CII (1885) 438–460.

Bourde, André J., *Agronomie et Agronomes en France au XVIIIè Siècle*. Paris, 1967.

Bourgin, G., "Notes sur l'Administration de l'Agriculture et de la Législation Rurale de 1788 à l'An VIII," France. Commission de Recherche et de Publication des Documents Relatifs à la Vie Economique de la Révolution, *Bulletin Trimestriel* (1907) 248–260.

Boutteville, Roger, *L'Eclairage Public à Paris*. Paris, 1925.

Bouvet, Marcel, "La Législation des Remèdes Secrets de 1778 à 1803," *Bulletin de la Société d'Histoire de la Pharmacie*, IV (1923) 204–216.

Boyer, Carl B., "Mathematicians of the French Revolution," *Scripta Mathematica*, XXV (1960) 11–31.

Bray, René, *La Formation de la Doctrine Classique en France*. Paris, 1921.

Brissot de Warville, Jacques-Pierre, *Correspondance et Papiers*, ed. Claude-Marie Perroud. Paris, [1912].

———, *De la Vérité, ou Méditations sur les Moyens de Parvenir à la Vérité dans Toutes les Connoissances Humaines.* Neufchâtel, 1782.

———, *Journal du Licée de Londres, ou Tableau de l'Etat Présent des Sciences et des Arts en Angleterre.* Paris, 1784.

———, *Licée de Londres,* Prospectus. [London, 1784.]

———, *Mémoires 1754–1793,* ed. Claude-Marie Perroud. Paris, [1910], 2 vols.

[———,] *Un Mot à l'Oreille des Académiciens de Paris.* n.p., [1786].
   Pamphlet in Paris, Bibliothèque de l'Arsenal, 8°S.6877 and BHVP, 8°939204, and 12030(II), no. 1.

———, *Le Pot Pourri. Etrennes aux Gens de Lettres.* London, 1777.

Broc, Numa, "Voyages et Géographie au XVIIIè Siècle," *Revue d'Histoire des Sciences,* XXII (1969) 137–154.

Brombert, Victor, *The Intellectual Hero. Studies in the French Novel 1880–1955.* Philadelphia, 1961.

Brown, A. W., "Some Political and Scientific Attitudes to Literature and the Arts in the Years following the French Revolution," *Forum for Modern Language Studies,* II (1966) 230–252.

Brown, Harcourt, *Scientific Organizations in Seventeenth Century France (1620–1680).* Baltimore, 1934.

———, "The Utilitarian Motive in the Age of Descartes," *Annals of Science,* I (1936) 182–193.

Brunet, Pierre, "La Méthodologie de Mariotte," *Archives Internationales d'Histoire des Sciences,* I (1947) 26–59.

———, *Les Physiciens Hollandais et la Méthode Expérimentale en France au XVIIIè Siècle.* Paris, 1926.

———, *La Vie et l'Œuvre de Clairaut, 1713–1765.* Paris, 1952.

Bry, Ilse, "To Foster Human Greatness. History and Review of the Literature of the Concept of Genius," *Insight,* I (1962) 2–15, 36.

Bucquet, Jean-Baptiste-Michel, *Mémoire sur la Manière dont les Animaux Sont Affectés par Différens Fluides Aériformes, Méphitiques.* Paris, 1778.

Buffon. See Roger, ed.

Buisson, François-Albert, "L'Institut de France," *Revue des Travaux de l'Académie des Sciences Morales et Politiques,* CVII (1954) 177–187.

Bulletin de Littérature, des Sciences et des Arts. Paris, an III–V [1795–1797].

Burke, John G., *Origins of the Science of Crystals.* Berkeley, 1966.

Burr, Charles W., "Jean Paul Marat, Physician, Revolutionary, Paranoic," *Annals of Medical History*, XI (1919) 248–261.

Busson, Henri, *La Religion des Classiques (1660–1685)*. Paris, 1948.

Cabanès, Augustin, *Marat Inconnu*. Paris, 1891.

———, *Marat Inconnu*, 2nd ed. Paris, 1911.

Cabanis, Pierre-Jean-Georges, *Œuvres Complètes*. Paris, 1823–1825, 5 vols.

Cahen, Léon, *Condorcet et la Révolution Française*. Paris, 1904.

Cain, A. J., "Deductive and Inductive Methods in Post-Linnaean Taxonomy," *Proceedings of the Linnean Society of London*, CLXX (1957–1958) 185–217.

Campardon, Emile, *Les Spectacles de la Foire*. Paris, 1877, 2 vols.

Cap, Paul-Antoine, *Etudes Biographiques pour Servir à l'Histoire des Sciences*. Paris, 1857–1864, 2 vols.

Cardwell, Donald Stephen Lowell, *The Organisation of Science in England*. London, 1957.

Cassini, Jacques-Dominique, *Mémoires pour Servir à l'Histoire des Sciences et à celle de l'Observatoire Royal de Paris*. Paris, 1810.

Cassirer, Ernst, *The Philosophy of the Enlightenment*, tr. F. C. A. Koelln and J. P. Pettegrove. Boston, 1955.

Cazé, Michel, *Le Collège de Pharmacie de Paris (1777–1796)*. Fontenay-aux-Roses, [1943].

Challamel, Augustin, *Les Clubs Contre-Révolutionnaires*. Paris, 1885.

Chambon, Félix, "Une Page Inconnue de l'Histoire du Collège de France (1774–1807)," *Revue Internationale de l'Enseignement*, XLVI (1903) 208–217.

Chamfort, Sébastien, *Des Académies*. Paris, 1791.

Champeix, Robert, "Marat, ou la Science Agressive," *Savants Méconnus. Inventions Oubliées*, pp. 234–264. Paris, 1966.

Chapin, Seymour L., "The Academy of Sciences during the Eighteenth Century: An Astronomical Appraisal," *French Historical Studies*, V (1968) 371–404.

———, "Les Associés Libres de l'Académie Royale des Sciences. Un Projet Inédit pour la Modification de leurs Statuts (1788)," *Revue d'Histoire des Sciences*, XVIII (1965) 7–13.

———, "Astronomy and the Paris Academy of Sciences during the Eighteenth Century," unpublished Ph.D. dissertation, University of California, Los Angeles, 1964.

———, "Expeditions of the French Academy of Sciences, 1735," *Navigation*, III (1952) 120–122.

Chaptal, Jean-Antoine, *Mes Souvenirs sur Napoléon*, ed. A. Chaptal. Paris, 1893.

Charavay, Etienne, "Autographes et Documents Révolutionnaires," *La Révolution Française*, VI (1884) 726–728.

Chardon, and Jean Perny de Villeneuve, *Académie des Arts Proposée par une Société d'Artistes à l'Assemblée Nationale.* [Paris], n.d.
>   Pamphlet in BN, Imprimés, 8°Vp. 19097.

Charliat, P. J., "L'Académie Royale de Marine et la Révolution Nautique au XVIIIè Siècle," *Thalès*, I (1934) 71–82.

Chronique de Paris. Paris, 1789–1793.

La Chronique du Mois. Paris, 1791–1793.

Chuquet, Arthur, *L'Ecole de Mars (1794).* Paris, 1899.

Clark, Terry N., "Institutionalization of Innovations in Higher Education: Four Models," *Administrative Science Quarterly*, XIII (1968) 1–25.

Clément, Pierre, *Lettres, Instructions et Mémoires de Colbert.* Paris, 1861–1882, 8 vols.

Clifford-Vaughan, Michalina, "Enlightenment and Education," *British Journal of Sociology*, XIV (1963) 135–143.

———, "Some French Concepts of Elites," *British Journal of Sociology*, XI (1960) 319–331.
>   See also Vaughan, Michalina.

Cochrane, Eric W., *Tradition and Enlightenment in the Tuscan Academies 1690–1800.* Chicago, 1961.

Cohen, I. Bernard, *Franklin and Newton.* Philadelphia, 1956.

Cole, Arthur H., and George B. Watts, *The Handicrafts of France as Recorded in the Descriptions des Arts et Métiers 1761–1788.* Cambridge, Mass., 1952.

Collas, Georges, *Jean Chapelain: 1595–1674.* Paris, 1912.

Compayré, Gabriel, "Condorcet," *Dictionnaire de Pédagogie et d'Instruction Primaire*, part I, 462–464. Paris, 1882–1887.

Condorcet, Marie-Jean-Antoine-Nicolas Caritat de, *Eloges des Académiciens de l'Académie Royale des Sciences, Morts depuis 1666, jusqu'en 1699.* Paris, 1773.

———, *Œuvres de Condorcet*, ed. A. C. O'Connor and F. Arago. Paris, 1847–1849, 12 vols.

———, *Rapport et Projet de Décret sur l'Organisation Générale de l'Instruction Publique*, ed. Gabriel Compayré. Paris, 1883.

Connaissance des Temps. Paris, 1680–1803.

Le Conservateur, ed. Dominique-Joseph Garat, Pierre-Claude-François Daunou, and Marie-Joseph-Blaise Chénier. Paris, an V—an VI [1797–1798].

Constitution des Sciences, Arts, et Métiers, avec le Projet de Décret Présenté a l'Assemblée Nationale et Rédigé par la Société du

Point Central des Arts-et-Métiers. [Paris], an IV de la Liberté
[1793].

This item is sometimes listed under the title *Nouvelle Constitution . . .*, or the author, Desaudray, Charles.

Contant, Jean-Paul, *L'Enseignement de la Chimie au Jardin Royal des Plantes de Paris*. Cahors, 1952.

Corlieu, A., *L'Ancienne Faculté de Médecine de Paris*. Paris, 1877.

Correspondance des Contrôleurs Généraux des Finances, avec les Intendants des Provinces, 1685–1699, ed. A. M. de Boislisle. Paris, 1874–1897, 3 vols.

Correspondance des Propriétaires et Entrepreneurs de la Manufacture de Monseigneur Comte d'Artois, pour les Acides et Sels Minéraux, Etablie à Javel, près Paris, sur la Fabrication en Grand de l'Alun-Factice, n.p., n.d.

Pamphlet in BN, Imprimés, 8°Vz. 1733.

Corson, David W., "Pierre Polinière, Francis Hauksbee, and Electroluminescence: A Case of Simultaneous Discovery," *Isis*, LIX (1968) 402–413.

Costaz, Claude-Anthelme, *Histoire de l'Administration en France de l'Agriculture, des Arts Utiles, du Commerce, des Manufactures, des Subsistances, des Mines et des Usines*. Paris, 1832, 2 vols.

Costaz, Louis, *Discours sur les Sciences Naturelles et Mathématiques*. [Paris, 1798.]

Courrier Français. Paris, 1789–1796.

Cousin, Jacques-Antoine-Joseph, *Traité Elémentaire de Physique*. Paris, an III [1795].

Cousin, Jean, *L'Académie des Sciences, Belles-Lettres et Arts de Besançon. Deux Cent Ans de Vie Comtoise, 1752–1952. Essai de Synthèse*. n.p., 1954.

Cousin, Jules, *Le Comte de Clermont. Sa Cour et ses Maitresses*. Paris, 1867, 2 vols.

Couturat, Louis, *La Logique de Leibniz*. Paris, 1901.

Crestois, Paul, *L'Enseignement de la Botanique au Jardin Royal des Plantes de Paris*. Cahors, 1953.

Critical Problems in the History of Science, ed. Marshall Clagett. Madison, 1959.

Crosland, Maurice P., "The Congress on Definitive Metric Standards, 1798–1799: The First International Scientific Conference?" *Isis*, LX (1969) 226–231.

————, ed., *Science in France in the Revolutionary Era*. Cambridge, Mass., 1969.

————, *The Society of Arcueil. A View of French Science at the Time of Napoleon I*. London, 1967.

Crouzet, François, "Angleterre et France au XVIIIè Siècle: Essai d'Analyse Comparée de Deux Croissances Economiques," *Annales*, XXI (1966) 254–291.

Croÿ, Emmanuel, *Journal Inédit du Duc de Croÿ 1718–1784*, ed. Emmanuel-H. de Grouchy and Paul Cottin. Paris, 1906–1907, 4 vols.

Crozier, Michel, *The Bureaucratic Phenomenon*. Chicago, 1964.

Curtius, Ernst Robert, *Essai sur la France*, tr. Jacques Benoist-Méchin. Paris, 1932.

Cuvier, Jean-Léopold-Nicolas-Frédéric, *Rapport Historique sur les Progrès des Sciences Naturelles depuis 1789, et sur leur Etat Actuel*. Paris, 1810.

———, *Recueil des Eloges Historiques*, new ed. Paris, 1861, 3 vols.

Dainville, François de, "Enseignement des 'Géographes' et des 'Géomètres'," *Enseignement et Diffusion*, pp. 481–491.

Daoust, Joseph, "Encyclopédistes et Jésuites de Trévoux (1751–1752)," *Etudes*, CCLXXII (1952) 179–191.

Darcet, Jean, and Bertrand Pelletier, *Instruction sur l'Art de Séparer le Cuivre du Métal des Cloches*. Paris, an II [1794].

Darnton, Robert, *Mesmerism and the End of the Enlightenment in France*. Cambridge, Mass., 1968.

Daudin, Henri, *Cuvier et Lamarck. Les Classes Zoologiques et l'Idée de Série Animale, 1790–1830*. Paris, 1926, 2 vols.

Daumas, Maurice, *Les Cabinets de Physique au XVIIIè Siècle*. Paris, 1951.

———, "Le Corps des Ingénieurs Brevetés en Instruments Scientifiques (1787)," *Archives Internationales d'Histoire des Sciences*, V (1952) 86–96.

———, "Esquisse d'une Histoire de la Vie Scientifique," *Histoire de la Science*, ed. Maurice Daumas, pp. 27–159. Paris, 1952.

———, *Les Instruments Scientifiques aux XVIIè et XVIIIè Siècles*. Paris, 1953.

———, *Lavoisier. Théoricien et Expérimentateur*. Paris, 1955.

[Daunou, Pierre-Claude-François], *Discours Prononcé par le Citoyen Daunou, Vice-Président de l'Institut dans la Séance du Cinquième Jour Complémentaire, an 5.* [Paris], an VI [1798].

Davy de Virville, Adrien, *Histoire de la Botanique en France*. Paris, 1954.

Dawson, Warren R., ed., *Catalogue of the Manuscripts in the Library of the Linnean Society of London, part I. The Smith Papers*. London, 1934.

Debidour, Antonin, ed., *Recueil des Actes du Directoire Exécutif*. Paris, 1910–1917, 4 vols.

La Décade Philosophique, Littéraire et Politique. Paris, 1794–1807.

Decremps, Henri, *La Science Sanculotisée. Premier Essai sur les Moyens de Faciliter l'Etude de l'Astronomie, tant aux Amateurs et aux Gens de Lettres, qu'aux Marins de la République Française, et d'Opérer une Révolution dans l'Enseignement.* Paris, an II [1794].

Dehérain, Henri, "Lettres à G. Cuvier sur l'Organisation de l'Institut en l'An XI," *Journal des Savants* (1916) 368–376.

Dejob, Charles, *L'Instruction Publique en France et en Italie au Dix-Neuvième Siècle.* Paris, 1894.

Delambre, Jean-Baptiste-Joseph, *Grandeur et Figure de la Terre,* ed. G. Bigourdan. Paris, 1912.

———, *Histoire de l'Astronomie au Dix-Huitième Siècle.* Paris, 1827.

———, *Histoire de l'Astronomie Moderne.* Paris, 1821.

———, *Rapport Historique sur les Progrès des Sciences Mathématiques depuis 1789 et sur leur Etat Actuel.* Paris, 1810.

Delandine, Antoine-François, *Courrones Académiques ou Recueil des Prix Proposés par les Sociétés Savantes.* Paris, 1787.

The copy examined at BI has manuscript corrections and additions.

———, *De la Conservation et de l'Utilité Politique des Sociétés Littéraires dans les Départemens.* Paris, 1791.

Delattre, François-Pascal. *Rapport sur la Recherche à Faire de M. de La Pérouse . . . Précédé de la Pétition de la Société d'Histoire Naturelle de Paris, sur le Même Sujet, 9 Février 1791.* Paris, 1791.

Delaunay, Paul, *D'une Révolution à l'Autre 1789–1848. L'Evolution des Théories et de la Pratique Médicales.* Paris, 1949.

———, *Le Monde Médical Parisien au 18è Siècle.* Paris, 1906.

Delisle de Sales, Jean-Baptiste-Claude, *Mémoire pour les Académies.* Paris, an VIII-1800.

Delorme, Suzanne, "L'Académie Royale des Sciences: Ses Correspondants en Suisse," *Revue d'Histoire des Sciences,* IV (1951) 159–170.

———, "Académies et Salons," *La Synthèse. Idée-Force dans l'Evolution de la Pensée,* pp. 115–133. Paris, 1951.

———, "La Vie Scientifique à l'Époque de Fontenelle d'après les 'Eloges des Savants'," *Archeion,* XIX (1937) 217–235.

Delsaux, Hélène, *Condorcet Journaliste (1790–1794).* Paris, 1931.

Demachy, Jacques-François, *Histoire et Contes,* ed. Léon G. Toraude. Paris, 1907.

Derrouch, André, *L'Institut de France. Son Statut Juridique.* Paris, 1938.

Desautels, Alfred-R., *Les Mémoires de Trévoux et le Mouvement des Idées au XVIIIè Siècle.* Rome, 1956.

Descriptions des Arts et Métiers, Faites ou Approuvées par Messieurs de l'Académie Royale des Sciences, avec Figures. Paris, 1761–1789, 76 vols.

[Desgenettes, René], R. D. G., *Souvenirs de la Fin du XVIIIè Siècle et du Commencement du XIXè.* Paris, 1835–1836, 2 vols.

Despois, Eugène, *Le Vandalisme Révolutionnaire,* 2nd ed. Paris, 1885.

Devérité, Louis-A., *Notice pour Servir à l'Histoire de la Vie et des Ecrits de S. N. H. Linguet.* Liège, 1781.

Devic, Jean-François Schlisteur, *Histoire de la Vie et des Travaux Scientifiques et Littéraires de J.-D. Cassini IV.* Clermont, 1851.

Dictionnaire de Biographie Française, ed. J. Balteau, Marius Barroux, and Michel Prévost. Paris, 1933–present.

Dictionnaire de Pédagogie et d'Instruction Primaire, ed. F. Buisson. Paris, 1880–1887, 2 parts in 4 vols.

Dieckmann, Herbert, "The Concept of Knowledge in the *Encyclopédie,*" *Essays in Comparative Literature,* pp. 73–107. St. Louis, 1961.

———, "Diderot's Conception of Genius," *Journal of the History of Ideas,* II (1941) 151–182.

———, ed., *Le Philosophe. Texts and Interpretation.* St. Louis, 1948.

Dircks, Henry, *Perpetuum Mobile, or, Search for Self-Motive Power During the 17th., 18th., and 19th. Centuries.* London, 1861.

Discours sur les Découvertes en Général et Particulièrement sur Deux des Principales Découvertes de ce Siècle, Lu au Musée de Paris, dans son Assemblée Publique du Jeudi 11 Décembre 1783. Paris, 1784.

Divers Ouvrages de Mathématique et de Physique par Messieurs de l'Académie Royale des Sciences. Paris, 1693.

Dixnard and Rubi, *Le Point Central des Arts et Métiers à la Convention Nationale.* [Paris], 1793.

Pamphlet in AN, AD VIII 40, vol. 2, pièce 18.

"Documents sur le Collège de France à la Veille de la Révolution . . . ," *Revue Internationale de l'Enseignement,* V (1883) 406–412.

Dolffus, Charles, *The Orion Book of Balloons*, tr. Carter Mason. New York, 1961.

Doneaud du Plan, Alfred, "L'Académie Royale de Marine, de 1752 à 1793," *Revue Maritime et Coloniale*, LVIII-LXXV (1878–1882).

Dorveaux, Paul, "Pilâtre de Rozier et l'Académie des Sciences," *Les Cahiers Lorrains*, VIII (1929) 162–166.

Dowd, David L., *Pageant-Master of the Republic. Jacques-Louis David and the French Revolution*. Lincoln,[1948].

Doyon, André, and Lucien Liaigre, *Jacques Vaucanson, Mécanicien de Génie*. Paris, 1966.

Dubois, Fréderic, "Documents pour Servir à l'Histoire de l'Académie de Chirurgie," *Mémoires de l'Académie Nationale* [or *Impériale*] *de Médecine*, XVI (1852) l–lxxviii, XX (1856) i–xxxiii.

Duclos, Charles Pinot, *Considérations sur les Mœurs de ce Siècle*. Paris, 1751.

——, "Honoraire," *Encyclopédie*, VIII, 291–292.

Du Clos, Samuel Cottereau, *Observations sur les Eaux Minérales de Plusieurs Provinces de France, Faites en l'Académie Royale des Sciences en l'Année 1670 et 1671*. Paris, 1675.

[Dufé-Lafrainaye, Louis], *Avis aux Plus Puissants Génies de l'Univers*. n.p., n.d.

[——], *Démonstration de la Quadrature Définie du Cercle*. La Haye, 1774.

Dufour, Léon, *Souvenirs d'un Savant Français à Travers un Siècle (1780–1865)*. Paris, 1888.

Dufresne, Hélène, *Le Bibliothécaire Hubert-Pascal Ameilhon (1730–1811)*. Paris, 1962.

Dulieu, Louis, "La Contribution Montpelliéraine aux Recueils de l'Académie Royale des Sciences," *Revue d'Histoire des Sciences*, XI (1958) 250–262.

——, "Le Mouvement Scientifique Montpelliérain au XVIIIè Siècle," *Revue d'Histoire des Sciences*, XI (1958) 227–249.

Dumas, Gustave, *Histoire du Journal de Trévoux depuis 1701 jusqu'en 1762*. Paris, 1936.

Dupin, Charles, "Notice Biographique sur Madame de Prony," *Mercure du Dix-Neuvième Siècle*, II (1823) 203–215.

Dupree, A. Hunter, "Nationalism and Science—Sir Joseph Banks and the Wars with France," *A Festschrift for Frederick B. Artz*, ed. David H. Pinkney and Theodore Ropp, pp. 37–51. Durham, N. C., 1964.

Dupuy, Paul, *Le Centenaire de l'Ecole Normale 1795-1895*. Paris, 1895.

Durry, Marie-Jeanne, *Autographes de Mariemont*. Paris, 1955, 2 vols.

Dussaud, René, *La Nouvelle Académie des Inscriptions et Belles-Lettres (1795–1914)*. Paris, 1946–1947, 2 vols.

[Dutens, Louis], *Mémoires d'un Voyageur qui se Repose*. London, 1806, 3 vols.

Duveen, Denis I., "Antoine-Laurent Lavoisier and the French Revolution," *Journal of Chemical Education*, XXXI (1954) 60–65, XXXIV (1957) 502–503, XXXV (1958) 233–234.

———, "Augustin François Silvestre and the Société Philomatique," *Annals of Science*, X (1954) 339–341.

———, *Supplement to A Bibliography of the Works of Antoine Laurent Lavoisier, 1743–1794*. London, 1965.

———, and Roger Hahn, "Deux Lettres de Laplace à Lavoisier," *Revue d'Histoire des Sciences*, XI (1958) 337–342.

———, and ———, "Laplace's Succession to Bézout's Post of Examinateur des Elèves de l'Artillerie," *Isis*, XLVIII (1957) 416–427.

———, and ———, "A Note on Some Lavoisiereana in the 'Journal de Paris'," *Isis*, LI (1960) 64–66.

———, and Herbert S. Klickstein, "Antoine Laurent Lavoisier's Contributions to Medicine and Public Health," *Bulletin of the History of Medecine*, XXIX (1955) 164–179.

———, and ———, "Benjamin Franklin (1706–1790) and Antoine Laurent Lavoisier (1743–1794)," *Annals of Science*, XI (1955) 284–301.

———, and ———, *A Bibliography of the Works of Antoine Laurent Lavoisier 1743–1794*. London, 1954.

Ehrard, Jean, *L'Idée de Nature en France dans la Première Moitié du XVIIIè Siècle*. Paris, 1963, 2 vols.

———, and Jacques Roger, "Deux Périodiques Français du 18è Siècle: 'Le Journal des Savants' et 'Les Mémoires de Trévoux'. Essai d'une Etude Quantitative," *Livre et Société dans la France du XVIIIè Siècle*, pp. 33–59. Paris, 1965.

Elias, Norbert, *Über den Prozess der Zivilisation; Soziogenetische und Psychogenetische Untersuchungen*. Basel, 1939.

Encyclopédie, ou Dictionnaire Raisonné des Sciences, des Arts et des Métiers, ed. Diderot and D'Alembert. Paris, 1751–1765.

L'Encyclopédie et le Progrès des Sciences et des Techniques, ed. Centre International de Synthèse. Paris, 1952.

Enseignement et Diffusion des Sciences en France au XVIIIè Siècle, ed. René Taton. Paris, 1964.

Estivals, Robert, *Le Dépot Légal sous l'Ancien Régime de 1537 à 1791.* Paris, 1961.

——, *La Statistique Bibliographique de la France sous la Monarchie au XVIIIè Siècle.* Paris, 1965. Monograph in series, *Livre et Sociétés,* vol. 2.

Fabian, Bernhard, "Der Naturwissenschaftler als Original-genie," *Europäische Aufklärung,* ed. Friedrich Hugo and Fritz Schalk, pp. 47–68. Munich, 1967.

Fabry, Jean-Baptiste-Germain, *Le Génie de la Révolution Considéré dans l'Éducation, ou Mémoires pour Servir à l'Histoire de l'Instruction Publique depuis 1789 jusqu'à Nos Jours.* Paris, 1817–1818, 3 vols.

Faivre, Jean-Paul, "Savants et Navigateurs: Un Aspect de la Coopération Internationale entre 1750 et 1840," *Cahiers d'Histoire Mondiale,* X (1966) 98–124.

Faujas de Saint-Fond, Barthélemy, *Voyage en Angleterre, en Ecosse et aux Iles Hébrides; Ayant pour Objet les Sciences, les Arts, l'Histoire Naturelle et les Mœurs.* Paris, 1797, 2 vols.

Fauré-Fremiet, E., "Les Origines de l'Académie des Sciences de Paris," *Notes and Records of the Royal Society of London,* XX (1966) 20–31.

Favre, Adrien, *Les Origines du Système Métrique.* Paris, 1931.

Fay, Bernard, "Learned Societies in Europe and America in the Eighteenth Century," *American Historical Review,* XXXVII (1932) 255–266.

Fayet, Joseph, *La Révolution Française et la Science, 1789–1795.* Paris, 1960.

Febvre, Lucien, "Civilisation. Evolution d'un Mot et d'un Groupe d'Idées," *Civilisation. Le Mot et l'Idée,* ed. Centre International de Synthèse, pp. 1–55. Paris, 1930.

Félibien, Jean-François, *Conférences de l'Académie Royale de Peinture et de Sculpture pendant l'Année 1667.* Paris, 1668.

Fellows, Otis E., "The Theme of Genius in Diderot's *Neveu de Rameau*," *Diderot Studies,* II (1952) 168–199.

Féry, F.-A.-R., *Plan des Ecoles de Mathématiques Pratiques et de Dessein.* Reims, 1748.

Festy, Octave, "L'Agriculture pendant la Révolution Française— Les Journaux d'Agriculture et le Progrès Agricole (1789–an VIII)," *Revue d'Histoire Economique et Sociale,* XXVIII (1950) 35–53, 148–168.

Feuille Critique et Littéraire. Paris, an III [1795].

Filon, François, "L'Ingénieur Lamblardie," *Revue de Géographie*, XLI (1897) 116–127, 172–185, 245–255.

Fontenelle, Bernard le Bovier de, *Entretiens sur la Pluralité des Mondes. Digression sur les Anciens et les Modernes*, ed. Robert Shackleton. Oxford, 1955.

Fonvielle, Wilfrid de, *La Mesure du Mètre, Dangers et Aventures des Savants qui l'Ont Déterminé*. Paris, 1886.

Ford, Franklin, *The Robe and the Sword : The Regrouping of the French Aristocracy after Louis XIV*. New York, 1962.

Formey, Jean-Henri-Samuel, "Considérations sur ce qu'On Peut Regarder Aujourd'hui comme le But Principal des Académies, et comme leur Effet le Plus Avantageux," *Histoire de l'Académie Royale des Sciences et Belles-Lettres [de Berlin]*, XXIII (1767) 367–381, XXIV (1768) 357–366.

　　Republished in *Encyclopédie*, Supplément I, 93–98.

Fortescue, George Knottesford, *List of the Contents of Three Collections of Books, Pamphlets and Journals in the British Museum Relating to the French Revolution*. London, 1899.

Foucault, Michel, *Naissance de la Clinique: Une Archéologie du Regard Médical*. Paris, 1963.

Fouchy, Jean-Paul Grandjean de, *Eloges des Académiciens de l'Académie Royale des Sciences Morts depuis l'An 1744*. Paris, 1761.

Fourcroy, Antoine-François, ed., *La Médecine Eclairée par les Sciences Physiques*. Paris, 1791–1792, 4 vols.

——, *Rapport Fait au Nom du Comité de Salut Public, sur les Arts qui Ont Servi à la Défense de la République, et sur le Nouveau Procédé de Tannage Découvert par le Citoyen Armand Séguin. Séance de la Convention, 14 Nivôse An III*. Paris, an III [1795].

Fourcy, A., *Histoire de l'Ecole Polytechnique*. Paris, 1828.

Fournier, Paul, *Les Voyageurs Naturalistes du Clergé Français avant la Révolution*. Paris, 1932.

Fréron, Alexandre, *Les Académies Provinciales, Salons ou Sociétés Savantes?* Rouen, 1934.

Furet, François, "La 'Librairie' du Royaume de France au 18è Siècle," *Livre et Société dans la France du XVIIIè Siècle*, pp. 3–32. Paris, 1965.

Gadroys, Claude, *Le Système du Monde selon Trois Hypothèses*. Paris, 1675.

Gage, Andrew Thomas, *A History of the Linnean Society of London*. London, 1938.

Gaillard, Gabriel-Henri, *De la Fraternité des Corps Littéraires ou Académies.* Paris, 1789.

Galante-Garrone, Alessandro, *Gilbert Romme, Storia de un Rivoluzionario.* [Turin], 1959.

Gauja, Pierre, "Les Origines de l'Académie des Sciences de Paris," Institut de France, Académie des Sciences, *Troisième Centenaire, 1666–1966*, pp. 1–51. Paris, 1967.

Gazette d'Agriculture, Commerce, Arts et Finances. Paris, 1769–1783.

Gazette de Paris. Paris, 1789–1792.

Gazette Littéraire de l'Europe. Paris, 1764–1784.

Genty, Maurice, "Les Dernières Années de la Société Royale de Médecine 1789–1793," *Progrès Médical. Supplément Illustré*, XIV (1937) 49–56.

———, "La Médecine sous la Révolution. Le Journalisme Médical," *Progrès Médical. Supplément Illustré*, XI (1934) 33–37.

———, "La Société Royale de Médecine et les Remèdes Secrets avant la Révolution," *Progrès Médical. Supplément Illustré*, XIV (1937) 17–21.

Geoffroy Saint-Hilaire, Isidore, *Vie, Travaux et Doctrines Scientifiques d'Etienne Geoffroy Saint-Hilaire.* Paris, 1847.

George, Albert J., "The Genesis of the Academy of Sciences," *Annals of Science*, III (1938) 372–401.

Gerbaux, Ferdinand, and Charles Schmidt, eds., *Procès-Verbaux des Comités d'Agriculture et de Commerce de la Constituante, de la Législative et de la Convention.* Paris, 1906–1937, 5 vols.

Gershoy, Leo, *Bertrand Barère. A Reluctant Terrorist.* Princeton, 1962.

Gillispie, Charles Coulston, "The Discovery of the Leblanc Process," *Isis*, XLVIII (1957) 152–170.

———, *The Edge of Objectivity.* Princeton, 1960.

———, "The *Encyclopédie* and the Jacobin Philosophy of Science: A Study of Ideas and Consequences," *Critical Problems in the History of Science*, ed. Marshall Clagett, pp. 255–289. Madison, 1959.

———, "The Natural History of Industry," *Isis*, XLVIII (1957) 398–407.

———, "Science and Technology," *War and Peace in an Age of Upheaval 1793–1830*, ed. C. W. Crawley, pp. 118–145. Cambridge, 1965.

　　In *The New Cambridge Modern History*, IX.

———, "Science in the French Revolution," *Behavioral Science*, IV (1959) 67–73.

Gilpin, Robert, *France in the Age of the Scientific State*. Princeton, 1968.

Girdlestone, Cuthbert, *Louis-François Ramond (1755–1827). Sa Vie, son Œuvre Littéraire et Politique*. Paris, 1968.

Glanvill, Joseph, *Scepsis Scientifica*. London, 1665.

Godechot, Jacques, "L'Aérostation Militaire sous le Directoire," *Annales Historiques de la Révolution Française*, VIII (1931) 213–228.

Goetz, Helmut, *Marc-Antoine Jullien de Paris (1775–1848): L'Evolution Spirituelle d'un Révolutionnaire*, tr. C. Cuénot. Paris, 1962.

Gottschalk, Louis R., *Jean Paul Marat, a Study in Radicalism*. New York, 1927.

Gourcy, François-Antoine-Etienne de, *Discours sur les Académies. Est-Il à Propos de les Multiplier?* Nancy, 1768.

Grand-Carteret, John, *Les Almanachs Français*. Paris, 1896.

Granger, Gilles-Gaston, *La Mathématique Sociale du Marquis de Condorcet*. Paris, 1956.

Grappin, Pierre, *La Théorie du Génie dans le Préclassicisme Allemand*. Paris, 1952.

Green, F. C., "Sir James Hall's Impressions of France in 1791," *French Studies*, XVIII (1964) 236–243.

Grégoire, Henri-Baptiste, *Mémoires de Grégoire, Ancien Evêque de Blois*. Paris, 1837, 2 vols.

Grétry, André-Ernest-Modeste, *De la Vérité, ce que Nous Fûmes, ce que Nous Sommes, ce que Nous Devrions Etre*. Paris, an IX [1801], 3 vols.

Grimaux, Edouard, *Lavoisier, 1743–1794*, 3rd ed. Paris, 1899.

Grosclaude, Pierre, *Malesherbes, Témoin et Interprète de son Temps*. Paris, 1961.

Guéhenno, Jean, *Jean-Jacques. En Marge des "Confessions." 1712–1750*. Paris, 1948.

Guerlac, Henry, "Commentary on the Papers of Charles Coulston Gillispie and L. Pearce Williams," *Critical Problems in the History of Science*, ed. Marshall Clagett, pp. 317–320. Madison, 1959.

——, "Science and French National Strength," *Modern France*, ed. E. M. Earle, pp. 81–105. Princeton, 1951.

——, "Science as a Social and Historical Phenomenon," *The Scientific Monthly*, LXXX (1955) 93–101.

——, "Some French Antecedents of the Chemical Revolution," *Chymia*, V (1959) 73–112.

——, "Three Eighteenth-Century Social Philosophers: Scientific Influences on their Thought," *Daedalus*, LXXXVII (1958) 7–24.

Guiffrey, Jules-Joseph, ed., *Comptes des Bâtiments du Roi sous le Règne de Louis XIV*. Paris, 1881–1901, 5 vols.

Guillaume, James, *Etudes Révolutionnaires*. Paris, 1908–1909, 2 vols.

Guittard, E.-H., "Berthollet et la Politique," *Revue d'Histoire de la Pharmacie*, L (1962) 238–239.

Gusdorf, Georges, *Introduction aux Sciences Humaines*. Paris, 1960.

——, *Les Sciences Humaines et la Pensée Occidentale*. Paris, 1966–1969, 3 vols in 4.

Haag, Eugène and Emile, *La France Protestante, ou Vies des Protestants Français*. Paris and Geneva, 1846–1859, 10 vols.

Hagstrom, Warren O., *The Scientific Community*. New York, 1965.

Hahn, Roger, "L'Académie Royale des Sciences et la Réforme de ses Statuts en 1789," *Revue d'Histoire des Sciences*, XVIII (1965) 15–28.

——, "The Application of Science to Society: The Societies of Arts," *Studies on Voltaire and the Eighteenth Century*, XXV (1963) 829–836.

——, "The Chair of Hydrodynamics in Paris 1775–1791: A Creation of Turgot," *Acts of the Tenth International Congress of the History of Science*, pp. 751–754. Paris, 1964.

——, "Elite Scientifique et Démocratie Politique dans la France Révolutionnaire," *Dix-Huitième Siècle*, I (1969) 229–235.

——, "L'Enseignement Scientifique aux Ecoles Militaires et d'Artillerie," *Enseignement et Diffusion*, pp. 513–545.

——, "L'Enseignement Scientifique des Gardes de la Marine au XVIIIè Siècle," *Enseignment et Diffusion*, pp. 547–558.

——, "The Fall of the Paris Academy of Sciences during the French Revolution," unpublished Ph.D. dissertation, Cornell University, 1962.

——, "Fourcroy, Advocate of Lavoisier?" *Archives Internationales d'Histoire des Sciences*, XII (1959) 285–288.

——, "Les Observatoires en France au XVIIIè Siècle." *Enseignement et Diffusion*, pp. 653–658.

——, "The Problems of the French Scientific Community, 1793–1795," *Actes du XIIè Congrès International d'Histoire des Sciences* (in press).

————, "Quelques Nouveaux Documents sur Jean-Sylvain Bailly," *Revue d'Histoire des Sciences*, VIII (1955) 338–353.

————, "Review of Joseph Fayet, *La Révolution Française et la Science*," *Isis*, LI (1960) 605–607.

Halem, Gerhard Anton von, *Paris en 1790, Voyage de Halem*, tr. A. Chuquet. Paris, 1896.

Hall, Alfred Rupert, "The Scientific Movement," *The Ascendancy of France, 1648–1688*, ed. F. L. Carsten, pp. 47–72. Cambridge, 1961.
    In *The New Cambridge Modern History*, V.

Hamy, Ernest-Théodore, *Les Débuts de Lamarck*. Paris, 1909.

————, *Les Derniers Jours du Jardin du Roi*. Paris, 1893.

————, *Joseph Dombey*. Paris, 1905.

Hanks, Leslie, *Buffon avant l' "Histoire Naturelle."* Paris, 1966.

Hans, Nicholas, "UNESCO of the Eighteenth Century. La Loge des Neuf Soeurs and Its Venerable Master, Benjamin Franklin," *Proceedings of the American Philosophical Society*, XCVII (1953) 513–524.

Hansard, Thomas C., ed., *Parliamentary History of England*. London, 1812–1820, 41 vols.

Hassenfratz, Jean-Henri, *La Sidérotechnie, ou l'Art de Traiter les Minerais de Fer pour en Obtenir de la Fonte, du Fer ou de l'Acier*. Paris, 1812, 4 vols.

Hautecœur, Louis, "Pourquoi les Académies Furent-Elles Supprimées en 1793?" *Revue des Deux Mondes*, XXIV (1959) 593–604.

Hazard, Paul, *La Crise de la Conscience Européenne 1680–1715*. Paris, 1933.

Helvétius, Claude-Adrien, *Œuvres Complettes*, new ed. Paris, 1795, 14 vols.

Henry, Charles, ed., *Correspondance Inédite de Condorcet et de Turgot 1770–1779*. Paris, 1882.

————, "Correspondance Inédite de D'Alembert avec Cramer, Lesage, Clairaut, Turgot, Castillon, Beguelin, etc.," *Bullettino di Bibliografia e di Storia delle Scienze Matematiche e Fisiche*, XVIII (1885) 507–570, 605–649.

Henry, Louis, "The Population of France in the Eighteenth Century," *Population in History: Essays in Historical Demography*, ed. D. V. Glass and D. E. C. Eversley, pp. 434–456. Chicago, 1965.

Hérissant, Louis-Antoine-Prosper, *Bibliothèque Physique de la France ou Liste de Tous les Ouvrages, tant Imprimés que Ma-*

nuscrits, Qui Traitent de l'Histoire Naturelle de ce Royaume. Paris, 1771.

"Discours sur l'utilité de l'histoire naturelle de la France, et la manière de l'étudier," pp. 17–40.

Heubaum, Alfred, *Geschichte des Deutschen Bildungswesens seit der Mitte des Siebzehten Jahrhunderts*. Berlin, 1905.

Hirschfield, John Milton, "The Académie Royale des Sciences (1666–1683): Inauguration and Initial Problems of Method," unpublished Ph.D. dissertation, University of Chicago, 1957.

Histoire de l'Académie Royale des Sciences (1666–1699). Paris, 1733, 2 vols.

Histoire de l'Académie Royale des Sciences, avec les Mémoires de Mathématique et de Physique. Tirés des Registres de cette Académie (1699–1790). Paris, 1702–1797, 92 vols.

Histoire de la Fondation de la Société d'Encouragement pour l'Industrie Nationale, ou Recueil des Procès-Verbaux des Séances de cette Société. Paris, 1850.

[Hourcastremé, Pierre], *Les Aventures de Messire Anselme, Chevalier des Loix*, 2nd ed. Paris, an IV [1796], 4 vols.

———, *Dissertation sur les Causes qu'Ont Produit l'Espèce de Contradiction que l'On Remarque entre un Premier Décret de l'Assemblée Nationale, sous la Date du 8 Mai 1790, et un Second Décret, sous celle du 26 Mars 1791, quoique Tendant l'Un et l'Autre au Même But d'Utilité Générale*. [Paris], 1791.

Pamphlet in BN, Imprimés, Vz. 1695.

Hours, Henri, *La Lutte contre les Epizooties et l'Ecole Vétérinaire de Lyon au XVIIIè Siècle*. Paris, 1957.

Huard, Pierre, *L'Académie Royale de Chirurgie*. Paris, 1967.

———, "L'Enseignement Médico-Chirurgical," *Enseignement et Diffusion*, pp. 171–236.

———, and M. Wong, "Les Enquêtes Scientifiques Françaises et l'Exploration du Monde Exotique aux XVIIè et XVIIIè Siècles," *Bulletin de l'Ecole Française d'Extrême-Orient*, LII (1964) 143–155.

Huygens, Christiaan, *Œuvres Complètes*. La Haye, 1888–1950, 22 vols.

Iggers, Georg G., *The Cult of Authority. The Political Philosophy of the Saint-Simonians, a Chapter in the Intellectual History of Totalitarianism*. The Hague, 1958.

Ingrand, Henry, *Le Comité de Salubrité de l'Assemblée Nationale Constituante (1790–1791)*. Paris, 1936.

Isambert, Gustave, *La Vie à Paris pendant une Année de la Révolution (1791–1792)*. Paris, 1896.

Isoré, Jacques, "De l'Existence des Brevets d'Invention en Droit Français avant 1791," *Revue Historique du Droit Français et Etranger*, ser. 4, XVI (1937) 94–130.

Jacquot, Jean, *Le Naturaliste Sir Hans Sloane (1660–1753) et les Echanges Scientifiques entre la France et l'Angleterre*. Paris, 1953.

———, "Sir Hans Sloane and French Men of Science," *Notes and Records of the Royal Society of London*, X (1952) 85–98.

Jancigny, L. Dubois de, *Tableau Annuel du Progrès de la Physique*. Paris, 1772.

Jaucourt, Louis de, "Sciences," *Encyclopédie*, XIV, 788–789.

Jodin, Jean, *Examen des Dernières Observations de M. de la Lande Insérées dans le "Mercure" de Juillet Dernier*. Paris, 1755.

Jones, Richard F., *Ancients and Moderns*, 2nd ed. St. Louis, 1961.

Jones, Tom Bard, *The Figure of the Earth*. Lawrence, 1967.

Journal de l'Ecole Polytechnique, ou Bulletin du Travail Fait à cette Ecole. Paris, an III–VIII [1795–1800].

Journal de Paris. Paris, 1777–1803.

Journal de Physique, ed. François Rozier, Jean-André Mongez, and Jean-Claude Delamétherie. Paris, 1773–1803, 57 vols.
    Variously known as *Observations et Mémoires sur la Physique* . . . and *Observations sur la Physique* . . . .

Journal des Etats Généraux. Paris, 1789–1791.

Journal des Savants. Paris, 1666–an V [1797].

Journal des Sciences, Arts et Métiers. Paris, 1792.

Journal des Sciences Utiles, ed. Pierre Bertholon. Paris, 1791.

Journal d'Histoire Naturelle, ed. Lamarck et al. Paris, 1792.

Journal du Point Central des Arts et Métiers. Paris, September 1791.
    BN, Imprimés, Vp. 8303.

Journal Gratuit. Physique. Paris, 1790.

Jouvenel, Bertrand de, "The Republic of Science," *The Logic of Personal Knowledge; Essays Presented to Michael Polanyi on his 70th Birthday, 11th March 1961*, pp. 131–141. Glencoe, 1961.

[Jussieu, Antoine-Laurent de], *Discours Prononcé par Jussieu, Président de l'Institut National en Présentant au Corps Législatif les Travaux de l'Institut pendant l'An 7*. [Paris], an VIII [1800].

Justin, Emile, *Les Sociétés Royales d'Agriculture au XVIIIè Siècle (1757–1793)*. St. Lô, 1935.

Kafker, Frank A., "Les Encyclopédistes et la Terreur," *Revue d'Histoire Moderne et Contemporaine*, XIV (1967) 284–295.

Kersaint, Georges, "Antoine-François de Fourcroy (1755–1809). Sa Vie et son Œuvre," *Mémoires du Muséum National d'Histoire Naturelle*, n.s., ser. D, II (1966) 1–296.

King, James E., *Science and Rationalism in the Government of Louis XIV, 1661–1683.* Baltimore, 1949.

Kirschstein, Max, *Klopstocks Deutsche Gelehrtenrepublik.* Berlin, 1929.

Kitchin, Joanna, *Un Journal 'Philosophique': La Décade (1794–1807).* Paris, 1965.

Kortum, Hans, "Die Hintergründe einer Akademiesitzung im Jahre 1687," *Neue Beiträge zur Literatur der Aufklärung,* ed. W. Krauss and W. Dietze, pp. 13–36. Berlin, 1964.

Krauss, Werner, "Entwicklungstendenz der Akademien im Zeitalter der Aufklärung," *Studien zur Deutschen und Französischen Aufklärung,* pp. 41–71. Berlin, 1963.

Kristeller, Paul O., "The Modern System of the Arts," *Journal of the History of Ideas,* XII (1951) 496–527, XIII (1952) 17–46.

Kronick, David A., *A History of Scientific and Technical Periodicals: The Origins and Development of the Scientific and Technological Press, 1665–1790.* New York, 1962.

Laboulbène, Alexandre, "L'Ancienne Académie de Chirurgie," *Revue Scientifique,* XLII (1888) 728–738.

Lacoarret, Marie, and Ter-Menassian, "Les Universités," *Enseignement et Diffusion,* pp. 125–163.

La Coste, Bertrand de, *Ce n'Est pas la Mort aux Rats ny aux Souris, mais c'Est la Mort des Mathematichiens [sic] de Paris.* Hamburg, 1676.

———, *Démonstration de la Quadrature du Cercle qui Est l'Unique Couronne et Principal Sujet de Toutes les Mathématiques.* Hamburg, 1677.

———, *Le Monde Désabusé ou La Démonstration de Deux Lignes Moyennes Proportionelles, Matin des Prétendus Savans Matématiciens de l'Académie Royale de Paris.* Hamburg, 1675.

———, *Ne Trompez Plus Personne ou Suite de Reveil-Matin des Pretendus Savans Matematiciens de l'Academie Royale de Paris.* Hamburg, 1675.

———, *Le Reveil Matin Fait par Monsieur Bertrand pour Reveiller les Pretendus Sçavans Matematiciens de l'Academie Royale de Paris.* Hamburg, 1674.

———, *Scheda de Inventa Quadratur a Circuli.* Hamburg, 1663.

Lacour-Gayet, Georges, *Bonaparte, Membre de l'Institut.* Paris, 1921.

———, *Talleyrand (1754–1838).* Paris, 1930–1934, 4 vols.

Lacroix, Alfred, "Une Famille de Bons Serviteurs de l'Académie des Sciences et du Jardin des Plantes," *Bulletin du Muséum d'Histoire Naturelle,* ser. 2, X (1938) 446–471.

Lacroix, Paul, *XVIIIè Siècle. Sciences et Arts*. Paris, 1878.

Lacroix, Sigismund, *Actes de la Commune de Paris pendant la Révolution*. Paris, 1894–1914, 2 ser., 15 vols.

Lacroix, Silvestre-François, *Essais sur l'Enseignement en Général et sur celui des Mathématiques en Particulier*. Paris, an XIV–1805.

La Fontainerie, François de, ed. and tr., *French Liberalism and Education in the Eighteenth Century*. New York, 1932.

La Harpe, Jean-François, *De la Guerre Déclarée par nos Derniers Tyrans à la Raison, à la Morale, aux Lettres et aux Arts*. Paris, an IV [1796].

Laissus, Yves, "Les Cabinets d'Histoire Naturelle," *Enseignement et Diffusion*, pp. 659–712.

——, "Gaspard Monge et l'Expédition d'Egypte (1798–1799)," *Revue de Synthèse*, ser. 3, LXXXI (1960) 309–336.

——, "Le Jardin du Roi," *Enseignement et Diffusion*, pp. 287–341.

Lakanal, Joseph, *Exposé Sommaire des Travaux de Joseph Lakanal*. Paris, 1838.

Lalande, Joseph-Jérôme, *Art de Faire le Papier*. Paris, [1761].

——, *Art du Tanneur*. Paris, 1764.

——, *Bibliographie Astronomique; avec l'Histoire de l'Astronomie depuis 1781 jusqu'à 1802*. Paris, 1803.

——, "Eloge de Louis Dupuy," *Décade Philosophique*, VI, 65–78.

——, "Histoire de l'Astronomie en 1794," *Magasin Encyclopédique* (1795) tome I, 1–35.

——, "Histoire de l'Astronomie pour l'Année 3 de l'Ere Française," *Magasin Encyclopédique* (1795) tome V, 433–452.

Lalanne, Ludovic, "Lettre de Séguier, de Nîmes, à Condorcet," *Revue des Sociétés Savantes des Départements*, ser. 7, II (1880) 145–148.

Lamy, Edouard, *Les Cabinets d'Histoire Naturelle en France au XVIIIè Siècle*. Paris, [1931].

Lancelin, P. F., *Introduction à l'Analyse des Sciences ou De la Génération, des Fondemens, et des Instruments de nos Connoissances*. Paris, an IX–XI [1801–1803], 3 vols.

Landes, David S., "Technological Change and Development in Western Europe, 1750–1914," *The Cambridge Economic History of Europe*, VI, part I, pp. 274–601. Cambridge, 1965.

Langlès, Louis-Mathieu, "Coup d'Œil Général sur l'Etat des Lettres, des Sciences et des Arts en Europe, au Commencement de la Cinquième Année de la République Française," *Journal des Savants* (1797) 24–29.

[Laplace, Pierre-Simon], *Discours Prononcé par le Citoyen Laplace au Nom de l'Institut National des Sciences et des Arts dans la Séance du Premier Jour Complémentaire, An 4.* [Paris], an IV [1796].

La Revellière-Lépeaux, Louis-Marie, *Mémoires.* Paris, 1895, 3 vols.

Las Vergnas, Raymond, *Le Chevalier Rutlidge, "Gentilhomme Anglais" 1742–1794.* Paris, 1932.

Laukhard, Friedrich Christian, *Un Allemand en France sous la Terreur. Souvenirs de Fr.-Chr. Laukhard (1792–94),* tr. W. Bauer. Paris, 1915.

Launay, Louis de, *Une Grande Famille de Savants. Les Brongniart.* Paris, 1940.

Laurent, Gustave, "Mémoire Historique du Chimiste Hassenfratz," *Annales Historiques de la Révolution Française,* I (1924) 163–164.

Lauzun, Philippe, "Histoire de la Société Académique d'Agen (1776–1900)," *Recueil des Travaux de la Société d'Agriculture, Sciences et Arts d'Agen,* ser. 2, XIV (1900) 1–359.

Lavoisier, Antoine-Laurent, *Correspondance,* ed. René Fric. Paris, 1955–present.

———, *Œuvres,* ed. Jean-Baptiste Dumas and Edouard Grimaux. Paris, 1862–1893, 6 vols.

Lefranc, Abel, *Histoire du Collège de France depuis ses Origines jusqu'à la Fin du Premier Empire.* Paris, 1893.

Le Franc de Pompignan, Georges, *Essai Critique sur l'Etat Présent de la République des Lettres.* Paris, 1744.

Le Guin, Charles A., "The Continuing Education of Jean-Marie Roland (1734–1793)," *History of Education Quarterly,* III (1963) 123–133.

Leith, James A., "French Republican Pedagogy in the Year II," *Canadian Journal of History,* III (1968) 52–67.

Lemonnier, Henry, "Les Derniers Jours de l'Académie Royale d'Architecture, 1789–1793," *Revue de France,* VII (1927) 276–292.

———, ed., *Procès-Verbaux de l'Académie Royale d'Architecture 1671–1793.* Paris, 1911–1929, 10 vols.

Lenormant, Charles, "L'Académie Royale de Chirurgie et la Révolution," *Progrès Médical. Supplément Illustré,* XVI (1939) 53–60.

Léon, Antoine, *La Révolution Française et l'Education Technique.* Paris, 1968.

Lepaute, Jean-André, *Traité d'Horlogerie.* Paris, 1755.

Lepreux, Georges, *Gallia Typographica*. Paris, 1909–1914, 7 vols.

[Le Roy, Pierre], *Mémoires sur l'Horlogerie Contenant Diverses Remarques sur les Ouvrages et les Pretentions de M. R[ivaz]*. [Paris], 1750.

Lesuire, Robert M., *Histoire de la République des Lettres et Arts en France. Année 1780*. Amsterdam, 1781.

Lettsom, John Coakley, *Le Voyageur Naturaliste, ou Instructions sur les Moyens de Ramasser les Objets d'Histoire Naturelle*. Amsterdam, 1775.

Linguet, Simon-Nicolas-Henri, *Le Fanatisme des Philosophes*. Geneva, 1764.

Luxembourg, Lilo K., *Francis Bacon and Denis Diderot: Philosophers of Science*. Copenhagen, 1967.

Mac-Auliffe, Léon, *La Révolution et les Hôpitaux (Années 1789, 1790, 1791)*. Paris, 1901.

Machines et Inventions Approuvées par l'Académie Royale des Sciences..., ed. Jean-Gaffin Gallon. Paris, 1735–1777, 7 vols.

Magasin Encyclopédique, ou Journal des Sciences, des Lettres et des Arts, ed. A. L. Millin. Paris, 1792–1803.

Maindron, Ernest, *L'Académie des Sciences*. Paris, 1888.

―――, *Les Fondations de Prix à l'Académie des Sciences: Les Lauréats de l'Académie, 1714–1880*. Paris, 1881.

Malesherbes, Chrétien-Guillaume de Lamoignon de, *Mémoire sur les Moyens d'Accéler les Progrès de l'Economie Rurale en France*. Paris, 1790.

Manuel, Frank E., *The Prophets of Paris*. New York, 1962.

Maras, Raymond Joseph, "Napoleon: Patron of the Arts and Sciences," unpublished Ph. D. dissertation, University of California, Berkeley, 1955.

Marat, Jean-Paul, *L'Ami du Peuple*. Paris, 1789–1792.

―――, *Les Charlatans Modernes, ou Lettres sur le Charlatanisme Académique*. Paris, 1791.

―――, *La Correspondance de Marat*, ed. Charles Vellay. Paris, 1908.

―――, *Journal de la République Française*. Paris, 1792–1793.

Maréchal, Henri, *L'Eclairage à Paris*. Paris, 1894.

Marguet, Frédéric, *Histoire de la Longitude à la Mer au XVIIIè Siècle en France*. Paris, 1917.

Marin, Louis, "Le Role de l'Abbé Grégoire dans la Fondation de l'Institut et l'Académie des Sciences Morales et Politiques," *Revue des Travaux de l'Académie des Sciences Morales et Politiques*, ser. 4, CIX (1956) 2nd sem., 38–69.

Marion, Marcel, "Un Révolutionnaire Très Conservateur: Creuzé-Latouche," *Revue d'Histoire Moderne*, XI (1936) 101–134.

Marsak, Leonard M., *Bernard de Fontenelle: The Idea of Science in the French Enlightenment*. Philadelphia, 1959.

———, "The Humanist as Scientist: Nicolas Claude Fabri de Peiresc," *Journal of World History*, VIII (1964) 93–99.

Martin, Geneviève, "Documents de l'Académie de Rouen Concernant l'Enseignement des Sciences au XVIIIè Siècle," *Revue d'Histoire des Sciences*, XI (1958) 207–226.

Martin-Allanic, Jean-Etienne, *Bougainville Navigateur et les Découvertes de son Temps*. Paris, 1964.

Masson, Frédéric, *L'Académie Française 1629–1793*. Paris, 1912.

Masuyer, Claude-Louis, *Discours sur l'Organisation de l'Instruction Publique et de l'Education Nationale en France*. Paris, 1793.

Mathiez, Albert, "L'Institut et la Liberté Scientifique," *Annales Révolutionnaires*, IX (1917) 577–582.

———, "La Mobilisation des Savants en l'An II," *Revue de Paris*, XXIV (1917) 524–565.

Matoré, Georges, and A. J. Greimas, "La Naissance du 'Génie' au XVIIIè Siècle. Etude Lexicologique," *Le Français Moderne*, XXV (1957) 256–272.

Maugras, Gaston, ed., *Journal d'un Etudiant pendant la Révolution, 1789–1793*. Paris, 1910.

Maupertuis, Pierre-Louis Moreau de, "Des Devoirs de l'Académicien," *Histoire de l'Académie Royale des Sciences et Belles-Lettres de Berlin depuis son Origine jusqu'à Présent (1750)*, pp. 137–147. Berlin, 1752.

Maupin, Louis, *Supplément Nécessaire à la Science des Académies, ou des Physiciens et Chimistes de Tous les Pays*. Paris, 1784.

Maury, L.-F.-Alfred, *L'Ancienne Académie des Inscriptions et Belles-Lettres*. Paris, 1864.

———, *L'Ancienne Académie des Sciences*. Paris, 1864.

McCloy, Shelby T., *French Inventions of the Eighteenth Century*. Lexington, 1952.

McKeon, Robert M., "Une Lettre de Melchisédech Thévenot sur les Débuts de l'Académie Royale des Sciences," *Revue d'Histoire des Sciences*, XVIII (1965) 1–6.

McKie, Douglas, *Antoine Lavoisier, Scientist, Economist, Social Reformer*. New York, 1952.

———, " 'Observations' of the Abbé François Rozier-I," *Annals of Science*, XIII (1957) 73–89.

———, "The Scientific Periodical from 1665 to 1798," *The Philo-*

sophical Magazine, Commemoration Number, pp. 122–132. London, 1948.

————, "Scientific Societies to the End of the Eighteenth Century," *The Philosophical Magazine*, Commemoration Number, pp. 133–143. London, 1948.

McRae, Robert, *The Problem of the Unity of the Sciences: Bacon to Kant*. Toronto, 1961.

Meister, Paul, *Charles Duclos (1704–1772)*. Geneva, 1956.

Mémoire pour Pierre de Rivaz; Demandeur aux Fins de l'Enregistrement des Lettres-Patentes à lui Accordées par le Roi, pour la Construction et Vente de ses Ouvrages d'Horlogerie pendant Quinze Années, avec Privilège Exclusif pour les Horloges de son Invention. n.p., n.d.

Mémoires de l'Académie d'Asnières. Neufchatel, 1783.
  The copy consulted at the University of California, Berkeley, has marginal comments.

Mémoires de l'Académie Royale des Sciences depuis 1666 jusqu'à 1699. Paris, 1729–1732, 9 vols.

Mémoires de l'Institut National des Sciences et Arts. Littérature et Beaux-Arts (an IV–1803). Paris, an VI–XII [1798–1804], 5 vols.

Mémoires de l'Institut National des Sciences et Arts. Sciences Mathématiques et Physiques (an IV–1805). Paris, an VI[1798]–1806, 6 vols.

Mémoires de l'Institut National des Sciences et Arts. Sciences Morales et Politiques (an IV–1803). Paris, an VI–XII[1798–1804], 5 vols.

Mémoires de Mathématique et de Physique Tirez des Registres de l'Académie Royale des Sciences. Paris, 1692–1693, 2 vols.

Mémoires de Mathématique et Physique Présentés à l'Académie Royale des Sciences par Divers Sçavans, et Lus dans ses Assemblées. Paris, 1750–1786, 11 vols.

Mémoires pour Servir à l'Histoire des Sciences et des Beaux-Arts. Lyons, February, 1733.
  Known as *Mémoires de Trévoux*.

Mémoires Présentés à l'Institut des Sciences Lettres et Arts par Divers Savans et Lus dans ses Assemblées. Sciences Mathématiques et Physiques. Paris, 1805–1811, 2 vols.

Mendelsohn, Everett, "The Context of Nineteenth-Century Science," *The Golden Age of Science*, ed. B. Z. Jones, pp. xiii–xviii. New York, 1966.

Mercier, Louis-Sébastien, *Tableau de Paris*, new ed. Amsterdam, 1783–1788, 12 vols.

Mercure de France. Paris, 1672–1803.
    Known as *Mercure Français* during the Revolution.

Merle, Louis, "La Vie et l'Œuvre du Dr. Jean-Gabriel Gallot (1774–1794)," *Mémoires de la Société des Antiquaires de l'Ouest*, ser. 4, V (1961).

Merton, Robert K., "Bureaucratic Structure and Personality," *Social Theory and Social Structure*, rev. ed., pp. 195–206. Glencoe, 1957.

Merz, John T., *A History of European Thought in the Nineteenth Century*. London, 1896–1914, 4 vols.

Mesnard, Jean, "Pascal à l'Académie Le Pailleur," *Revue d'Histoire des Sciences*, XVI (1963) 1–10.

Meyer, Friedrich Johann Lorenz, *Fragments sur Paris*, tr. Charles-François Dumouriez. Hamburg, 1798, 2 vols.

Meyer, Rudolf W., *Leibnitz and the Seventeenth Century Revolution*, tr. J. P. Stern. Cambridge, 1952.

Millerson, Geoffrey, *The Qualifying Associations*. London, 1964.

Millin, Aubin-Louis, *Discours sur l'Origine et les Progrès de l'Histoire Naturelle en France, Servant d'Introduction aux Mémoires de la Société d'Histoire Naturelle*. Paris, 1792.

Mirault, Claude-François, "Notice Nécrologique sur Charles de Saudray," *Procès-Verbaux de la 100è Séance Publique de l'Athenée des Arts*, pp. 20–31. Paris, 1834.

Mols, Roger, *Introduction à la Démographie Historique des Villes d'Europe du 14è au 18è Siècle*. Gembloux, 1954–1956, 3 vols.

Monglond, André, *Le Préromantisme Français*. Grenoble, 1930, 2 vols.

Monin, Hippolyte, "Les Papiers de Boissy d'Anglas," *Revue Historique de la Révolution Française*, VI (1914) 288–293.

Moniteur Universel. Paris, 1789–an VIII [1800].
    Known also as the *Gazette Nationale*.

Montaiglon, Anatole de, ed., *Procès-Verbaux de l'Académie Royale de Peinture et de Sculpture, 1648–1792*. Paris, 1875–1892, 10 vols.

Montbas, Hugues de, *La Police Parisienne sous Louis XVI*. Paris, 1949.

Montucla, Jean-Etienne, *Histoire des Mathématiques*, ed. J.-J. Lalande, 2nd ed. Paris, an VII–an X [1799–1802], 4 vols.

———, *Histoire des Recherches sur la Quadrature du Cercle*, ed. S.-F. Lacroix, 2nd ed. Paris, 1831.

Moravia, Sergio, *Il Tramonto dell'Illuminismo. Filosofia e Politica nella Società Francese (1770–1810)*. Bari, 1968.

Morazé, Charles, "Le Siècle de la Curiosité," *Histoire Générale des Sciences*, ed. René Taton, II, 425–431. Paris, 1958.

Moreau de Saint-Méry, Médéric-Louis-Elie, *Discours sur l'Utilité des Assemblées Publiques Littéraires*. Parma, 1805.

————, *Discours sur l'Utilité du Musée Etabli à Paris, Prononcé dans sa Séance Publique du 1er Décembre 1784*. Parma, 1805.

Morgan, Betty T., *Histoire du Journal des Sçavans depuis 1665 jusqu'en 1701*. Paris, 1928.

Morin-Lavallée, François-M., *Essai de Bibliographie Viroise*. Caen, 1879.

Mornet, Daniel, "Les Enseignements des Bibliothèques Privées (1750–1780)," *Revue d'Histoire Littéraire de la France*, XVII (1910) 449–496.

————, *Les Origines Intellectuelles de la Révolution Française 1715–1787*, 5th ed. Paris, 1954.

————, *Les Sciences de la Nature en France, au XVIIIè Siècle*. Paris, 1911.

Mousnier, Roland, *Progrès Scientifique et Technique au XVIIIè Siècle*. Paris, 1958.

Mouy, Paul, *Le Développement de la Physique Cartésienne, 1646–1712*. Paris, 1934.

Muller, Johannes, *Epistola, Bertrando de la Coste, Colonello rei tormentariae Brandeburgio, de Inventa se Circuli Quadratura, Scheda peculiari gloriato, Opposita*. Hamburg, 1663.

Mulot, François-Valentin, *Réflexions sur l'Etat Actuel de l'Instruction Publique en France*. Paris, an V [1797].

Muratori, Ludovico Antonio, *Opere*, ed. G. Falco and F. Fort. Milan, 1965.
   Letteratura Italiana. Storia e Testi, XLVI, 1.

Nielsen, Niels, *Géomètres Français sous la Révolution*. Copenhagen, 1929.

Notice sur l'Institution de la Société Philomatique. Paris, n.d.
   Pamphlet in AdS, Lav 162.

Nouvelle Biographie Générale, ed. Jean C.-F. Hoefer. Paris, 1842–1877, 46 vols.

Nouvelle et Seule Manufacture d'Encre d'Argonne Inaltérable et Toujours Fluide. Composée par le Sieur Vicq, Marchand Epicier-Droguiste, Cirier et Chandelier, Rue Saint-Honoré, No. 463. n.p., n.d.
   Handbill in BN, Imprimés, Vz. 1837, fol. 38.

Observations Physiques et Mathématiques pour Servir à l'Histoire Naturelle et à la Perfection de l'Astronomie et de la Géographie:

Envoyées des Indes et de la Chine à l'Académie des Sciences à Paris, par les Pères Jésuites, ed. Thomas Gouye. Paris, 1692.

Oldenburg, Henry, *The Correspondence of Henry Oldenburg,* ed. and tr. A. Rupert Hall and Marie Boas Hall. Madison, 1965–1969, 6 vols.

Olivier-Martin, François-Jean-Marie, *L'Organisation Corporative de la France d'Ancien Régime.* Paris, 1938.

Olmsted, John W., "The Scientific Expedition of Jean Richer to Cayenne," *Isis,* XXXIV (1942) 117–128.

———, "The Voyage of Jean Richer to Arcadia in 1670: A Study in the Relations of Science and Navigation under Colbert," *Proceedings of the American Philosophical Society,* CIV (1960) 612–634.

Ornstein, Martha, *The Role of Scientific Societies in the Seventeenth Century,* rev. ed. Chicago, 1938.

Oursel, Noémi-N., *Nouvelle Biographie Normande.* Paris, 1886, 2 vols.

Palmer, Robert R., *The Age of the Democratic Revolution.* Princeton, 1959–1964, 2 vols.

Papillon, Fernand, "De la Rivalité de l'Esprit Leibnizien et l'Esprit Cartésien au 17è Siècle," *Revue des Travaux de l'Académie des Sciences Morales et Politiques,* XCVIII (1872) 689–720.

Pappas, John N., "Berthier's Journal de Trévoux and the Philosophes," *Studies on Voltaire and the Eighteenth Century,* III (1957).

Paris, Bibliothèque Nationale, Département des Imprimés, *Catalogue de l'Histoire de la Révolution Française,* ed. André Martin and Gérard Walter. Paris, 1936–1955, 5 vols.

Paris, Musée du Conservatoire National des Arts et Métiers, *Histoire et Prestige de l'Académie des Sciences 1666–1966.* [Paris, 1966.]

Parker, Harold T., "French Administrators and French Scientists during the Old Regime and the Early Years of the Revolution," *Ideas in History,* ed. R. Herr and H. T. Parker, pp. 85–109. Durham, N.C., 1965.

———, "Two Administrative Bureaus under the Directory and Napoleon," *French Historical Studies,* IV (1965) 150–169.

The Parliamentary Register: or History of the Proceedings and Debates of the House of Commons (and House of Lords) Containing the Most Interesting Speeches. London, 1775–1813, 112 vols.

Parsons, Talcott, *Essays in Sociological Theory,* 2nd ed. Glencoe, 1958.

————, and Edward Shils, *Toward a General Theory of Action.* Cambridge, Mass., 1954.

Partington, James R., *A History of Chemistry.* London, 1961–1964, 3 vols.

Passy, Louis, *Histoire de la Société Nationale d'Agriculture de France. 1761–1793.* Paris, 1912.

Payen, Jacques, *Capital et Machine à Vapeur au XVIIIè Siècle. Les Frères Périer et l'Introduction en France de la Machine à Vapeur de Watt.* Paris, 1969.

Pelletier, Bertrand, and Jean Darcet, *Instruction sur l'Art de Séparer le Cuivre du Métal des Cloches.* Paris, an II [1794].

Penrose, Stephen B. L., Jr., *The Reputation and Influence of Francis Bacon in the Seventeenth Century.* New York, 1934.

Perkins, Merle L., "Late Seventeenth-Century Scientific Circles and the Abbé de Saint-Pierre," *Proceedings of the American Philosophical Society,* CII (1958) 404–412.

Perrault, Charles, *Mémoires de ma Vie,* ed. Paul Bonnefon. Paris, 1909.

————, *Parallèle des Anciens et des Modernes en ce qui Regarde les Arts et les Sciences,* ed. Hans R. Jauss. Munich, 1964.

————, *Recueil de Divers Ouvrages en Prose et en Vers.* Paris, 1675.

Perrault, Claude, *Mémoires pour Servir à l'Histoire Naturelle des Animaux.* Paris, 1671–1676, 2 vols.

————, *Recueil de Plusieurs Machines de Nouvelle Invention.* Paris, 1700.

Perronet, Jean-Rodolphe, *Description des Projets et de la Construction des Ponts de Neuilli, de Mantes, d'Orléans, de Louis XVI,* 2nd ed. Paris, 1788.

Pevsner, Nikolaus B. L., *Academies of Art, Past and Present.* Cambridge, 1940.

Pflaum, Georg Michael, *Geschichte des Wortes "Zivilisation."* Munich, 1961.

Pichois, Claude, "Un Concours sur la 'Légèreté Française' à l'Académie de Dijon, en 1809," *Société Française de Littérature Comparée, Actes du Troisième Congrès National,* pp. 87–97. Paris, 1960.

Pictet, Marc A., "Journal d'un Genevois à Paris sous le Consulat," *Mémoires et Documents Publiés par la Société d'Histoire et d'Archéologie de Genève,* ser. 2, V (1893–1901) 98–133.

[Pidansat de Mairobert, Mathieu-François], *L'Espion Anglais.* London, 1779–1783, 6 vols.

Pigeire, Jean, *La Vie et l'Œuvre de Chaptal (1756–1832)*. Paris, 1932.

Pire, Georges, *Stoicisme et Pédagogie: De Zénon à Marc-Aurèle, de Sénèque à Montaigne et à J.-J. Rousseau*. Liège, 1958.

Plantefol, Lucien, "L'Académie des Sciences durant les Trois Premiers Siècles de son Existence, ses Visages Successifs, ses Publications," Institut de France, Académie des Sciences, *Troisième Centenaire, 1666–1966*, pp. 56–139. Paris, 1967.

Le Point du Jour. Paris, 1789–1791.

Poisson, Charles, *Les Fournisseurs aux Armées sous la Révolution Française*. Paris, 1932.

Poissonnier, Pierre-Isaac, *Discours sur la Naissance de Mgr. le Dauphin, Prononcé au Collège Royal le 18 Février 1782*. Paris, 1782.

Polanyi, Michael, *Science, Faith and Society*. Chicago, 1964.

Popkin, Richard H., *The History of Skepticism from Erasmus to Descartes*, rev. ed. Assen, 1964.

Portal, Antoine, *Histoire de l'Anatomie et de la Chirurgie*. Paris, 1770–1772, 6 vols.

Pottinger, David T., *The French Book Trade in the Ancien Regime, 1500–1791*. Cambridge, Mass., 1958.

Pouchet, Georges, *Les Sciences pendant la Terreur, d'après les Documents du Temps et les Pièces des Archives Nationales*, ed. James Guillaume. Paris, 1896.

Prevet, François, *Contribution à l'Etude de l'Evolution Historique des Techniques d'Organisation Sociale Appliquées à la Pharmacie*. Paris, 1940.

Procès-Verbaux de l'Académie des Beaux-Arts, ed. M. Bonnaire. Paris, 1937–1943, 3 vols.

Procès-Verbaux des Séances de l'Académie Tenues depuis la Fondation de l'Institut, jusqu'au Mois d'Août 1835. Hendaye, 1910–1922, 10 vols. (cited in notes as *Institut, Reg*).

Procès-Verbaux du Comité d'Instruction Publique de la Convention Nationale, ed. James Guillaume. Paris, 1891–1959, 7 vols. (cited in notes as *CIPCN*).

Procès-Verbaux du Comité d'Instruction Publique de l'Assemblée Législative, ed. James Guillaume. Paris, 1889. (cited in notes as *CIPAL*).

Le Propagateur des Lois, des Evénements et des Arts. Paris an VI–an VIII [1798–1800].

Proust, Jacques, *Diderot et l'Encyclopédie*. Paris, 1962.

——, *L'Encyclopédisme dans le Bas-Languedoc au XVIIIè Siècle*. Montpellier, 1968.

Rabbe, Félix, "Pahin de la Blancherie et le Salon de la Correspondance," *Bulletin de la Société Historique du VIè Arrondissement de Paris*, II (1899) 30–52.

Ranc, Albert, "Les Débuts de la Presse Quotidienne à l'Académie des Sciences," *Etudes de Presse*, X (1958) 38–45.

——, "La Presse Scientifique sous la Révolution Française," *La Nature*, LXXVI (1948) 31–32.

Rapport Fait à l'Académie des Sciences le 7 Septembre 1790 des Moyens Hydrauliques Présentés par M. Detrouville, Publié par les Commissaires de l'Académie. n.p., [February, 1791].

Raspail, François-Vincent, *Nouvelles Etudes Scientifiques et Philosophiques, 1861–1864.* Paris, 1864.

[Raymond-Latour, Jean-Michel], *Souvenirs d'un Oisif.* Lyon, 1836, 2 vols.

Réaumur, René-Antoine Ferchault, *Memoirs on Steel and Iron*, tr. A. G. Sisco and ed. Cyril S. Smith. Chicago, 1956.

Recueil de Plusieurs Traitéz de Mathématique de l'Académie Royale des Sciences. Paris, 1676.

Recueil des Pièces qui Ont Remporté les Prix de l'Académie 1720–1772. Paris, 1721–1777, 9 vols.

Reglement de la Société des Arts. Paris, 1730.
    BN, Imprimés, F. 27662.

Règlements Intérieurs de l'Académie des Sciences. [1795–1886]. Paris, n.d.

Reichardt, Johann Friedrich, *Un Hiver à Paris sous le Consulat, 1802–1803*, ed. A. Laquiante. Paris, 1896.

Reinhard, Marcel, "Elite et Noblesse dans la Seconde Moitié du XVIIIè Siècle," *Revue d'Histoire Moderne et Contemporaine*, III (1956) 5–37.

Rémond, André, *John Holker, Manufacturier et Grand Fonctionnaire en France au XVIIIè Siècle 1719–1786.* Paris, 1946.

Remond de Sainte-Albine, Pierre de, *Mémoire sur le Laminage du Plomb*, 3rd ed. Paris, 1746.

Restout, Turcaty, and Dufournoy, *Pétition Motivée de la Commune des Arts à l'Assemblée Nationale.* n.p., n.d.
    Pamphlet in AN, AD VIII 21, and BHVP, 965227.

Retz, Noël, "Exposé Succinct à l'Assemblée Nationale, sur les Facultés et les Sociétés de Médecine," *Nouvelles ou Annales de l'Art de Guérir*, VII (1791) 108–117.

Révolutions de Paris. Paris, 1790–1793.

Rey, Auguste, "Le Naturaliste Bosc, un Girondin Herborisant," *Revue de l'Histoire de Versailles et de Seine-et-Oise* (1900) 241–277, (1901) 17–42.

Richard, Camille, *Le Comité de Salut Public et les Fabrications de Guerre sous la Terreur*. Paris, 1921.

Ricken, Ulrich, *"Gelehrter" und "Wissenschaft" im Französischen. Beiträge zu ihrer Bezeichnungsgeschichte vom 12.–17. Jahrhundert*. Berlin, 1961.

Ritterbush, Philip C., *Overtures to Biology. The Speculations of Eighteenth-Century Naturalists*. New Haven, 1964.

Rivaz, Pierre de, *Réponse du Sieur de Rivaz à un Mémoire Publié contre ses Découvertes en Horlogerie*. Paris, 1751.

Robespierre, Maximilien, *Œuvres Complètes*, ed. Victor Barbier and Charles Vellay. Paris, 1910–1967, 10 vols.

Robinet, Jean-F.-E., *Condorcet, sa Vie, son Œuvre, 1743–1794*. Paris, [1893].

Roche, Daniel, "La Diffusion des Lumières. Un Exemple: L'Académie de Châlons-sur-Marne," *Annales*, XIX (1964) 887–922.

——, "Milieux Académiques Provinciaux et Société des Lumières," *Livre et Société dans la France du XVIIIè Siècle*, ed. F. Furet, pp. 93–184. Paris, 1965.

Rochedieu, Charles Alfred, *Bibliography of French Translations of English Works*. Chicago, 1948.

Roddier, Henri, "Pour une Histoire des Académies," Société Française de Littérature Comparée, *Actes du Troisième Congrès National*, pp. 45–54. Paris, 1960.

Roger, Jacques, ed., Buffon, Georges-Louis Leclerc, *Les Epoques de la Nature*. Paris, 1962.

    Mémoires du Muséum National d'Histoire Naturelle, n.s., ser. C, X (1962).

——, *Les Sciences de la Vie dans la Pensée Française du XVIIIè Siècle*. Paris, 1963.

Roland de la Platière, Jean-Marie, *Compte Rendu à la Convention Nationale, par Jean-Marie Roland, Ministre de l'Intérieur, de Toutes les Parties de son Département, de ses Vues d'Amélioration et de Prosperité Publique, le 6 Janvier de l'An II de la République Française*. Paris, 1793.

Rolle, Michel, *Traité d'Algèbre ou Principes Généraux pour Résoudre les Questions de Mathématiques*. Paris, 1690.

Rose, Arnold M., "Voluntary Associations in France," *Theory and Method in the Social Sciences*, pp. 72–115. Minneapolis, 1954.

Rosenthal, Bronislaw, "Der Geniebegriff des Aufklärungszeitalters," *Germanische Studien*, CXXXVIII (1933) 1–215.

Ross, Sydney, "Scientist: The Story of a Word," *Annals of Science*, XVIII (1962) 65–85.

Rossi, Paolo, *Francis Bacon, from Magic to Science*, tr. Sacha Rabinovitch. London, 1968.

Rouland, *Tableau Historique des Propriétés et des Phénomènes de l'Air*. Paris, 1784.

Roule, Louis, *Bernardin de Saint-Pierre et l'Harmonie de la Nature*. Paris, 1930.

Rousseau, Jean-Jacques, *Discours*, ed. George R. Havens. New York, 1946.

———, *The First and Second Discourses*, ed. Roger D. Masters. New York, 1964.

———, *Œuvres Complètes*, ed. Bernard Gagnebin and Marcel Raymond. Paris, 1959–1969, 4 vols.

Roux, Claudius, "Documents Bio-biblio-iconographiques sur Marat," *Albums du Crocodile*, XXII (1954) 4–52.

Rozbroj, Hugo, *Jean Paul Marat (1743–1793), ein Naturforscher und Revolutionar, sein Zusammentreffen in der Geisteswelt mit Goethe, Lamarck, Rousseau, u. a.* Berlin, 1937.

Rozier, François, *Nouvelle Table des Articles Contenus dans les Volumes de l'Académie Royale des Sciences de Paris depuis 1666 jusqu'en 1770*. Paris, 1775–1776, 4 vols.

Ruplinger, André, *Charles Bordes, Membre de l'Académie de Lyon (1711–1781)*. Lyon, 1915.

Russo, François, "L'Hydrographie en France au XVIIè et XVIIIè Siècles," *Enseignement et Diffusion*, pp. 419–440.

Rutledge, James, *Nouvelle Théorie Astronomique pour Servir à la Détermination des Longitudes*. London, 1788.

Sage, Balthazar-Georges, *Mémoires Historiques et Physiques*. Paris, 1817.

———, *Opuscules de Physique*. Paris, 1813.

———, *Supplément aux Institutions de Physique*. Paris, 1812.

———, *Tableau Comparé de la Conduite qu'Ont Tenue envers Moi les Ministres de l'Ancien Régime, avec celles des Ministres du Nouveau Régime*. Paris, 1814.

Saigey, Emile, *Les Sciences au XVIIIè Siècle*. Paris, 1873.

Saint-Pierre, Jacques-Henri Bernardin de, *La Chaumière Indienne*. Paris, 1791.

———, *Etudes de la Nature*, 1st ed. Paris, 1784, 3 vols. 3rd ed. Paris, 1788, 4 vols.

———, *Vœux d'un Solitaire pour Servir de Suite aux Etudes de la Nature*. Paris, 1789.

Sanders, Heinrich, *Beschreibung seinen Reisen durch Frankreich, die Niederlande, Holland, Deutschland und Italien.* Leipzig, 1783–1784, 2 vols.

Saricks, Ambrose, *Pierre Samuel Du Pont de Nemours.* Lawrence, 1965.

Sas, Louis Furman, "Semantic Notes on the French Noun 'Savant'," *French Review,* XIV (1941) 195–201.

Saucerotte, Constant, *Les Médecins pendant la Révolution, 1789–1799.* Paris, 1887.

Saunders, Richard M., "The Provincial Academies of France during the Seventeenth and Eighteenth Centuries," unpublished Ph.D. dissertation, Cornell University, 1931.

[Saurin, Joseph], *Lettre Critique de Monsieur **** à Monsieur **** sur le Traité de Mathématique du P. C.[astel] et les Extraits qu'Il A Faits dans les Journaux des Mémoires de l'Académie des Sciences de l'Année 1725.* Paris, 1730.

Scarfe, Francis, *André Chénier. His Life and Work 1762–1794.* Oxford, 1965.

Schalk, Fritz, *Studien zur Französischen Aufklärung.* Munich, 1964.

Scheler, Lucien, *Lavoisier et la Révolution Française.* Paris, 1956–1960, 2 vols.

Schier, Donald S., *Louis Bertrand Castel, Anti-Newtonian Scientist.* Cedar Rapids, 1941.

Schiller, Joseph, "Les Laboratoires d'Anatomie et de Botanique à l'Académie des Sciences au XVIIè Siècle," *Revue d'Histoire des Sciences,* XVII (1964) 97–114.

———, "La Transfusion Sanguine et les Débuts de l'Académie des Sciences," *Clio Medica,* I (1965) 33–40.

Schmeisser, Johann Gottfried, *Beytraege zur Naeheren Kenntniss des Gegenwaertigen Zustandes der Wissenschaften im Frankreich.* Hamburg, 1797–1798, 2 vols.

Schneider, Emil, *Der Animale Magnetismus. Seine Geschichte und seine Beziehungen zur Heilkunst.* Zurich, 1950.

Schneider, René, *Quatremère de Quincy et son Intervention dans les Arts (1788–1830).* Paris, 1910.

Schofield, Robert, "Histories of Scientific Societies: Needs and Opportunities for Research," *History of Science,* II (1963) 70–83.

———, *The Lunar Society of Birmingham.* Oxford, 1963.

Sedillot, L.-A., *Les Professeurs de Mathématiques et de Physique Générale au Collège de France.* Rome 1869.

Reprinted from *Bullettino di Bibliografia e di Storia delle Scienze Matematiche e Fisiche,* II (1869).

Sénard, Jean, "L'Académie des Sciences Est-Elle Morte ou 'Récupérable'?" *Le Figaro Littéraire* (18 November 1965) 14.

Sergescu, Pierre, "La Littérature Mathématique dans la Première Période (1665–1701) du 'Journal des Savants'," *Archives Internationales d'Histoire des Sciences*, I (1947) 60–99.

[Servan, Joseph-Michel-Antoine], *Discours sur le Progrès des Connaissances Humaines en Général, de la Morale et de la Législation en Particulier.* [Lyon, 1781.]
   BN, Imprimés, Rz. 3231.

Sieffert, Charles-Henri, "Deux Lettres de Condorcet à Sabbathier Secrétaire de l'Académie des Sciences, Arts et Belles-Lettres de Châlons (1777)," *Mémoires de la Société d'Agriculture, Commerce, Sciences et Arts de la Marne*, ser. 2, XXIX (1955) 89–91.

Siégler-Pascal, S., *Les Projets de l'Abbé de Saint-Pierre 1658–1743.* Paris, 1900.

Sigaud de la Fond, Joseph-Aignan, *Essai sur Différentes Espèces d'Air-Fixe ou de Gas*, new ed., ed. Rouland. Paris, 1785.

Silvestre de Sacy, Jacques, *Alexandre-Théodore Brongniart, 1739–1813, sa Vie, son Œuvre.* Paris, 1940.

Simon, Jules, *Une Académie sous le Directoire.* Paris, 1885.

Smeaton, William A., "L'Avant-Coureur. The Journal in which Some of Lavoisier's Earliest Research Was Reported," *Annals of Science*, XIII (1957) 219–234.

———, "The Early Years of the Lycée and the Lycée des Arts. A Chapter in the Lives of A. L. Lavoisier and A. F. de Fourcroy," *Annals of Science*, XI (1955) 257–267, 309–319.

———, *Fourcroy, Chemist and Revolutionary 1755–1809.* Cambridge, 1962.

———, "Lavoisier's Membership of the Société Royale d'Agriculture and the Comité d'Agriculture," *Annals of Science*, XII (1956) 267–277.

———, "Lavoisier's Membership of the Société Royale de Médecine," *Annals of Science*, XII (1956) 228–244.

Smelser, Neil J., *Social Change in the Industrial Revolution.* Chicago, 1959.

Smith, Edwin Burrows, *Jean-Sylvain Bailly, Astronomer, Mystic, Revolutionary 1736–1793.* Philadelphia, 1954.

Smith, James Edward, *A Sketch of a Tour on the Continent in the Years 1786 and 1787.* London, 1793, 3 vols.

Smith, Preserved, *The Enlightenment, 1687–1776.* New York, 1962.

Sneath, Peter H. A., "Mathematics and Classification from Adanson to the Present," *Adanson*, pp. 471–498. Pittsburgh, 1964.

[Société du Point Central des Arts et Métiers], *Adresse Présentée à l'Assemblée Nationale Législative*. [Paris], 16 October 1791.
  Pamphlet in AN, AD VIII, 29.

[————], *A Tous les Artistes et Autres Citoyens*. Paris, 1792.
  Pamphlet in AN, AD VIII, 21.

[————], *Le Point Central des Arts et Métiers à la Convention Nationale*. [Paris], an II [1794].
  Pamphlet in BN, Imprimés, Vz. 620.

Société Philomatique de Paris, *Bulletin des Sciences*. 1791–an IX [1801], 2 vols.

————, *Rapport Général des Travaux de la Société Philomatique de Paris*, ed. C.-A. Riche and A.-F. Silvestre. Paris, [1795]– an VIII [1800], 4 vols.

Sprat, Thomas. *History of the Royal Society of London*. London, 1667.

Stafleu, Frans A., "Adanson and the 'Famille des Plantes'," *Adanson*, pp. 123–264. Pittsburgh, 1963.

————, "L'Héritier de Brutelle: The Man and his Work," L'Héritier, *Sertum Anglicum 1788*, pp. xiii–xliii. Pittsburgh, 1963.

Steele, Arthur R., *Flowers for the King. The Expedition of Ruiz and Pavon and the Flora of Peru*. Durham, N.C., 1964.

Stein, Jay W., *The Mind and the Sword*. New York, 1961.

Stocking, George W., Jr., "French Anthropology in 1800," *Isis*, LV (1964) 134–150.

Storrs, F. C., "Lavoisier's Technical Reports 1768–1794," *Annals of Science*, XXII (1966) 251–275, XXIV (1968) 179–197.

Sully, Henry, *Règle Artificielle du Temps*. Paris, 1737.

Suppression de Toutes les Académies du Royaume comme Onéreuses à l'Etat et Nuisibles aux Sciences, à la Littérature et aux Arts. n.p., [1789].
  Pamphlet in BN, Imprimés, pièce 8°Z. 1063.

Suremain de Missery, Antoine, "Mémoire Contenant la Réfutation de l'Opinion de M. Bernardin-Henri de Saint-Pierre, au Sujet de la Figure de la Terre," *Journal de Physique*, XLI (1792) 239–243.

Tachard, Gui, *Voyage de Siam des Pères Jésuites, Envoyés par le Roy, aux Indes et à la Chine*. Paris, 1686.

Taillandier, Alphonse-Honoré, *Documents Biographiques sur P. C. F. Daunou*, 2nd ed. Paris, 1847.

Taine, Hippolyte-Adolphe, *Les Origines de la France Contemporaine*. Paris, 1899–1910, 12 vols.

Talleyrand-Périgord, Charles-Maurice, *Mémoires*, ed. Paul-Louis and Jean-Paul Couchoud. Paris, 1958, 2 vols.

————, *Proposition Faite à l'Assemblée Nationale sur les Poids et Mesures.* Paris, 1790.

Tatarkiewicz, W., "A Note on the Modern System of the Arts," *Journal of the History of Ideas,* XIV (1963) 422.

Taton, René, "Les Conditions du Progrès Scientifique en Europe Occidentale," *Histoire Générale des Sciences,* III, 615–629. Paris, 1961.

————, "Condorcet et Sylvestre-François Lacroix," *Revue d'Histoire des Sciences,* XII (1959) 127–158, 243–262.

————, "L'Ecole du Génie de Mézières," *Enseignement et Diffusion,* pp. 559–615.

————, "L'Ecole Polytechnique et le Renouveau de la Géometrie Analytique," *Mélanges Alexandre Koyré,* I, 552–564. Paris, 1964.

————, "The French Revolution and the Progress of Science," *Centaurus,* III (1953) 73–89.

————, *Les Origines de l'Académie Royale des Sciences.* Paris, 1966.

————, "Projet de Publication des Registres des Séances de l'Académie Royale des Sciences de Paris (1666–1793)," *Actes du Dixième Congrès International d'Histoire des Sciences,* pp. 283–286. Paris, 1964.

Taylor, Eva G., *The Mathematical Practitioners of Hanoverian England, 1714–1840.* Cambridge, 1966.

Techtmeier, Wolfgang, "Die Rolle Brissot de Warvilles als Publizist und Philosoph," *Neue Beiträge zur Literatur der Aufklärung,* ed. W. Krauss and W. Dietze, pp. 239–257. Berlin, 1964.

Tisserand, Roger, *Au Temps de l'Encyclopédie: L'Académie de Dijon de 1740 à 1793.* Vesoul, 1936.

Tissier, André, *Collin d'Harleville Chantre de la Vertu Souriante (1755–1806).* Paris, 1963, 2 vols.

Tonelli, Giorgio, "Kant's Early Theory of Genius (1770–1779)," *Journal of the History of Philosophy,* IV (1966) 109–131, 209–224.

Torlais, Jean, *L'Abbé Nollet, un Physicien au Siècle des Lumières, 1700–1770.* Paris, 1954.

————, "L'Académie de La Rochelle et la Diffusion des Sciences au XVIIIè Siècle," *Revue d'Histoire des Sciences,* XII (1959) 111–125.

————, "La Physique Expérimentale," *Enseignement et Diffusion,* pp. 619–645.

————, "Un Prestidigitateur Célèbre, Chef de Service d'Electrothérapie au XVIIIè Siècle. Ledru dit Comus (1731–1807)," *Histoire de la Médecine,* V (1955) 13–25.

————, *Réaumur. Un Esprit Encyclopédique en dehors de l'Encyclopédie*. Paris, 1936.

Tournefort, Joseph Pitton de. *Elémens de Botanique*. Paris, 1694.

Tourneux, Maurice, *Bibliographie de l'Histoire de Paris pendant la Révolution Française*. Paris, 1890–1913, 5 vols.

————, "Journal Intime de l'Abbé Mulot (1777–1782)," *Mémoires de la Société de l'Histoire de Paris et de l'Ile-de-France*, XXIX (1902) 19–124.

————, "Un Projet d'Encouragement aux Lettres et aux Sciences sous Louis XVI," *Revue d'Histoire Littéraire de la France*, VIII (1901) 281–311.

Townsend, Joseph, *A Journey through Spain in the Years 1786 and 1787*, 2nd ed. London, 1792, 3 vols.

Trembley, Emilie, "Un Savant Genevois: Jean Trembley-Colladon 1749–1811. Son Jugement sur le Monde Scientifique de Paris en 1786," *Bulletin de la Société d'Histoire et d'Archéologie de Genève*, IX (1948) 103–117.

Trénard, Louis, "L'Académie de Lyon et ses Relations Etrangères au XVIIIè Siècle," Société Française de Littérature Comparée, *Actes du Troisième Congrès National*, pp. 65–86. Paris, 1960.

Tresse, René, "Un Atelier Pilote de la Révolution Française—l'Atelier de Perfectionnement des Armes Portatives, Mai–Décembre 1794," *Techniques et Civilisations*, V (1956) 54–64.

————, "Le Conservatoire des Arts et Métiers et la Société d'Encouragement pour l'Industrie Nationale au Début du XIXè Siècle," *Revue d'Histoire des Sciences*, V (1952) 246–264.

————, "Les Dessinateurs du Comité de Salut Public," *Techniques et Civilisations*, V (1956) 1–10.

Tudesq, André-Jean, *Les Grands Notables en France (1840–1849)*. Paris, 1964.

Tuetey, Louis, ed., *Procès-Verbaux de la Commission Temporaire des Arts*. Paris, 1912–1917, 2 vols.

Turgot, Etienne-François, *Mémoire Instructif sur la Manière de Rassembler, de Préparer, de Conserver et d'Employer les Diverses Curiosités d'Histoire Naturelle*. Paris, 1758.

Vadier, Marc-Guillaume-Albert, *Le Montagnard Vadier à Mr. Caritat, Ci-Devant Marquis de Condorcet, Académicien Auteur d'une Constitution à la Détrempe Rejetée par la Convention Nationale, et d'une Feuille Empoisonnée, Corruptrice de l'Esprit Public, Appellée la Chronique, Ci-Devant Président du Club Hermaphrodite, dit de 1789, etc.* Paris, n.d.

    BN, Imprimés, fol. Lb⁴¹ 3196.

Van Duzer, Charles Hunter, *Contribution of the Idéologues to French Revolutionary Thought.* Baltimore, 1935.

Van Leeuwen, Henry G., *The Problem of Certainty in English Thought 1630–1690.* The Hague, 1963.

Varignon, Pierre, *Nouvelles Conjectures sur la Pesanteur.* Paris, 1690.

Vaughan, Michalina, "The Grandes Ecoles," *Governing Elites,* ed. Rupert Wilkinson, pp. 74–107. New York, 1969.
    See also Clifford-Vaughan, Michalina.

Vausenville, Guillaume Roberger de, *Essai Physico-Géométrique.* Paris, 1778.

————, *Quadrature Définie du Cercle. Lettre de M. Le Rohbergherr-de-Vausenville . . . à M. D'Alembert.* n.p., [1774].

————, *Supplément à l'Ouvrage Intitulé "Essai Physico-Géométrique."* Paris, 1779.

Vergnaud, Marguerite, "Un Savant pendant la Révolution," *Cahiers Internationaux de Sociologie,* XVII (1954) 123–139.

Vernière, Paul, "L'Encyclopédie de Diderot et D'Alembert," *La Synthèse. Idée-Force dans l'Evolution de la Pensée,* ed. Centre International de Synthèse, pp. 134–149. Paris, 1951.

Vess, David M., "The Collapse and Revival of Medical Education in France: A Consequence of Revolution and War, 1789–1795," *History of Education Quarterly,* VII (1967) 71–92.

Vicq d'Azyr, Félix, "Nouveau Plan de Constitution pour la Médecine en France," *Histoire et Mémoire de la Société Royale de Médecine, 1787–1788,* pp. 1–201. Paris, 1790.

Vignery, Robert J., *The French Revolution and the Schools. Educational Policies of the Mountain, 1792–1794.* Madison, 1965.

Villeterque, Alexandre-Louis, *Quelques Doutes sur la Théorie des Marées par les Glaces Polaires, ou Lettre à M. B. de Saint-Pierre.* Paris, 1793.

Voltaire, François-Marie Arouet de, *Correspondance,* II, ed. Theodore Besterman. Geneva, 1953.

————, *Lettres Philosophiques.* Amsterdam, 1734.

————, *Le Siècle de Louis XIV.* Berlin, 1751, 2 vols.

Wachs, Morris, "A Study of D'Alembert's *Histoire des Membres de l'Académie Française,*" unpublished Ph.D. dissertation, University of Wisconsin, 1958.

Walter, Gérard, *Marat,* 2nd ed. Paris, 1960.

Watson, E. C., "The Early Days of the Académie des Sciences as Portrayed in the Engravings of Sébastien Le Clerc," *Osiris,* VII (1939) 556–587.

Watts, George B., "Panckoucke's Saving of Academic Mémoires," *Kentucky Foreign Language Quarterly*, V (1958) 212–216.

Weil, Hans, *Die Entstehung des Deutschen Bildungsprinzips*. Bonn, 1930.

Welvert, Eugène, "La Légende de Lakanal," *Revue des Questions Historiques*, LXXXIV (1908) 113–144.

White, Howard B., "The Influence of Bacon on the *Philosophes*," *Studies on Voltaire and the Eighteenth Century*, XXVII (1963) 1849–1869.

Willard, Nedd, *Le Génie et la Folie au Dix-Huitième Siècle*. Paris, 1963.

Williams, L. Pearce, "The Politics of Science in the French Revolution," *Critical Problems*, ed. M. Clagett, pp. 291–308. Madison, 1959.

———, "Science, Education and Napoleon I," *Isis*, XLVII (1956) 369–382.

———, "Science, Education and the French Revolution," *Isis*, XLIV (1953) 311–330.

Wolf, Charles-Joseph-Etienne, *Histoire de l'Observatoire de Paris de sa Fondation à 1793*. Paris, 1902.

Wood, Henry T., *A History of the Royal Society of Arts*. London, 1913.

Woolf, Harry, *The Transits of Venus. A Study of Eighteenth-Century Science*. Princeton, 1959.

Znaniecki, Florian, *The Social Role of the Man of Knowledge*. New York, 1940.

INDEX

Index of Names

General Index